DESIGN OF
FEEDBACK CONTROL SYSTEMS

DESIGN OF FEEDBACK CONTROL SYSTEMS

Douglas B. Miron
South Dakota State University

Harcourt Brace Jovanovich, Publishers
Technology Publications

San Diego   New York   Chicago   Austin   Washington, D.C.
London   Sydney   Tokyo   Toronto

ISBN 0-15-517368-5

Library of Congress Number 88-83244

Printed in the United States of America

To James Hayes: a teacher who opened my innermost ear to music, and my mind to engineering.

PREFACE

This book is intended to support a first course in control systems for senior students in engineering. It concentrates on feedback control systems describable by linear, continuous, time-invariant differential equations. Both single-input single-outpout (SISO) and multi-input multi-output (MIMO) systems are treated. The book in turn, should be supported by a background in mathematics through matrix algebra and Laplace Transforms. It is further assumed that the student has access to a software package, either on a mainframe computer or a microcomputer, that will do the various calculations and plots discussed and presented here.

Design is a major emphasis in this text, and the student is led through a design cycle from problem statement to component specification. In my course, we finish the SISO material about the middle of the semester, so that students can begin work on a paper design project, which is due a week before the end of the semester. In order to cover the design material in this schedule, I have to limit the topics covered and I cannot afford to spend time on review of previous coursework in mathematics. I must choose one approach to analysis and design, either the Nyquist–Bode plots or the root–locus plots, but not both. Most students come to this course with some knowledge of the use and meaning of Bode plots and s-plane pole and zero locations, so I make reference to both of these in most chapters. Other methods of displaying frequency–response information are omitted entirely. The Routh–Hurwitz Test and its extensions are given in an appendix. A teacher who does not wish to cover quite so much design and who is not going to assign a design project can spend time on alternative ways to proceed through a given stage of the cycle.

The presence of personal computers, terminals to mainframes, and work stations in both school and industry makes it even more necesary to emphasize ways to test the results of computation. We must be ever-vigilant against an unquestioning faith in a computing machine that is prey both to programming errors and to electromechanical malfunctions. The user must know the shape of the expected results and should exercise the computing system with test cases relevant to the particular problem at hand. Therefore, in addition to student use of a computing system for plots and calculations, I take many opportunities to point the way to checking results.

In the later chapters and in Appendix B, I give some algorithms which the student can code in any language and use as the basis of a personal software system for control system analysis and design. I have done this for myself in APL, which is, for reasons discussed in Appendix B, a language particularly suited to engineering work. In this text, when I describe calculations I've done or plots that I've generated, I sometimes refer to the tools I've used—APL and my LINEAR SYSTEMS PACKAGE. I do not mean to imply that these are either essential or the best possible tools for the particular job, just that these are the tools I actually used. The same things can be done with any language and graphics package or specialized software system. It is important that an author say something about the tools used, whether he is writing a journal article or a design text.

Many people have made great efforts to eliminate errors in the text, to refine the presentation, and redirect sections of the book. I must thank my colleagues and reviewers for their generous participation. Any errors that remain in the text, I must claim.

Especially, I want to extend appreciation to Dr. Robert Van Nuys of Grove City College and to Dr. Robert I. Egbert of the Wichita State University. Each reviewer's careful eye and incisive criticisms have contributed immeasurably to this book.

CONTENTS

INTRODUCTION

1.1   BASIC CONCEPTS AND PROPERTIES

Good definitions of the term "control system" are not easy to come by. Many are too general to be useful or don't convey the real essence of what a control engineer thinks he or she is working with. Part of this problem is caused by the fact that the modeling concepts and many analysis methods developed for control systems have been generalized and applied to systems that aren't really control systems. Consider the following definition: A control system is a collection of objects designed to produce a specified output function. Under this definition, the paperweight on my card pile is a control system, since its purpose is to maintain the cards in a specified position. But that's not what we mean. One might think of the system consisting of a baseball, a pitcher, and a catcher as a control system whose object is to keep the ball moving in an oscillatory fashion. In this mind frame, a batter would be viewed as an external disturbance source. Likewise, an economy nowadays is frequently described as a system whose purpose is to maximize production of goods and services, maximize total wealth, or maximize some social good. Even though both of these "systems" can be described in the language of the control engineer, they are not what we mean by "control systems." The essence of control systems is that they are designed for the purpose of control. This statement is not intended as a definition either, but to point the way to the following examples.

Consider a wood-burning heater. The basic heater has an adjustable opening to the firebox, the draft, which admits air. This heater is not a control system because it doesn't directly control the combustion rate, the heat rate, the exhaust temperature, or anything. It has a static influence on all these things, which the operator may think about when adjusting the draft. It can, however, be an element in a control system. Many heaters have bimetal coil springs connected to the draft by a lift chain. If the spring's temperature rises, it will lower the chain and reduce the air supply until the temperature and the draft reach an equilibrium. This arrangement is a control system whose operation is really to maintain the spring's temperature—and indirectly the air temperature around the heater. The final, most sophisticated, step is to use a remote thermostat to sense the air temperature to turn on a motor to open the draft when more heat is needed in the house. These last two are examples of control systems. A control system is usually considered to include the element being controlled as well as the elements that do the controlling. So the control engineer speaks of the entire system—the heater, draft, lifter, and thermostat—as a control system. Without the control elements though, the rest is just a system, a plant, or a process.

The thermostat-heater system is an example of a particular class of control systems, *regulators*. Their objective is to maintain some variable at or near a fixed value. They usually operate by sensing, or using, the variable to cause a motion to reduce the variable's current tendency which is due to other factors. The regulator also embodies another principle and classification, *feedback*. The output variable is converted to a signal which is used to control a system input. It is fed back to the input. Control system designs from the earliest records we have through the nineteenth century have all been regulators. In the twentieth century, we encounter another category, *servomechanisms* or *followers*. For these systems, the objective is to make an output variable follow a time-varying input signal. Such systems occur in steering systems and speed controllers for large and complex vehicles. Again, these systems usually include feedback so that the difference between the output and the desired function represented by the input signal is used to drive the plant, which may be a control surface (rudder, aileron), a motor, or an engine.

A common example of a control system which does not use feedback is a variable-speed electric drill. The user moves the trigger to turn on the electronic circuit which regulates the average voltage to the motor, but the shaft speed is strongly load-dependent, so the operator must observe and counter the speed drop without any help from the system. If a shunt-excited, rather than series, electric motor were used, this open-loop control system would be much less load-sensitive, but the need for high starting and stall torque dominates the motor choice. Take a few minutes now and think of some other control systems you use or commonly encounter. Are they regulators or servos? Do they use feedback or not?

Figures 1.1–1.3 show progressive levels of abstraction in representation of the heating system. In Fig. 1.1 we have a pictorial drawing of the system, with labels to identify the parts. In Fig. 1.2, the part labels are put in boxes and lines drawn to show the flow of cause and effect between the parts. This is a physical block diagram. The next level of abstraction is the block diagram in Fig. 1.3 showing the cause-and-effect variables, as time functions, and the functions that

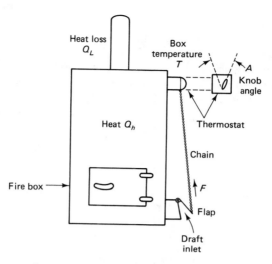

Figure 1.1 Wood-burning heater.

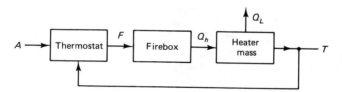

Figure 1.2 Physical block diagram for the heater in Fig. 1.1

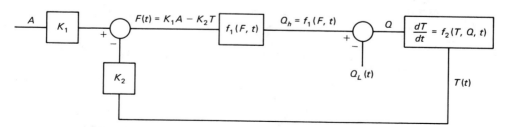

Figure 1.3 Signal-and-function block diagram for the heater in Fig. 1.1.

relate them. Of these, the control engineer spends the most time on variations and elaborations of the signal-and-function diagram, so that when the phrase "block diagram" is used, it is this kind that is meant.

Figure 1.4 shows a general and standard form of the block diagram, with commonly used symbols for the variables. R is for reference, E is for error, and C is for controlled output. The variables are shown as time functions and as transform functions of the complex frequency s. The G and H in the blocks are constants for the present discussion, so that one may write

$$c = Ge \quad \text{and} \quad C = GE \tag{1.1}$$

The circle with the plus and minus signs near it is called a *summing junction*. Its output is the algebraic sum of its inputs.

$$e = r - Hc \quad \text{and} \quad E = R - HC \tag{1.2}$$

Since the purpose of the system is to make the output like the input in some sense, we want an equation showing the relation between C and R. If you eliminate E from the two equations, you will find that

$$C = \frac{GR}{1 + GH} \tag{1.3}$$

Only if the block outputs are constants times their inputs will the time functions have the same relationship. If the GH product is large compared to unity, the dependence on G disappears:

$$C = \frac{R}{H} \quad GH \gg 1 \tag{1.4}$$

For the case of the heater, this has an obvious advantage. If G represents the combustion process and H represents the heat transfer and damper mechanics, G is not a constant over the long term, and for any short term it is a different constant. Equation (1.4) tells how to achieve a dependable, uniform input-output relationship.

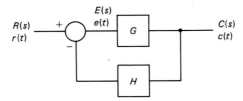

Figure 1.4 General block diagram.

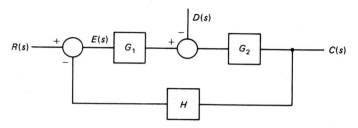

Figure 1.5 General block diagram including an external disturbance.

The heater and the electric drill are subject to other influences than their inputs and their internal processes. In the case of the heater, an open door will cause the house temperature to drop, cooling the box and the thermostat. In the case of the drill, load torque has a strong effect on the output speed. These influences are often considered disturbances to the system and modeled as additional inputs usually at or near the output. This is generally represented by the kind of diagram shown in Fig. 1.5. Again, for the present purpose, let the blocks be constant gains. Setting $R = 0$, you will find the transfer from the disturbance to the output is

$$C = \frac{G_2 D}{1 + G_1 G_2 H} \qquad (1.5)$$

and again,

$$C = \frac{D}{G_1 H} \qquad G_1 G_2 H \gg 1 \qquad (1.6)$$

To make the effect of the disturbance small compared to that of the reference, $G_1 H \gg H$, or $G_1 \gg 1$ for R and D values of the same order of magnitude.

Equations (1.4) and (1.6) express the properties that make feedback control so very attractive. By introducing a well-made control mechanism, represented by the block H, the crafter or designer has overcome two major problems at once. The variability of the main mover, the plant, the actuator has been brought to heel by controlling its input. The disturbance, very often a load variation, has been shut out, again by driving the input to the plant to act against the variation. What are the costs?

From the conditions in (1.4) and (1.6), you can see that more gain is required with feedback than without it. If feedback were not present, R and E would be identical. With feedback, E is a much smaller quantity than R. This is illustrated quite simply by the electronic amplifier shown in Fig. 1.6. Suppose that a gain of

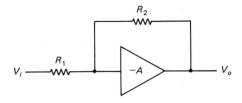

Figure 1.6 Amplifier with feedback.

-10 is required. Without feedback, one can make $A = 10$, which can be done with one transistor and a few passive parts. With feedback, (1.4) says that H should be 0.1. In this case, H is the resistor ratio and $G = A$. To meet the inequality, A must be at least 100. This may mean that a second transistor and another handful of parts are needed. This will cause a considerable increase in cost both for parts and assembly. If the amplifier is to operate at "low" frequencies, an operational amplifier (op amp) is probably the easiest solution anyway, so the part count won't change and feedback will be used whatever the gain requirement. At higher frequencies, discrete transistors must be used, and even at low frequencies stringent conditions may dictate a discrete-part design.

For smooth operation, the input must be capable of making fine motions or changes in a smooth manner. This raises the level of skill needed to make the system in many cases. Frequently, the plant has to be redesigned to accommodate the control mechanism. The heater draft without feedback is hand-operated and is either a sliding plate or a body on a threaded shaft which closes the inlet hole by turning the body. With feedback, the draft is a steel flap hinged at the top, which falls over a hole whose sides project outward to meet the plate and form a seal when the fire is to be shut down. The inlet is moved from the firebox door to either a side or the back of the heater. In this particular case, the net change in manufacturing cost may be zero, but a considerable cost is incurred in design and retooling.

Before the nineteenth century, the systems to which feedback was applied were slow-acting in comparison to the feedback mechanism. With the advent of steam engines and electric motors, actuators became capable of fast responses to input changes. This led, in some cases, to a situation in which the output oscillated and the signal delay through the system (phase shift in the frequency domain) was such that the correcting signal was just in time to push the plant in the wrong direction, reinforcing the motion instead of countering it. This caused the amplitude to increase until some limit was reached, set either by mechanical stops, available power, or destruction. This is called *dynamic instability* or the *stability problem*. This problem, and the corollary problem of meeting specifications for the response to dynamic reference signals, brought in the academic and analytical communities who are still active in making extensions of the known solutions to new problems and in finding new solutions to larger and less understandable systems. The description, analysis, and design of the dynamics of a system are the major focus of this and most other texts on control systems.

■ Example 1.1

Suppose I observe a wood heater of the sort in Fig. 1.1 and find the following. A 90° turn of the thermostat moves the flap chain $\frac{3}{4}$ in. Holding the flap wide open at maximum heat rate (a good 50-kW fire) raises the box temperature 200°F in 10 min., and when the flap is controlled by the thermostat, this same temperature change causes a $\frac{1}{2}$-in. drop in the flap. What estimates can I make about the steady-state behavior of this system?

The transfer from thermostat knob to draft opening is $\frac{3}{4}/90$ = 1/120 in./deg. The transfer from heat rate to box temperature rise is 200/50 = 4°F/kW. The transfer from box temperature to draft opening is $\frac{1}{2}$/200 = 1/400 in./°F. The transfer from draft opening to heat rate is 50/(1/2) = 100 kW/in. With these numbers, I have constructed Fig. 1.7 as a model.

An important difference between Fig. 1.7 and Figs. 1.5 and 1.6 is the presence of the block between the control input A and the summing junction. The thermostat is two transducers in one package. It converts knob angle to flap lift, and it also converts box temperature to flap lift. These transducer gains are each given their own block in Fig. 1.7. This diagram can be converted to the standard form, which we will discuss in Chap. 2. In the meantime, you should be able to verify the following directly from the figure and (1.3) and (1.5):

$$\frac{T}{A} = \frac{(1/120) \times 400}{1 + (4 \times 100/400)} = \frac{5}{3} = 1.67°\text{F/deg}$$

$$\frac{Q_h}{A} = \frac{5}{12} = 0.417 \text{ kW/deg}$$

$$\frac{T}{Q_L} = \frac{4}{12} = -2°\text{F/kW}$$

The gain around the loop is unity, so one can't use the approximate relations that apply for good performance. Suppose a fresh piece of wood is put in the heater and this causes the transfer from flap to heat rate to drop 20%. What will be the new knob-to-temperature gain?

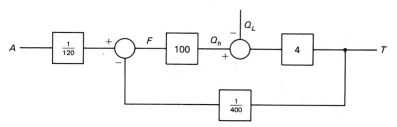

Figure 1.7 Block diagram for the wood heater using the numbers from Example 1.1.

$$\frac{T}{A} = \frac{(1/120) \times 4 \times 80}{1 + 0.8} = 1.48°\text{F/deg}$$

and the heat load will now pull the temperature down at the rate

$$\frac{T}{Q_L} = \frac{-4}{1.8} = -2.22°\text{F/kW}$$

A 20% change in the plant gain causes about a 10% change in the control and disturbance gains. This is due to the low loop gain. This sort of behavior, along with the loose connection between box and room temperatures, explains why people tend to fiddle with these thermostats.

Clearly, one could improve the system by raising the loop gain. If one makes the thermostat more sensitive, the capacity of the firebox to follow the flap changes may be exceeded. This may result in a low-frequency, on-off type of oscillation, which is the way gas, oil, and electric heaters and furnaces are generally operated. Another option, which can be used in combination with the first, is to use a larger firebox. If performance is a problem, this, in effect, tells the analyst that a bigger heater should have been used for the application. This is an example of the general problem of choosing the right size actuator for a specific requirement. ∎

1.2   HISTORICAL SUMMARY OF THE FIELD

Regulators have been devised and constructed by people of craft and ingenuity from the most ancient of times to the present day. I doubt that the records we have from antiquity represent fairly the equipment used in those days, but we do have records of an important type of device, the float valve. Water clocks were invented in the third century B.C. Their basic principle is that a small hole in a vessel will leak water at a constant rate if the water level is held constant. A float valve was used to keep the level constant. The constant-flow exit was used to raise another float with a pointer attached, and the pointer indicated the time on a fixed scale. Thus, the float valve converted a decreasing pressure source, the initial water supply vessel, into a constant pressure and flow source. Fig. 1.8 illustrates one such design. The Greeks used the float valve for other things as well, including Philon's famous oil lamp which was kept full from a valve-regulated reservoir, and an always-full goblet. Water clocks were made in Baghdad until 1258 A.D., when the Mongols captured the city.

During the Renaissance in Europe, temperature and speed regulators were invented. Cornelis Drebble, of Holland, invented the first temperature regulator in 1624. It was used for a chicken incubator and for small furnaces for chemistry. During the seventeenth and eighteenth centuries, millwrights built and improved windmills for grinding grain into flour. In particular, they developed the fantail,

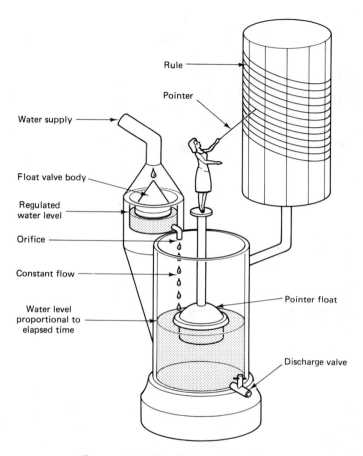

Figure 1.8   Kresibian waterclock.

an auxiliary propellor, to keep the main rotor directly across the wind. They also invented several mechanisms which acted to decrease the main rotor sail area as the wind picked up, so that the shaft speed would remain more nearly constant. The most important invention of the millwrights was not used at first as a speed regulator. It was used to control the distance between moving and stationary grindstones. A sketch of this mechanism is shown in Fig. 1.9. Two lead balls are on the lower ends of two levers. The levers pivot on horizontal axles which are fixed in a plate which turns with the drive shaft. When the plate turns, the weight of the lead balls swings the levers out by centrifugal force, and this lifts the levers. The levers lift a horizontal rod which slides through collars fixed to the levers. This rod in turn lifts a collar through which it passes. The rod and its collar turn at shaft speed, so the turning collar is used to lift another one that slides on it.

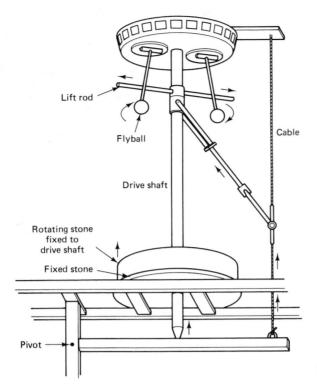

Lift rod

Flyball

Drive shaft

Rotating stone
fixed to
drive shaft

Fixed stone

Pivot

Cable

Figure 1.9   Mead's lift-tenter.

The last collar moves only vertically and pulls a cable which lifts the lower stone. There are many possible variations on this principle, one of which is illustrated in a problem in Chap. 2. A flyball was used to roll up the sails on the mill in a 1787 patent drawing by Matthew Mead. In 1788, James Watt borrowed the idea to control the steam throttle valve on his steam engines. This adoption, along with the general development of steam-powered industry, brought the flyball governor to the attention of the scholarly community and the governor became a subject of analysis for most of the nineteenth century.

Governors were trouble almost from their beginning. Sometimes they worked wonderfully well, but on other units they caused oscillations in the speed. Sir George Airy, royal astronomer, used a governor to help make his telescope track a star in spite of the earth's motion. One design misbehaved so violently that he described it as being "perfectly wild." He devised a damper for the governor and did some analysis of its operation. J. C. Maxwell, famous today for his contributions to electromagnetics, took an interest in the problem as an example of a dynamic system. He linearized the equations for the governor, derived the differential equations to describe the system, and established conditions for the stability of second- and third-order systems. He published this work in a paper entitled, "On Governors," in 1868. Edward J. Routh, a Canadian, derived a test on the coefficients of the differential equation for a system of any order, while he was associated with Cambridge University. He published his work in an 1876 paper, "A Treatise on the Stability of a Given State of Motion." In 1893, a Swiss professor tried to apply a graphical method presented in Paris, in 1876, by a Russian engineer. The method was good for a third-order system, but extensions of the problem drove the order to 7. He gave the problem to the German mathematician Adolf Hurwitz, who found another general test on the equation coefficients and published it in 1895. Although the European scientific community made a good effort to keep up with each other's work, Hurwitz was unaware of Routh's paper. In 1911, an Italian demonstrated the equivalence of their results. This test is still used occasionally today, and it is presented in Appendix A with some extensions.

The preceding story seems to be a largely British tale, but you should recall that the analytical work took place in the context of the general European development of dynamic analysis. France, Germany, Italy, and Switzerland had especially strong groups of scientists and mathematicians making great progress in differential and integral equations, complex variables, harmonic analysis, calculus of variations (What is the function that minimizes this cost integral?) and other fields which were drawn on then and would be drawn on later to solve problems in control engineering. From the nineteenth century to the present, Russia has always had a strong school of mathematicians. Unfortunately, their work did not receive the prompt attention it often deserved because of slow or indifferent translation efforts by the West. In the last 30 years, this communication gap has been reduced, so that we have been able to use Lyapunov's work on nonlinear systems (1890) and Pontryagin's maximum principle (optimization again, 1956), for example. These two subjects have been particularly important in our space programs.

The electrical age began in the nineteenth century. By the turn of the century a multitude of regulators were being applied in power and lighting systems. Until the 1930s, the Routh-Hurwitz test remained the only analytical tool used in control systems. The invention of the feedback amplifier (1927) by H. S. Black shifted the focus of analysis to the frequency domain and the locus of leadership to the United States. Black applied negative feedback to reduce the distortion in telephone repeater amplifiers. Like the engine governor, it worked well sometimes, but caused the amplifier to oscillate in other cases. Unlike the governor and engine, the amplifier couldn't be adequately modeled with a low-order differential equation. Since the order of a differential equation model is equal to the number of independent energy storage elements in the system, an amplifier of a few stages could generate a 40- or 50-degree equation. This made the coefficient test hard to use. By that time, communication engineers were used to transforming circuit problems to the frequency domain, so it was natural to seek a stability criterion expressed in complex frequency. Nyquist and Bode, of Bell Laboratories, developed such a criterion and design methods in the 1930s. This approach formed the dominant frame of reference for control system engineering until the 1960s.

In 1948, W. R. Evans introduced the root locus method. He was working with aircraft guidance and control systems in which the plants often were inherently unstable. In the frequency domain, this means that the transfer function of the plant might have poles in the right half of the s plane. This condition makes the Bode diagram approach trickier to use, so Evans developed a method for directly displaying the location of the closed-loop poles as one varied a single parameter. This method became quite popular and is the major alternative today for the design of single-input single-output systems.

Frequency domain concepts were extended to optimization by Wiener (1942), sampled-data systems by Hurewicz (1947), and nonlinearities by Kochenberger (1950). The space program and the development of industrial process controls in the 1950s and 1960s introduced new problems which were less amenable to frequency response methods. New optimization needs, multiple inputs and outputs, and less tractable nonlinearities were the major problems that caused a return to time domain methods. The classical work on calculus of variations, Poincare's use of state variables in celestial mechanics (France, 1892), and old and new Russian work were among the resources brought to bear on these problems. The state variable model, representing a system by N first-order differential equations instead of one of Nth order, is probably the most significant change from the work of Maxwell and Routh. This new point of view allowed the development of numerous original results and solutions and has led to a degree of unification of notions from the frequency and time domains. At this writing, most new developments in control engineering are taking place in the context of these ideas.

I have written this section to show you several things, the most important of which is the international nature of the development of knowledge. Even people who have ideas that solve major problems or introduce a fundamentally new point of view seldom develop the consequences of their inspiration. It takes knowledge of the work of many to lay the foundation for a new idea, and it takes the contributions of many more to develop its theoretical and practical consequences. The rate of progress is strongly related to the quality of communication between investigators in a given field and across many fields in all nations. In the exercise of our most distinctive faculty we have been and must continue to be one people of this world. A second point is the increased pace of progress due to the feedback between crafter and scholar. The engineer very often forms the bridge between these two communities and must understand them both. Some of us are more like the crafter, strong on intuition and physical understanding, and some of us are more like the scholar, strong on modeling and mathematical analysis—but we must not abandon one mode while working in the other. It is our unique position that requires one as a check on or a guide to the other to bring improvements to the human family from our area. My last point is—it's exciting! Control engineering is the field which makes one get into the dynamics of all engineering areas. The control engineer is sometimes required to dig out knowledge from specialists which the specialists don't know they have. No learning is ever wasted, especially in this field.

Bennett, S., *A History of Control Engineering: 1800–1930*, London, Peter Peregrinus, 1979.

Fuller, A. T., "The Early Development of Control Theory," *Trans. ASME J. Dyn. Sys. Meas. Control*, vol. 98, pp. 109–118, 224–235, 1976.

MacFarlane, A. G. J., *Frequency-Response Methods in Control Systems*, New York, IEEE Press, 1979.

Mayr, O., *The Origins of Feedback*, Cambridge, Mass., MIT Press, 1970.

PROBLEMS

1.1 For the system in Fig. 1.4, let $H = 1$ and G lie in the range $[G_m, 3G_m]$. If C/R is to be in [0.99, 1.01], what must be the minimum value of G_m?

1.2 For the system in Prob. 1.1, $C/R < 1$. Find H and G_m so that the variation in C/R fills the permitted interval.

1.3 The unit-to-unit variation in gain for a batch of operational amplifiers is $4:1$. Suppose that the process parameters can be adjusted to move this range above a desired minimum value. Consider the circuit in Fig. 1.6. Let $R_1 = 10\text{K}\Omega$ and choose R_2 and the minimum value of A so that the circuit gain is -10 ± 0.01.

1.4 Figure P1.4 shows a system in which the plant transfer is represented by a graph. This plant has saturation; its output is limited after the input exceeds certain values. Find the range on w corresponding to the linear range of the plant.

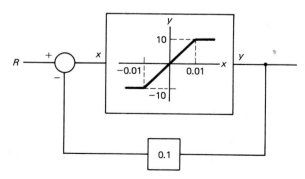

Figure P1.4 Plant with saturation.

1.5 Figure P1.5 shows a two-amplifier system. G_1 is a high-gain low-power preamp, and G_2 is a low-gain power amp. G_2 will saturate at ± 30 V, while G_1 saturates at $\pm V_{cc}$. If $G_2 = 5$ and $H = 1$,

(a) What is the minimum value of V_{cc} needed to avoid saturation of the preamp before it occurs in the power amp?

(b) If $G_1 = 50$, what will be the maximum values of R, E, and x short of saturation?

(c) Repeat (b) for $G_1 = 2000$.

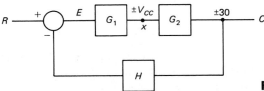

Figure P1.5   Two amplifiers with feedback.

Chapter 2

MODELS OF COMPONENTS AND SYSTEMS

2.1 EQUATIONS AND DIAGRAMS

In Chap. 1 you saw three levels of abstraction used to depict a system: the pictorial sketch, the physical block diagram, and the functional block diagram. The functional block diagram is a description of the input-output relation of each element in the physical block diagram. This close correspondence between the physical components and the functional blocks is not necessary and often not present. Functional blocks are also used to represent equations which, in turn, represent portions of the system. If a mathematical description of the signal transfer of several components in the system is obtained, this description may be represented by one block, and the input-output properties of the individual elements will be buried (included) in the block. As you know, the mathematical description of several components connected together is not unique. For example, an electrical network's behavior can be represented by any of the following: branch currents, branch voltages, mesh currents, or node voltages. The assignment of labels, the ordering of subscripts, and the choice of input and output variables are all external to the elements themselves and are almost arbitrary. Likewise, the description of a mechanical system depends on the choice of references, position, or velocity as the main variable type, and labeling. Thus the physics of the elements determines the order of the equations, but the details depend on the analyst's choices. Since this is so, a block diagram which represents the equations for portions of the system is not a unique representation. In fact, it is often as useful to manipulate a diagram to produce another form as it is to manipulate equations.

A block diagram is a partitioning of a system into nominally independent subsystems or components. This partitioning is only as valid as the independence of the subsystems. This idea translates into the requirement that a block's output is determined only by its input, is not affected by loading from succeeding blocks, and does not load preceding blocks. We will see examples of how to get a valid partition later in the chapter during the discussion of component models.

In linear systems, there are four elementary operations possible on signals: multiplication by a constant, differentiation, integration, and algebraic addition of signals. These operations are represented in the time and frequency domains by block diagram elements as shown in Fig. 2.1a and b. There is another fairly popular way to represent signal flow in the frequency domain. Signals are represented by labeled nodes, and the transfer between them is shown by a joining line segment, with an arrow to show the input-output direction and the transfer

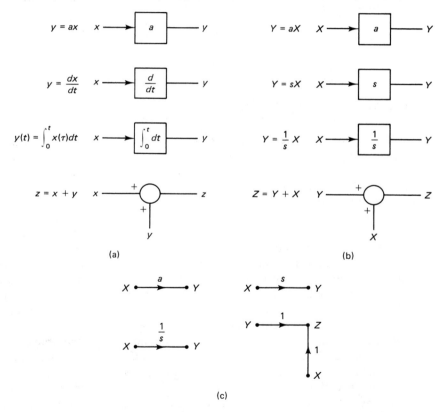

Figure 2.1   Basic linear operations and diagram elements (a) Time domain. (b) Frequency domain, zero initial conditions. (c) Signal flow graphs for (b). Arrows are used when needed to work signal flow direction. There are none on the summing junction because each input is already marked for its effect (+ or −) on the output (sum).

value shown alongside. Figure 2.1c shows the signal flow graph equivalents of Fig. 2.1a and b. The basic elements in these figures can be cascaded, looped, and paralleled in many ways, and the resulting diagrams can be quite complex. The next two sections present methods for simplifying these diagrams and obtaining their overall transfer functions.

2.2   SOME BLOCK DIAGRAM MANIPULATIONS

There are two different objectives which may motivate one to manipulate a block diagram. The most obvious is to obtain the overall transfer function. In the design process, though, the objective is more often to obtain a simpler form, such as a single-loop system. A single-loop system is the basic form for performance analysis and compensator design, as you will see in later chapters. Figure 2.2 shows

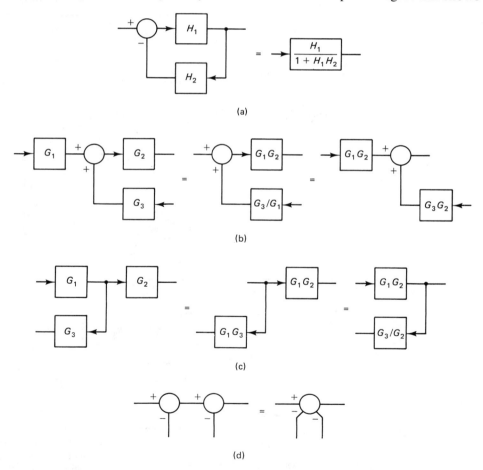

Figure 2.2   Equivalent block diagram segments.

some equivalent block arrangements. The first row is the single-loop close-up in which a single loop is replaced by a single forward block. You have already seen the equation for this block several times. In the rest of these operations, a summing point or a pickoff point is moved from the interior to one end of a line of blocks. Mathematically, the terminal relations in a figure row must be the same. From a physical point of view, this is equivalent to requiring that terminal signal levels be preserved. Thus, in Fig. 2.3, the summing junction is moved back to precede G_1. If nothing else were done, the contribution to the input of G_2 from $A(s)$ would be larger by G_1. To restore the proper level, H_2 must be divided by G_1. In Fig. 2.4, the minor loop involving H_2 can't be closed up because of the feedforward signal entering its interior. An alternative is to move the pickoff point forward to the output. In order to preserve the signal level at the output of H_2, H_2 must be divided by G_3.

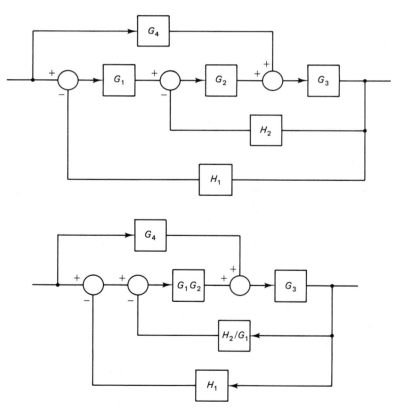

Figure 2.3 A reduction step: moving a summing junction back.

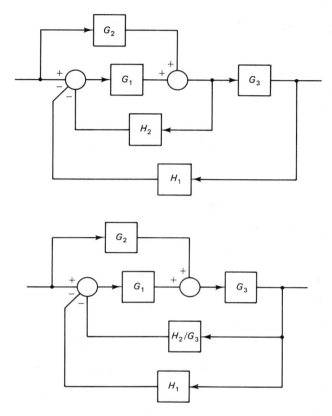

Figure 2.4 A reduction step: moving a pickoff point ahead.

2.3   MASON'S GAIN RULE

Mason developed his gain rule for signal flow graphs, but it is almost as easy to use on block diagrams. Before reading the rule, you must learn some definitions and ideas about graphs.

Figure 2.5 illustrates the following discussion. A *branch* is a line segment joining two nodes, and it has gain and direction. A *loop* is a sequence of connected branches which, when their directions are followed, closes on itself. If, in the course of traversing a loop, one finds parallel paths between nodes, each path counts as part of a different loop. Thus, if there are two parallel branches between two succesive nodes and there is one return branch between them, two loops are formed. A *forward path* is a sequence of branches leading from an input node to an output node. If a loop and a forward path have at least one common node, they are said to be *touching*. Likewise, if a loop has at least one node in common

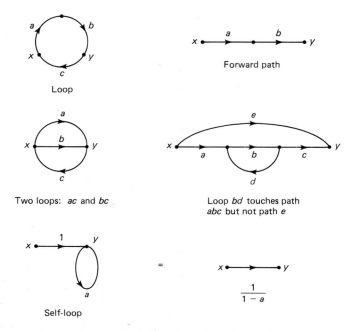

Figure 2.5   Some definitions for signal flow graphs.

with another loop, they are also said to be touching. Nontouching, having not even one node in common, is an important property for the gain rule. A *self-loop* is a branch which starts and ends on the same node. It is, in effect, equivalent to a control loop with unity forward gain.

Mason's gain rule is a generalization of the single-loop transfer formula illustrated by the first entry in Fig. 2.2 and derived in Chap. 1. Let $P_j(k)$ be the gain product of the jth combination of k nontouching loops, as illustrated in Fig. 2.6. There are four loops. For $k = 1$, this will give four terms each of which is the gain of one loop: a_2b_1, a_4b_2, a_6b_3, and $a_3a_4a_5b_4$. Now count every possible pair of nontouching loops. Observe that the big loop touches each of the little loops, but that they do not touch each other. There are three pairs, $a_2b_1a_4b_2$, $a_2b_1a_6b_3$, and $a_4b_2a_6b_3$. There is only one group of three nontouching loops, and there are no groups with more nontouching loops. The triplet is $a_2b_1a_4b_2a_6b_3$. Therefore, these are all the $P_j(k)$ for this graph. Let

$$\Delta = 1 + \sum_{k=1}^{\infty} (-1)^k \sum_{j=1}^{\infty} P_j(k) \tag{2.1}$$

and

$$\Delta_i = 1 + \sum_{k=1}^{\infty} (-1)^k \sum_{j=1}^{\infty} P_{ij}(k) \tag{2.2}$$

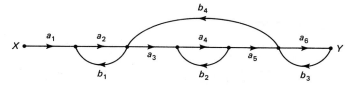

Figure 2.6 A signal flow graph example with nontouching loops.

where $P_{ij}(k)$ contains only loops that do not touch the ith forward path. Now, the overall gain can be written as

$$G = \frac{\sum\limits_{i=1}^{\infty} G_i \Delta_i}{\Delta} \qquad (2.3)$$

This is Mason's gain formula.

■ Example 2.1

Since we already have the $P_j(k)$ for Fig. 2.6, let us finish the work. There is only one forward path, $a_1 a_2 a_3 a_4 a_5 a_6$, and there are no loops which do not touch it. The numerator sum has only one nonzero term, $\Delta_1 = 1$, so

$$\frac{Y}{X} = \frac{a_1 a_2 a_3 a_4 a_5 a_6}{1 - (a_2 b_1 + a_4 b_2 + a_6 b_3 + a_3 a_4 a_5 b_4)} \qquad (2.4)$$
$$+ (a_2 b_1 a_4 b_2 + a_2 b_1 a_6 b_3 + a_4 b_2 a_6 b_3) - a_2 b_1 a_4 b_2 a_6 b_3$$

■

■ Example 2.2

As stated earlier, one does not always wish simply to know the overall transfer function or to reduce a system diagram to a single block. Consider the system in Fig. 2.7a. It may be that the only design flexibility lies in the choice of G_1. To be able to study the effects of this choice, we must not allow it to be absorbed in a reduction process. We will convert Fig. 2.7a to a single-loop system by the following steps:

1. The pickoff point for H_2 is moved to the output. This clears the minor loop and yields two parallel feedback paths as shown in Fig. 2.7b.
2. The inner loop is closed up and the outer feedback signals added to give Fig. 2.7c.

Now we have a single-loop system to which the powerful methods of the next few chapters can be applied.

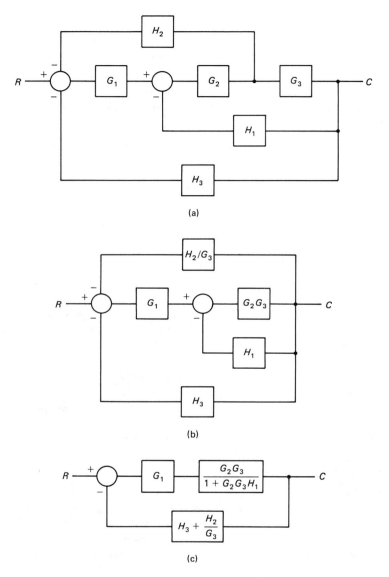

(a)

(b)

(c)

Figure 2.7 System for Example 2.2.

Mason's gain rule can be applied to Fig. 2.7a directly. There is one forward path, and there are three loops, all of which are touching.

$$\Delta = 1 - (-G_2G_3H_1 - G_2G_1H_2 - G_1G_2G_3H_3)$$

$$= 1 + G_2[G_3(H_1 + G_1H_3) + G_1H_2] \quad (2.5)$$

and

$$\frac{C}{R} = \frac{G_1G_2G_3}{\Delta} \quad (2.6)$$

You should not only verify the veracity of these results but also show that Fig. 2.7c leads to the same finding.   ∎

2.4   STATE VARIABLES AND EQUATIONS

As mentioned in Chap. 1, any Nth-order differential equation in one unknown can be transformed into N first-order equations in N unknowns. In your previous studies of electric circuits or dynamic mechanical systems you assigned names to several important variables, wrote differential equations of various orders, and in some cases eliminated all but one variable. This process led you from several low-order equations to one whose order was the sum of the orders of the original set. Enough examples of this sort should teach you several things:

1. Any variable in a system can be chosen and solved for, yielding an equation whose order is not greater than the number of energy-storing elements in the system.
2. The choice of variables is not unique.
3. Any variable can be written in terms of the others, so that once the system is solved, it is really solved for all the variables of interest.

If we can combine equations and eliminate variables by differentiation and substitution, it seems reasonable that we should be able to go in the opposite direction: We should be able, by defining more variables, to write more and simpler equations. The limit is N first-order equations in N variables. A set of such variables, providing they completely represent the behavior of the system, is called a *set of state variables*. Once a set is found, any other variable of interest can be expressed as a combination of the state variables, as in principle 3. The association between the order of a system's equations and its energy-storing elements suggests that the variables representing the energy stored ought to be a set of state variables. This is indeed the case, and I will give examples in the following sections using these energy variables to write the system equations. Such variables are voltage for a capacitor, current for an inductor, velocity for a mass, displacement for a spring, and temperature for a heat system.

Another way to generate a set of state variables and equations is to start from a high-order equation in one variable that also represents the system. One can define a set of variables, each of which is equal to a derivative of the original, from the zeroth up to the $(N - 1)$st. Consider the following second-order equation:

$$a\ddot{c} + b\dot{c} + dc = qr(t) \tag{2.7}$$

Define

$$x_1 = C \tag{2.8}$$

$$x_2 = \dot{c} = \dot{x}_1 \tag{2.9}$$

$$\dot{x}_2 = \ddot{c} \tag{2.10}$$

$$a\dot{x}_2 + bx_2 + dx_1 = qr(t) \tag{2.11}$$

Equations (2.9) and (2.11) are two first-order equations that can replace (2.7), but there is a standard form for the state equations that makes them particularly useful later on. It is an arrangement in which each equation has but one derivative, which is on the left with unity coefficient; the terms on the right are in subscript order with the forcing function(s) on the far right. Thus,

$$\dot{x}_1 = x_2 \tag{2.12}$$

$$\dot{x}_2 = -\frac{d}{a}x_1 - \frac{b}{a}x_2 + \frac{q}{a}r(t) \tag{2.13}$$

One can now represent the system state in vector-matrix form as

$$\begin{bmatrix} \dot{x}_1 \\ \dot{x}_2 \end{bmatrix} = \begin{bmatrix} 0 & 1 \\ -d/a & -b/a \end{bmatrix} \begin{bmatrix} x_1 \\ x_2 \end{bmatrix} + \begin{bmatrix} 0 \\ q/a \end{bmatrix} r(t) \tag{2.14}$$

which fits the general form

$$\dot{\mathbf{x}} = \mathbf{A}\mathbf{x} + \mathbf{B}\mathbf{u} \tag{2.15}$$

Whatever one wishes to choose as the output variables, they can generally be expressed as a linear combination of the state variables and forcing functions. These equations can also be put into a standard vector-matrix form as

$$\mathbf{y} = \mathbf{C}\mathbf{x} \tag{2.16}$$

Methods of deriving state equations from a high-order equation will be treated more fully in Chap. 10.

Resistors, capacitors, and inductors are the three basic passive electrical components. Ideally, they are linear and can be very well approximated by practical products in the frequency range of interest in most control systems. Since the advent of semiconductor operational amplifiers, inductors have been largely replaced by combinations of resistors, capacitors, and op amps (active *RC* networks) in circuits operating below 100 kHz. The trade-off might appear to be increased complexity for smaller size and weight, since low-frequency inductors usually have iron cores and are large compared to integrated circuit (IC) packages. However, the complexity of control circuits is actually about the same because an inductor would not be used in isolation, but together with other components to produce a required transfer function. An op amp would likely be included to impedance-isolate sections and provide gain. In many cases of interest, the transfer function can be produced by about the same number of components whether designed around indicators or active *RC* networks. Inductance is a property of many actuators used in control systems, so that their models include inductors. This will be illustrated by the dc motor in Sec. 2.7. Networks involving op amps will be discussed in Chap. 7.

Figure 2.8 shows the symbols and describing equations for the three elements and for a pair of coupled coils (a transformer). The frequency domain equations

(a) $v = Ri$ $i = Gv$

(b) $v = \dfrac{1}{C}\displaystyle\int_0^t i\,d\tau$ $i = C\dfrac{dv}{dt}$
$V = \dfrac{I}{Cs}$

(c) $v = L\dfrac{di}{dt}$ $i = \dfrac{1}{L}\displaystyle\int_0^t v\,d\tau$
$V = sL\,I$

(d) $v_k = L_k\dfrac{di_k}{dt} + M\dfrac{di_j}{dt}$ $\begin{array}{l} k = 1 \text{ or } 2 \\ j \neq k \end{array}$
$V_k = sL_kI_k + sMI_j$

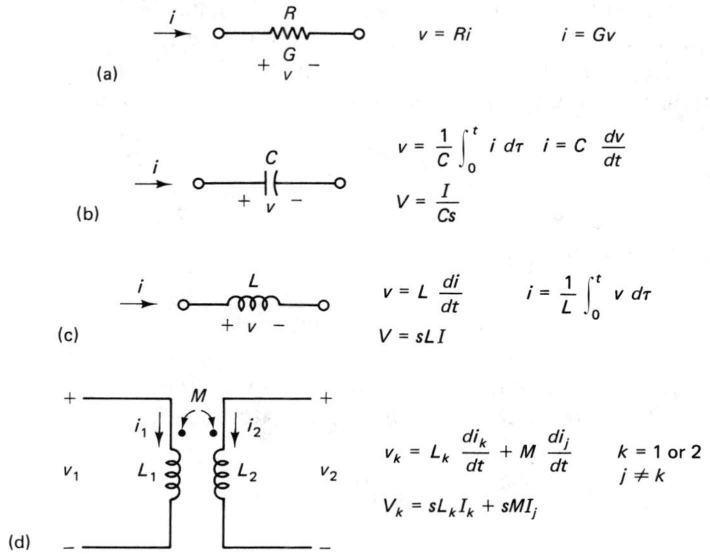

Figure 2.8 Passive electrical components with their time and frequency domain equations. (a) Resistor. (b) Capacitor. (c) Inductor. (d) Transformer.

are Laplace transforms of the time domain equations, assuming zero initial energy storage. The inductor and the capacitor are energy-storing devices. The inductor stores energy in a magnetic field which is proportional to the current through it, and the capacitor stores energy in an electric field which is proportional to the voltage drop across it. Therefore, these are the variables to choose when writing state equations directly from circuits.

The coupled coils shown in Fig. 2.8d are a special and somewhat tricky case. Physically, the magnetic field produced by a current through a coil induces a voltage not only in the current-carrying coil but in any coil in its neighborhood. The dot convention shows whether the cross-induced voltage is in or out of phase with the self-induced voltage in a coil. If the two currents are flowing the same way with respect to their dots, their magnetic fields will aid each other and the mutually induced voltage will be in the same direction as the self-induced voltage in each coil. If one current flows into its dot and the other current flows out of its dot, then the mutual voltage will have the opposite sign of the self-induced voltage. This matter of same or opposite sign is important to distinguish from plus or minus signs, because the plus or minus sign on the self-induction voltage depends on the reference direction for the circuit voltage with respect to the current. Thus, if the polarity marks for V_1 are reversed in Fig. 2.8d, one has $V_1 = -L_1sI_1 - MsI_2$. If, in addition, the I_2 reference direction is reversed, one then has $V_1 = -L_1sI_1 + MsI_2$. There will be other instances in which you will have to take account of both reference direction and dot conventions.

Consider the circuits in Fig. 2.9. They may be analyzed by the usual node or mesh equations to produce the transfer functions, or we may write state equations for them. In Fig. 2.9a, the output variable is R times the inductor current, which may be chosen as the state variable. We may write the state equation as

$$v_s = L \frac{di_L}{dt} + Ri_L \qquad \frac{di_L}{dt} = -\frac{R}{L} i_L + \frac{1}{L} v_s \qquad (2.17)$$

and the output equation as

$$v_o = Ri_L \qquad (2.18)$$

In Fig. 2.9b, the output is the state variable, and the charging current is G times the voltage drop across it. Thus, the state equation is

$$C \frac{dv_o}{dt} = G(v_s - v_o)$$

$$\frac{dv_o}{dt} = -\frac{G}{C} v_o + \frac{G}{C} v_s \qquad (2.19)$$

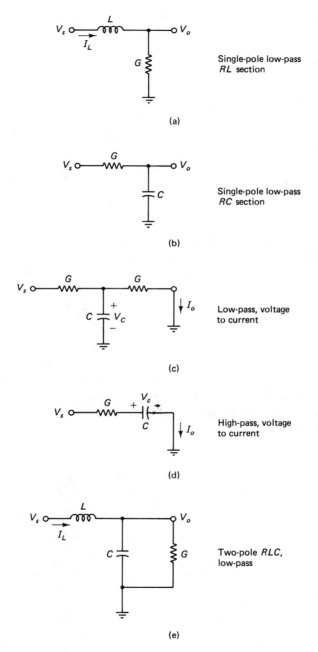

Figure 2.9 Some simple filters.

In some operational amplifier arrangements, we are interested in the transfer from voltage to current. A simple network of this kind is shown in Fig. 2.9c. Here the output variable is G times the state variable V_C, and the capacitor charging current must be written in terms of the voltages:

$$C \frac{dv_C}{dt} = G(v_s - v_C) - Gv_C$$

$$\frac{dv_C}{dt} = -\frac{2G}{C} v_C + \frac{G}{C} v_s \tag{2.20}$$

and the output equation is

$$i_o = Gv_C \tag{2.21}$$

In Fig. 2.9d, we have a simple case in which the forcing function directly drives the output variable as well as the state.

$$\frac{dv_C}{dt} = -\frac{G}{C} v_C + \frac{G}{C} v_s \tag{2.22}$$

$$i_o = -Gv_C + Gv_s \tag{2.23}$$

Did you notice that the state equation is the same as in an earlier case? The circuit in Fig. 2.9e involves two energy-storing elements, so it requires two state variables. The voltage across the inductor needs to be expressed in terms of the input and states, as well as the charging current of the capacitor. The state and output equations are

$$\frac{di_L}{dt} = -\frac{1}{L} v_o + \frac{1}{L} v_s \tag{2.24}$$

$$\frac{dv_o}{dt} = \frac{1}{C} i_L - \frac{G}{C} v_o \tag{2.25}$$

$$v_o = v_o \tag{2.26}$$

In vector-matrix form, these are

$$\begin{bmatrix} i_L \\ \dot{v}_o \end{bmatrix} = \begin{bmatrix} 0 & -1/L \\ 1/C & -G/C \end{bmatrix} \begin{bmatrix} i_L \\ v_o \end{bmatrix} + \begin{bmatrix} 1/L \\ 0 \end{bmatrix} v_s \tag{2.27}$$

$$v_o = \begin{bmatrix} 0 & 1 \end{bmatrix} \begin{bmatrix} i_L \\ v_o \end{bmatrix} \tag{2.28}$$

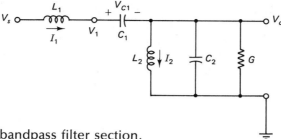

Figure 2.10 A bandpass filter section.

We have a more complex-looking circuit in Fig. 2.10. However, writing the state equations is about as simple as in the previous cases. One must remember to express the inductor voltages and the capacitor currents exclusively in terms of the state variables and the input. In this case, we should choose the two inductor currents and the two capacitor voltages as the state variables.

$$\frac{di_1}{dt} = -\frac{1}{L_1} v_{C1} - \frac{1}{L_1} v_o + \frac{1}{L_1} v_s \qquad (2.29)$$

$$\frac{di_2}{dt} = \frac{1}{L_2} v_o \qquad (2.30)$$

$$\frac{dv_{C1}}{dt} = \frac{1}{C} i_1 \qquad (2.31)$$

$$\frac{dv_o}{dt} = \frac{1}{C_2} i_1 - \frac{1}{C_2} i_2 - \frac{G}{C_2} v_o \qquad (2.32)$$

2.6   PASSIVE MECHANICAL ELEMENTS AND SYSTEMS

2.6.1   Element Models

Mechanical elements generally do not exist in pure forms. That is, there are no massless springs or completely rigid shafts. In many applications, though, the secondary properties of an element can be neglected in comparison to the same property of the system to which the element is connected. A common example is the automobile shock absorber. This element's primary property is viscous damping, and its secondary properties are mass and compliance (spring action). The mass is certainly negligible compared to that of the automobile, and the spring constant of the spring connected in parallel with the shock absorber is much greater. Even if the secondary properties are not negligible, it is useful to model them as separate elements, as is the case with electrical components at high radio

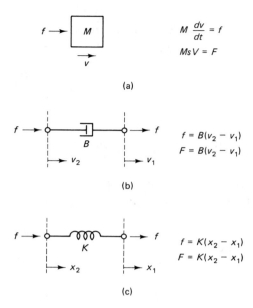

$$M \frac{dv}{dt} = f$$

$$Ms V = F$$

(a)

$$f = B(v_2 - v_1)$$

$$F = B(v_2 - v_1)$$

(b)

$$f = K(x_2 - x_1)$$

$$F = K(x_2 - x_1)$$

(c)

Figure 2.11 Linear motion mechanical elements. (a) Mass, M. (b) Damper, B. (c) Spring, K.

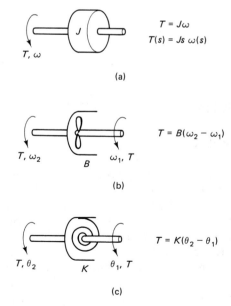

$$T = J\omega$$

$$T(s) = Js\, \omega(s)$$

(a)

$$T = B(\omega_2 - \omega_1)$$

(b)

$$T = K(\theta_2 - \theta_1)$$

(c)

Figure 2.12 Rotational mechanical elements. (a) Inertia, J. (b) Damper, B. (c) Spring, K.

30

frequencies. Figures 2.11–2.13 show the ideal models and describing equations for basic mechanical elements. They display a wider variety than electrical elements. There are two types of energy-storing devices, inertia and spring, and one type of dissipator, the damper. These components exist in forms for both linear motion and rotary motion. In addition, mechanical transformers exist in the forms of levers for small linear motions, and pulley, sprocket, and gear systems for rotary motion. Unlike electrical transformers, these systems work for signals down to zero frequency.

Consider the variables used in these figures. The mass and damper are described in terms of force and velocity. No fixed reference is needed, since displacement is not used. Because the reaction force of the damper is proportional (ideally) to the difference in the velocities of its two parts, a separate velocity is defined for each end. The spring's reaction force is proportional to the displacement of its length from its rest value, so each of the displacement variables must

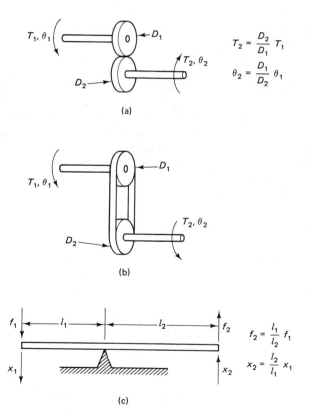

$$T_2 = \frac{D_2}{D_1} T_1$$

$$\theta_2 = \frac{D_1}{D_2} \theta_1$$

(a)

(b)

$$f_2 = \frac{l_1}{l_2} f_1$$

$$x_2 = \frac{l_2}{l_1} x_1$$

(c)

Figure 2.13   Mechanical transformers. (a) Gears. (b) Pulleys and belt or sprockets and chain. (c) Lever.

have its origin at the rest position of its end of the spring. This same pattern is followed for rotary components in Fig. 2.12. An important point to remember is that the force applied to a spring or a damper is passed through and applied to the connection at the other end. A mass, on the other hand, acts more like a summing junction. If an applied force and a spring attached to a mass are in equilibrium, the mass is not in motion and no force is applied to anything else that might be connected to it. The mass moves in response to the vector sum of the applied and reaction forces acting on it. The reaction forces of the connected elements, in turn, must be expressed in terms of the motion imparted to these elements by the mass. A discussion of Fig. 2.14 will illustrate this process.

The equations describing the transformers in Fig. 2.13 idealize them as lossless and without energy storage. Spring effects, mass, and friction may be added as external elements. As they are shown then, the transformers have conservation of power. That is, power in on one side is equal to power out on the other. The torque-speed product is the same on both sides of the rotary transformers. I use this fact along with the obvious one that the small wheel has to turn faster than the large one to check that I have the gear or diameter ratio written correctly.

Finally, I must insert a note of gravity into this discussion of models and equations. In some mechanical systems, a mass may move vertically, or have a vertical component to its motion, so that it stores energy as potential energy proportional to its vertical displacement. In these cases a position variable must be defined for the mass, thereby associating two state variables with it. Another situation in which position must be made a state variable arises when it is important either to the observer or to a connected system. In effect, this makes position an output. If the only energy variables are velocities and derivatives thereof, position will have to be defined as a state in order to prevent an output from having an integral relation to a state. This preserves the standard form at the price of one more, usually simple, state equation.

2.6.2 Examples of Simple Networks

Let us consider the linear mechanical networks in Fig. 2.14. Since, as a student of either mechanical or electrical engineering, you are probably not used to writing transfer functions for them, we will derive both transfer functions and state equations for each item in the figure. In Fig. 2.14a, the applied force compresses the spring and is passed through to accelerate the mass. Thus

$$f = K(x_1 - x_2) = M\ddot{x}_2 \tag{2.33}$$

and

$$F(s) = K[X_1(s) - X_2(s)] = Ms^2X_2(s) \tag{2.34}$$

For the purpose of writing transfer functions and designing in the frequency domain we will treat X_1 as an input and X_2 as an output and will be interested in

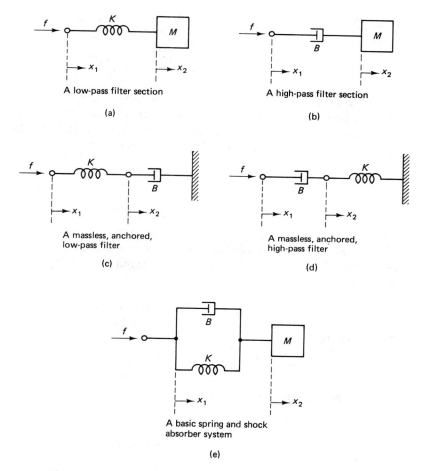

Figure 2.14   Some simple mechanical filter sections.

$F(s)$ as the force required from the driving system. If we are concerned with a sensor, we want F to be small compared to the load forces. To this end rearrange the second part of (2.34) as

$$KX_1 = (Ms^2 + K)X_2 \tag{2.35}$$

and then divide to obtain

$$\frac{X_2}{X_1} = \frac{1}{1 + (M/K)s^2} \tag{2.36}$$

From (2.34) and (2.36) you can find the cost in force to sense X_1 as

$$F(s) = \frac{Ms^2}{1 + (M/K)s^2} X_1(s) \qquad (2.37)$$

The ratio of X_1 to F may be thought of as an input impedance for the network.

The energy storage variables are the net displacement on the spring and the mass velocity. Let $x = x_1 - x_2$ and $v = \dot{x}_2$. Since x_1 is to be the input, replace x_2 by $x_1 - x$. With this and (2.33) the state equations are

$$\dot{v} = \frac{K}{M} x$$

$$\dot{x} = -v + \dot{x}_i \qquad (2.38)$$

In Fig. 2.14b, the applied force causes the damper to close at a velocity proportional to the force, and the force is again passed through to accelerate the mass. Thus

$$f = B(\dot{x}_1 - \dot{x}_2) = M\ddot{x}_2 \qquad (2.39)$$

Transforming,

$$F(s) = Bs[X_1(s) - X_2(s)] = Ms^2 X_2(s) \qquad (2.40)$$

The second equality again yields a transfer function:

$$\frac{X_2}{X_1} = \frac{s}{1 + (M/B)s} \qquad (2.41)$$

From the ends of (2.40) and the transfer function above, the required force is

$$F(s) = \frac{Ms^3}{1 + (M/B)s} X_1(s) \qquad (2.42)$$

If you look at (2.41) for frequencies below B/M, what time relation is implied? If this is part of a sensor network, what does x_2 measure?

Since there is only one energy-storing element in Fig. 2.14b, only one state variable and equation are needed. Again, choose the velocity of the mass as the state variable. Letting $v = \dot{x}_2$ in (2.39) gives

$$\dot{v} = -\frac{B}{M} v + \frac{B}{M} \dot{x}_1 \qquad (2.43)$$

In Figs. 2.14c and d we have two massless systems which appear to be inverse to each other. In both cases, the force is applied to one element and passed through to the other. It causes the spring to compress and the damper to close. For Fig. 2.14c, the time domain equations are

$$f = K(x_1 - x_2) = B\dot{x}_2 \tag{2.44}$$

The transform is

$$F = K(X_1 - X_2) = BsX_2 \tag{2.45}$$

and the transfer function and input force are

$$\frac{X_2}{X_1} = \frac{1}{1 + (B/K)s} \tag{2.46}$$

$$F = \frac{Bs}{1 + (B/K)s} X_1 \tag{2.47}$$

Since the energy storage is in the spring compression and x_1 is the input, choose x_2 as the state variable:

$$\dot{x}_2 = -\frac{K}{B} x_2 + \frac{K}{B} x_1 \tag{2.48}$$

For Fig. 2.14d, in the time domain,

$$f = B(\dot{x}_1 - \dot{x}_2) = Kx_2 \tag{2.49}$$

Transforming,

$$F = Bs(X_1 - X_2) = KX_2 \tag{2.50}$$

The transfer function and input force are

$$\frac{X_2}{X_1} = \frac{(B/K)s}{1 + (B/K)s} \tag{2.51}$$

$$F = \frac{Bs}{1 + (B/K)s} X_1 \tag{2.52}$$

Again, the energy-storing variable is the spring displacement, x_2 in this case. The state equation is

$$\dot{x}_2 = -\frac{K}{B} x_2 + \dot{x}_1 \tag{2.53}$$

Consider the transfer functions for these two cases. As shown Fig. 2.14c is low pass and Fig. 2.14d is high pass. For both functions, the frequency K/B is a boundary which divides a region of approximately unity gain from one of decreasing gain for values further away from the boundary. This agrees with the physics of the two elements involved. At low frequencies, the ends of the spring will track each other, while the damper will offer little resistance to being moved. At high frequencies, which correspond to high speeds, the damper becomes stiff, forcing the spring to do all the displacing.

Finally, look at the simplified automotive suspension model in Fig. 2.14e. Here the force is divided between the spring and damper and passed through to the mass. In the time domain,

$$f = K(x_1 - x_2) + B(\dot{x}_1 - \dot{x}_2) = M\ddot{x}_2 \tag{2.54}$$

Transforming,

$$F = (K + Bs)(X_1 - X_2) = Ms^2 X_2 \tag{2.55}$$

The transfer function and input force are

$$\frac{X_2}{X_1} = \frac{1 + (B/K)s}{1 + (B/K)s + (M/K)s^2} \tag{2.56}$$

$$F = \frac{Ms^2[1 + (B/K)s]}{1 + (B/K)s + (M/K)s^2} X_1 \tag{2.57}$$

If we follow the practice of previous examples, we choose $x = x_1 - x_2$ as one state variable and $v = \dot{x}_2$ as another state variable and let x_1 be the input. Then

$$\dot{x} = -V + \dot{x}_1$$
$$\dot{v} = \frac{K}{M} x - \frac{B}{M} v + \frac{B}{M} \dot{x}_1 \tag{2.58}$$

If the force is considered the input, with the same state definitions, the equations are

$$\dot{x} = -\frac{K}{B} x + \frac{1}{B} f$$
$$\dot{v} = \frac{1}{M} f \tag{2.59}$$

One's choice of representation depends on what's important. If this were a vehicle suspension problem, the known input would be the road (or terrain) contour, not the force. In fact, for this application one must find the force in order to assess the strength required of the system for a given roughness of road.

2.6.3   Toward a General Network Theory

Electrical engineering has an elaborate theory for analysis and synthesis of networks. It is the foundation of the discipline and an indispensible tool for the engineer's daily work. The spring and damper have a force-displacement property like the current-voltage property of electrical elements; the force is passed through the mechanical element like current through an electrical element, and the displacement is a linear operation on the applied force. As suggested in the last subsection, the displacement-to-force ratio may be regarded as an impedance, in which case its inverse is an admittance. Figure 2.15a shows a schematic symbol for such an element, and Fig. 2.16 shows the series addition of impedances and the parallel addition of admittances. For the series case, the total displacement must be the sum of the individual displacements, and since each displacement is an impedance times the force, the impedance of the whole must be the sum of the element impedances. For the parallel case, the displacement is the same for both elements, and the total force is the sum of the forces taken by the individual elements. Since the force through each element is the admittance of the element times the displacement, the admittance of the whole must be the sum of the admittances of the elements. Remember that we are dealing with one-dimensional motion in all of this discussion. The series case is the simplest example of the generally obvious proposition that the sum of the forces at a (massless) junction is zero. The parallel case is the simplest example of the proposition that the sum of the displacements around a closed loop must be zero. You should try a few more elaborate networks to test these ideas.

The properties of springs and dampers connected in one-motion-dimensional networks are also the basis of the theory of electrical networks. Since we have only two kinds of elements though, only that part of the theory which applies to *RC* networks can be borrowed. We can synthesize rational polynomial functions

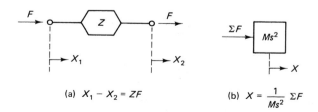

(a) $X_1 - X_2 = ZF$ (b) $X = \dfrac{1}{Ms^2} \Sigma F$

Figure 2.15 Diagrams for linear network elements. (a) Impedance. (b) Mass.

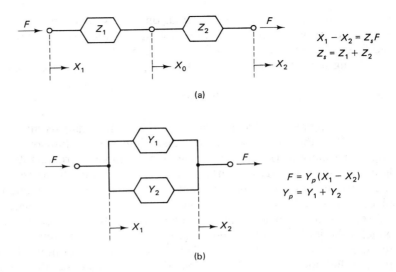

(a)

(b)

Figure 2.16 (a) Series and (b) parallel mechanical impedances.

of s either as impedance functions or transfer functions, with the limitation that the power of the numerator cannot differ from that of the denominator by more than 1. Also, inertia must be taken into account sooner or later. Figure 2.15b shows a schematic symbol and Newton's law for a mass. A mass is a junction point in a linear network because it is rigid; all points displace together.

Let us consider the network in Fig. 2.17. Assume at first that $M = 0$. In that case, you can show that

$$\frac{X_2}{X_1} = \frac{Z_2}{Z_1 + Z_2} \tag{2.60}$$

This is the general form that applies to Fig. 2.14a and b. If the Zs are formed from more than a single element, they will produce more interesting transfer functions. Suppose each is formed from a parallel spring and damper. Then

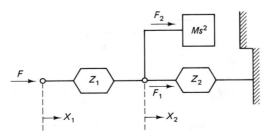

Figure 2.17 Displacement divider.

$$Z_1 = \frac{1}{K_1 + B_1 s} \qquad Z_2 = \frac{1}{K_2 + B_2 s} \tag{2.61}$$

and the transfer function becomes

$$\frac{X_2}{X_1} = \frac{Y_1}{Y_1 + Y_2}$$

$$= \frac{K_1}{K_1 + K_2} \cdot \frac{1 + (B_1/K_1)s}{1 + [(B_1 + B_2)/(K_1 + K_2)]s} \tag{2.62}$$

This is a very useful transfer function form, as you will see when we present compensation design.

Now let M be nonzero. F will equal $F_1 + F_2$, and you may wish to find the transfer function by writing branch equations. I prefer to write node equations, representing the elements by their admittances and recognizing that the displacements X_i are node variables:

$$Y_1 X_1 = (Y_1 + Y_2 + Ms^2)X_2 \tag{2.63}$$

$$\frac{X_2}{X_1} = \frac{Y_1}{Y_1 + Y_2 + Ms^2} \tag{2.64}$$

This form is low pass and is an illustration of the general truth that all real systems are ultimately low pass. Note that the mass was treated as an admittance to ground. Evidently, since every junction involving a mass can be treated as a junction to which the mass is attached, every mass can be likewise treated as an admittance to ground. In this way, we can incorporate inertia into the general network rules. This adds another dimension to the synthesis possibilities in that two storage elements are available, but we are limited to low-pass and incomplete bandpass structures since the inertia is a grounded element.

Suppose now that F_2 is an input force to another network with input admittance Y_i. Then the transfer function becomes

$$\frac{X_2}{X_1} = \frac{Y_1}{Y_1 + Y_2 + Y_i} \tag{2.65}$$

A mass connected to a following network is shown in Fig. 2.18a. Its grounded-mass equivalent is shown in Fig. 2.18b. The following network is represented by its input admittance Y_{i+1}, and a node equation for X_i gives the net input admittance:

$$F_i = (Ms^2 + Y_{i+1})X_i \tag{2.66}$$

$$Y_i = Ms^2 + Y_{i+1} \tag{2.67}$$

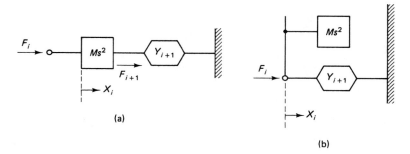

Figure 2.18 A mass in series with a grounded network.

The theory given to this point is sufficient for the purposes of this book. Interested students are invited to examine other points of view and backgrounds as presented in the books on the reading list at the end of this chapter, and to see what extensions can be made by studying these books and exercising their own analytical imagination.

2.6.4 Rotational Systems

The rotary elements in Fig. 2.12 are described by equations involving torque, angular displacement, and angular velocity which have the same form as those describing the linear elements in Fig. 2.11. It follows that the discussions in the previous two subsections apply equally well to networks of rotating elements. Unfortunately, rotating systems almost never look like the pictorial schematics in Fig. 2.12, so we must use some care in modeling real systems to decide what is in parallel and what is in series. For example, the sliding friction and wind drag of a generator armature act in parallel on the same shaft as the inertia, so the applied torque is equal to the sum of the torques taken by each of these components. On the other hand, if a motor is driving a load through a long shaft, that shaft is a torsion spring in series with the load. Like a linear spring, it passes the applied torque through, and its ends displace (in angle) with respect to each other. Figures 2.19 and 2.20 illustrate these situations.

2.6.5 Mechanical Transformers

You are surrounded by mechanical transformers in your daily life. In most cases, they serve the purpose of reducing the speed of a motor or engine to that desired for a load. Generally, it is not economical to build motors or engines to run at the low speeds we need for such applications as vehicles, mixers, hand-held electric drills, etc. For some industrial-type cutting tools belt-and-pulley systems provide a range from speed reduction to speed increase. Lathes, drill presses, and table saws are familiar examples. So we use gear systems or pulley systems to

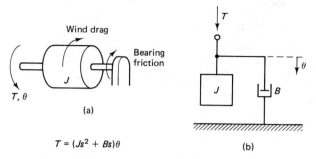

$$T = (Js^2 + Bs)\theta$$

Figure 2.19 A generator armature model. (a) Pictorial. (b) Schematic.

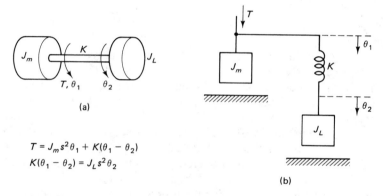

$$T = J_m s^2 \theta_1 + K(\theta_1 - \theta_2)$$
$$K(\theta_1 - \theta_2) = J_L s^2 \theta_2$$

Figure 2.20 Motor armature and load connected by a long shaft. (a) Pictorial. (b) Schematic.

change the speeds from movers to loads. For reasons which will be discussed in detail later, many control systems are made more efficient, and therefore less expensive, by the inclusion of speed reducers.

A mechanical transformer has an effect on the load seen by a mover similar to the action of an electrical transformer. Look at the gear train shown in Fig. 2.21. It drives a load represented by its admittance Y_L. For a speed reducer, $G > 1$. What is the admittance seen at the input shift? Writing the admittance equation for the load and then replacing torque and angle by the input values from the gear equations gives the following sequence:

$$T_2 = Y_L \theta_2 \tag{2.68}$$

$$GT_1 = \frac{Y_L \theta_1}{G} \tag{2.69}$$

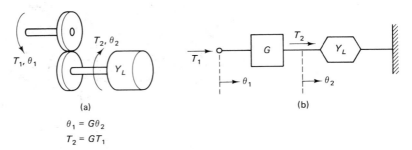

$$\theta_1 = G\theta_2$$
$$T_2 = GT_1$$

Figure 2.21   A mechanical transformer and load. (a) Pictorial. (b) Schematic.

$$T_1 = \left(\frac{Y_L}{G^2}\right)\theta_1 \qquad (2.70)$$

This shows that the mechanical admittance is reduced by G^2, an effect similar to that of a step-down electrical transformer. If the load is a simple parallel combination of inertia, spring, and damping, the constant for each of these is reduced by this factor. This same property is common to the other transformers shown in Fig. 2.13.

Mechanical transformers generally serve the purpose of directing power as well as changing speed. Pulley-and-sprocket systems are commonly used to connect parallel shafts, and beveled gears and worm gears are used to drive shafts at right angles to each other. A bent lever can be used to make an arbitrary change in the direction of small linear motions. Other devices, such as cams, rack-and-pinion assemblies, and screw-and-pinned-nut assemblies, are used to convert between linear and rotary motion. They also can be modeled as ideal transformers with loss and inertia externalized. Motion conversion has been a good area for the exercise of ingenuity.

2.7   STATIC AND DYNAMIC VARIABLES

At this point, I want to remind you of some things you know in specific cases but may not have thought about in a general way. In the previous section, the variables used were offsets from rest (equilibrium) values. Since we were dealing with displacements most of the time, this was easy to picture mentally and on paper. In control system analysis, we are often interested in the deviation of a variable from its steady-state value. This is true even if the variable is a velocity and the steady-state value is not zero. Thus, we generally divide a quantity into two parts—its static (steady-state) value and its dynamic (deviated or perturbed or displaced) value. This practice is familiar to you in the design of electronic amplifiers. The dc design sets the operating point (static values), and small-signal methods are used to analyze and design the ac gain and bandwidth of the stage. The electronic amplifier is also an example of a nonlinear device whose behavior is approximately linear for small deviations from the static values. This small-signal approach is one of the ways we can treat nonlinearities in general. Suppose that the drag term on a spinning load is $B\omega_s{}^3$, where ω_s is the angular velocity. If the system's equilibrium speed is ω_0, define the dynamic deviation as ω. Then the damping becomes

$$\begin{aligned} B\omega_s{}^3 &= B(\omega_0 + \omega)^3 \\ &= B(\omega_0{}^3 + 3\omega_0{}^2\omega + 3\omega_0\omega^2 + \omega^3) \end{aligned} \tag{2.71}$$

By assumption, $\omega \ll \omega_0$, so the third and fourth terms are small compared to the second term. The first term is the static value of the drag, and the second term is the linear term we need for dynamic analysis. In an analysis in which all quantities are divided into static and dynamic values, the static terms will cancel because they specify the operating point, and one will be left with equations involving the dynamic values. This will be illustrated and developed further in Sec. 2.12.

2.8   THE DC MOTOR

In order to provide precise control over a wide power range and be tractable in design, a mover should be capable of operating from rated to zero speed without stalling or abrupt changes in behavior and should be accurately represented by a linear model. The dc motor, in which the stator field is constant and the armature is driven from either a controlled voltage or current source, fits these needs very well. Figure 2.22 shows the linear model for a dc motor. The heart of the motor is an ideal armature which converts between electrical and mechanical power with no loss. To its left are the armature electrical properties, resistance R_a and inductance L_a. To its right are the armature mechanical properties, inertia J_m and friction B_m. The mechanical load is represented by its input admittance Y_L. For the reference directions shown, electrical power equal to $I_a E_b$ is converted to

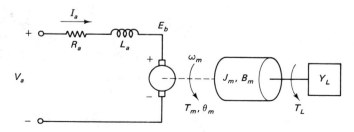

Figure 2.22   Schematic of an armature-controlled dc motor.

mechanical power equal to $T_m\omega_m$. The back voltage generated by the motor is $E_b = K_v\omega_m$, and the torque produced is $T_m = K_t I_a$. In a consistent set of units, such as SI or MKS, $K_v = K_t$, and I will use K_m for both hereafter. The basic equations are

$$V_a = (R_a + sL_a)I_a + E_b \tag{2.72}$$

$$T_m = (J_m s^2 + B_m s + Y_L)\theta_m \tag{2.73}$$

To find the transfer function from input voltage to motor shaft position, you can use the transducer (armature) relations in these equations to find

$$\frac{\theta_m}{V_a}$$

$$= \frac{K_m}{s[J_m L_a s^2 + (B_m L_a + R_a J_m)s + B_m R_a + (1/s)(R_a + sL_a)Y_L + K_m{}^2]} \tag{2.74}$$

Without the leading s, this is the transfer-to-motor-shaft speed. The simplest load to deal with in this expression is another inertia and friction. The load admittance for this case is

$$Y_L = J_L s^2 + B_L s \tag{2.75}$$

In effect, the inertias will add and the friction constants will add. Let $J = J_m + J_L$ and $B = B_m + B_L$. The transfer function becomes

$$\frac{\theta_m}{V_a} = \frac{K_m}{s(BR_a + K_m{}^2)\left(1 + \dfrac{BL_a + JR_a}{BR_a + K_m{}^2}s + \dfrac{JL_a s^2}{BR_a + K_m{}^2}\right)} \tag{2.76}$$

The denominator contains a quadratic, which can be very underdamped in the case of low losses (R_a and B small).

To write state equations for the motor, it is best to represent the load by its torque until its nature is known for a particular case. If the load is excluded then, energy is stored in the inductance and the inertia, and so current and speed should be the state variables. Making these changes in (2.72) and (2.73) yields

$$V_a = (R_a + sL_a)I_a + K_m\omega_m \tag{2.77}$$

$$K_m I_a = (J_m s + B_m)\omega_m + T_L \tag{2.78}$$

Converting to the time domain and rearranging yields the desired state equations:

$$\frac{di_a}{dt} = -\frac{R_a}{L_a} i_a - \frac{K_m}{L_a} \omega_m + \frac{1}{L_a} v_a$$

$$\frac{d\omega_m}{dt} = \frac{K_m}{J_m} i_a - \frac{B_m}{J_m} \omega_m - \frac{1}{J_m} T_L \tag{2.79}$$

Block diagrams are often used to show the electrical and mechanical transfer functions involved in describing a motor-load combination. Figure 2.23 shows a version directly from Fig. 2.22. If the desired output is some other displacement in the load network, one merely has to add a block with the network's transfer function in it. In a system with a gear train, it is often the practice to translate the motor constants to the load end. To generate the proper back voltage from the low-speed end of the gears, one must include a block with gain G in the speed feedback path as shown in Fig. 2.24.

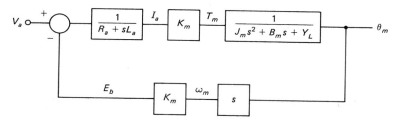

Figure 2.23   Block diagram for a general motor-load combination.

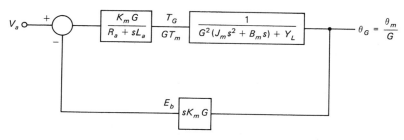

Figure 2.24   Motor-gear-load block diagram. G is the gear ratio.

■ Example 2.3 Some DC servomotor calculations

One of the more difficult things a control engineer has to do is extract information, from any source, for dynamic descriptions for the components of a proposed system. Even when the component in question is a product intended for control system use, the manufacturer may not provide anything more than maximum ratings or may supply data in nonstandard units. An example of the latter is Table 2.1 which lists servomotor data. The conversions were done by using 1 N − m = 141.612 oz-in and 1000 rpm = 104.72 rad/s.

Let's consider some steady-state values before we examine the dynamics of this motor and some possible loads. The maximum stall torque implies a maximum current:

$$I_{max} = \frac{T_{max}}{K_m} = 2.68 \text{ A}$$

This, in turn, implies a maximum applied voltage at stall of

$$V_{stall} = I_{max}R_a = 8.15 \text{ V}$$

Suppose we wish to run a load at 1 krpm from a 15-V supply. The back electromotive force (emf) will be

$$E_b = K_m\omega = 13.8 \text{ V}$$

If the semiconductor saturation voltage in the output stage of the driving amplifier is 0.2 V, the maximum steady-state armature current is

$$I_{ss} = 1 \text{ V}/3.04 \text{ }\Omega = 329 \text{ mA}$$

If the steady-state load is friction, its torque is

$$T_{ss} = I_{ss}K_m = 0.0434 \text{ N-m}$$

The total friction coefficient is

$$B = \frac{T_{ss}}{\omega} = 0.4141 \text{ mN-m-s}$$

Subtracting $B_m = 13.5$ μN-m-s leaves

$$\text{Max } B_L = 400.6 \text{ μN-m-s}$$

Table 2.1   Electro-Craft E-588-B-B motor-tachometer[a]

Parameter	Supplied	MKS Units
Inertia	0.0078 oz-in.-s^2	55×10^{-6} N-m-s^2
Damping	0.2 oz-in./krpm	13.49×10^{-6} N-m-s
Motor resistance	3.04 Ω	3.04 Ω
Motor inductance	8.41 mH	8.41 mH
Torque constant	18.66 oz-in./A	0.1318 N-m/A
Motor back emf	13.8 V/krpm	0.1318 V-s
Tachometer voltage	7 V/krpm	0.0668 V-s
Tachometer resistance	570 Ω	570 Ω
Tachometer inductance	33 mH	33 mH
Maximum stall torque	50 oz-in.	0.353 N-m
Maximum No-load speed	6 krpm	628.3 rad/s

[a] Tolerances on most of the above are 10–15%

Equation (2.76) tells a lot about the dynamic behavior of the motor and simple loads. The most important feature is the quadratic factor. Let us examine the relative importance of the motor properties in determining the properties of the quadratic. The resonant frequency is given by

$$\omega_0{}^2 = \frac{BR_a + K_m{}^2}{JL_a}$$

Let's suppose the load is purely friction with $B = 414$ μN-m-s as above and $J = J_m = 55$ μN-m-s^2:

$$BR_a = 0.00126 \qquad K_m{}^2 = 0.0174$$

Clearly, the motor constant dominates the loss terms in this case.

$$\omega_0 = 201 \text{ rad/s} \qquad f_0 = 32 \text{ Hz}$$

For the center term,

$$BL_a = 3.48 \times 10^{-6} \qquad JR_a = 167.2 \times 10^{-6}$$

Again, one term dominates the other. At resonance, the magnitude of the quadratic is 1.04. This shows that there will be no peak in the frequency response and no ringing in the transient response of this motor and load combination. Since both coefficients in the quadratic are nearly independent of B for smaller values, the motor's behavior will not change much for such load variations.   ∎

2.9.1   Heat

Many applications require the measure and control of temperature. Compared to mechanical and electromagnetic elements, heat flow is slow. Compared to electrical conduction, thermal conduction is not well confined because the differences in conductivity among materials are many orders of magnitude less than for electrical conductivity. For these reasons, thermal elements are not designed in as part of the controller but must be dealt with as part of the plant to be controlled.

Heat is mechanical energy on an atomic scale. It travels by diffusion in the same way an added liquid spreads through and mixes with the original liquid in a container. Temperature is a measure of the heat energy density in a material. Heat flows from a higher-temperature region to a lower-temperature region because a diffusion process occurs and all diffusion processes flow from higher to lower density to try to equalize densities. If heat power is flowing through a

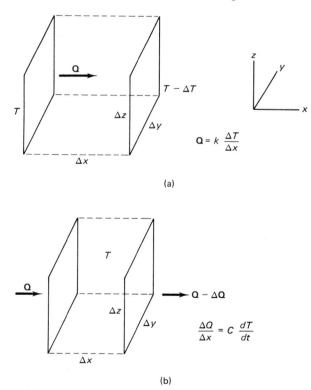

(a)

(b)

Figure 2.25   Basic heat flow. (a) Heat flow caused by a temperature difference. (b) Temperature rise and heat flow change.

differential volume and more flows in than out, the energy of that volume will rise at a rate equal to the difference. The temperature rise is related to the energy rise by a parameter called the *specific heat*. Figure 2.25 shows these two relations between heat flow and temperature. Q is the heat power density in W/m², C is the value heat capacity (specific heat times density) in J/(m³ − K), k is the thermal conductivity in W/(m − K), and T is temperature in K (kelvins). Both cells are differential volumes in which it is assumed that the interior variable is uniform through the cell and the exterior variables are uniform over their respective cell faces. In each case, one of possibly three directions of variation is shown. The general form for the cell in Fig. 2.25a is

$$\mathbf{Q} = k\mathbf{\nabla}T \tag{2.80}$$

and that for the cell in Fig. 2.25b is

$$\mathbf{\nabla}{\cdot}\mathbf{Q} = -C\frac{\partial T}{\partial t} \tag{2.81}$$

Substituting to eliminate $\mathbf{Q}$ yields the diffusion equation

$$\mathbf{\nabla}{\cdot}\mathbf{\nabla}T - \frac{C}{k}\frac{\partial T}{\partial t} = 0 \tag{2.82}$$

The main point of the previous paragraph is that thermal problems are not easily solved because they depend on the geometry of the elements involved. In special cases, a heat problem can be simplified from the differential form to a bulk difference form in the space variables. Figure 2.26 shows two of these cases. Figure 2.26a shows a slab of material, length l into the page, whose thickness d is small compared to the width w and length l. Further, if we have good reason to believe that the temperatures on the broad faces are uniform, we may assume that the heat flows straight from one broad face to the other with negligible leakage out the narrow edges. The temperature gradient is given by the difference in face temperatures divided by the slab thickness:

$$Q = \frac{k}{d}(T_U - T_L) \tag{2.83}$$

The total power H flowing through the slab is Q times the face area:

$$H = Qwl = \frac{k}{d}wl(T_U - T_L) \tag{2.84}$$

Straight-line flow may also be assumed if the body in question is long in the flow direction but has much higher conductivities than the surrounding medium.

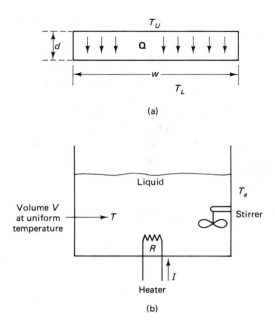

Figure 2.26   Simplified heat problems. (a) Thin slab. (b) Stirred liquid.

Figure 2.26b pictures a tank of liquid which is being heated and stirred. In this case, one of the design objectives was to keep the temperature uniform by forcing mixture at a much higher rate than diffusion would have given. By design then, T is the same everywhere in the tank. The input electric power goes to raise the temperature of the liquid and supply the loss to the ambient (surroundings), which is at temperature T_a, through the total thermal conductance G. This is expressed as

$$H = I^2R = CV\frac{dT}{dt} + G(T - T_a) \tag{2.85}$$

This equation can be rearranged as a state equation with T as the state variable and H and T_a as inputs. It also lends itself to the block diagram in Fig. 2.27. The assumption of uniform temperature can also be made if the material in question has a much higher conductivity than the surroundings and the diffusion time is much shorter than the period of the highest frequency component of the input heat power. Reciprocally, $k/C \gg f_{max}$.

Material properties, units and conversions, and heat transfer values for many practical situations are given in two of the books in the reading list for this chapter.

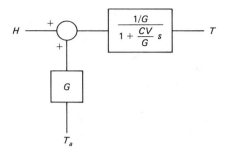

Figure 2.27 Block diagram for Fig. 2.26b.

2.9.2 Hydraulics

Control engineers are mainly concerned with hydraulics in two situations:

1. A system exists in which liquid transport must be controlled. In this case the system has been designed for some purpose in which the liquid is involved in the product.
2. A hydraulic system is designed to apply control force to a mechanical load, such as an aircraft flight control surface.

The principal physical variables in hydraulics are pressure and flow, either mass flow or volume flow. As you would expect, it takes a pressure difference to cause flow, elements exist which absorb power by requiring a significant pressure drop for flow, and there are storage elements in which liquid is accumulated as the pressure is raised. Unfortunately, the pressure-flow relations for common elements such as pipes, nozzles, orifices (holes), and valves of all sorts are nonlinear. Not only is the relation between pressure drop and flow nonlinear for a given valve, but it changes as a function of valve position. For process control, the engineer either has to work with small-signal linearization, work with large-signal approximations, or use a computer in the system, with an algorithm that deals with the nonlinearity directly.

Hydraulic actuators have a high power-to-weight ratio. This allows the designer to put the main power converter in a place where its weight is not so important and use a small, light mover at a point where added weight and bulk are less tolerable. The elements of a typical hydraulic control system are shown in Fig. 2.28. It is a mixture of electric and mechanical parts. The electric motor, pump, and accumulator usually supply more than one actuator, and the actuators may be pistons for linear motion or hydraulic motors for rotary motion. The pump and accumulator act together to maintain a nearly constant pressure to the system. At times of peak load, the accumulator supplies liquid with a small pressure drop, and at times of light load the pump refills the accumulator and brings the system pressure back to maximum. The piston is usually controlled by a spool valve,

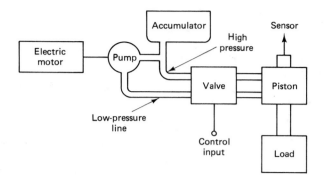

Figure 2.28   Hydraulic control system.

which may or may not be near the piston. The spool valve is in turn driven by a small electric solenoid. The valve-piston combination is a mechanical amplifier which works on the same conversion principle as an electronic amplifier. A small signal modulates the unidirectional power supply (a liquid under system pressure) to reproduce itself in the large power applied to the load. If the valve is placed next to, and in line with, the piston, an opportunity for mechanical feedback is created. A pictorial schematic of such a system appears in Fig. 2.29.

The valve and piston are shown in their neutral position. If the valve moves down, the high-pressure line will force liquid in below the piston, forcing the piston up. The piston will pass a force through the admittance Y_2 to reclose the valve. This is a negative feedback connection. Note that, for the reference directions shown, a positive X_1 will produce a negative flow F. This is the point where the negative feedback will enter the analysis. I will proceed on the following assumptions:

1. A linear relation can be found between the flow F, the valve displacement X_1, and the back pressure P_b caused by the load on the piston.
2. The valve load is insignificant to the network.
3. The network load is small compared to the output load.

The node equation for the network is

$$Y_1 X = (Y_1 + Y_2)X_1 - Y_2 X_2 \tag{2.86}$$

A linearized flow equation for the valve is

$$F = -K_v X_1 - K_b P_b \tag{2.87}$$

where K_v and K_b can be determined from exact equations for the valve or from

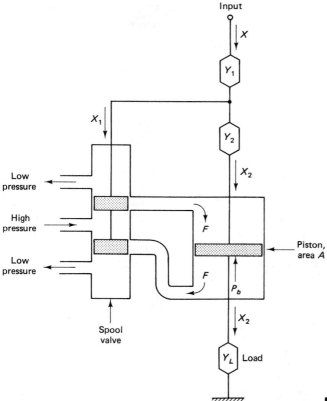

Figure 2.29 Hydraulic amplifier.

manufacturer's curves. The flow is volume flow and can occur only if the piston moves. It is equal to the piston area times the piston velocity:

$$F = sAX_2 \tag{2.88}$$

The back pressure times the piston area is the force needed to supply the output load:

$$AP_b = Y_L X_2 \qquad Z_L \ll Z_1 + Z_2 \tag{2.89}$$

You can combine the last three equations to get a relation between X_1 and X_2:

$$\left(s + \frac{K_b}{A^2} Y_L\right)X_2 = \frac{K_v}{A} X_1 \tag{2.90}$$

Their ratio is the forward gain of the amplifier. You can now combine this result with the node equation to find the overall transfer function:

$$\frac{X_2}{X} = \frac{-Y_1}{Y_2 + (Y_1 + Y_2)\dfrac{A}{K_v}\left(s + \dfrac{K_b}{A^2}Y_L\right)} \tag{2.91}$$

To be similar to an electronic operational amplifier, K_v/A should be large compared to 1 so that the transfer will reduce to

$$\frac{X_2}{X} = -\frac{Y_1}{Y_2} = -\frac{Z_2}{Z_1} \qquad \frac{K_v}{A} \gg 1 \tag{2.92}$$

This approximation can hold only for some finite range of frequency, since there is explicitly an s and Y_L probably increases with frequency. Let's examine a simple case to get an idea of what the frequency response might be.

Suppose each admittance is just a spring, $Y_1 = K_1$ and $Y_2 = K_2$. Suppose also that the load is inertia and friction, $Y_L = s(Js + B)$.

$$\frac{X_2}{X} = \frac{-K_1}{K_2\left[1 + \left(1 + \dfrac{K_1}{K_2}\right)\dfrac{A}{K_v}\left(1 + \dfrac{K_b}{A^2}B\right)s + \left(1 + \dfrac{K_1}{K_2}\right)\dfrac{JK_b}{K_vA}s^2\right]} \tag{2.93}$$

The effect of an admittance which goes to zero at zero frequency is to make the system look like an operational amplifier even without the above assumption on K_v/A. Another way of looking at it is that the piston integrates flow and the load integrates pressure, so there are two integrators in the system and integrators have infinite gain at zero frequency. However, like other real amplifiers, the frequency response does go down. The denominator is a quadratic whose parameters may be adjusted by the valve and piston choices or, if these have already been made, by choosing an appropriate mechanical feedback network. The importance of these matters will be made plain in Chap. 6.

2.10 TIME OR TRANSPORT DELAY

In some process control systems, we cannot measure certain quantities directly at the points where they are being affected. For example, if a second liquid is being added to a first in a pipe through which it is flowing, the mixture cannot be tested at the entry point because thorough mixing will not occur until some point further downstream. If we measure the concentration and use this value to control the inlet valve setting, the concentration response will be delayed with respect to the valve adjustment by the time it takes the liquid to move from the valve to the sensor. This is expressed in the time domain by writing

$$g(t) = f\left(t - \frac{d}{v}\right) \tag{2.94}$$

where $f(t)$ is the average concentration at the valve, d is the distance between the valve and the sensor, $g(t)$ is the concentration at the sensor, and v is the liquid's speed. The transformation to the frequency domain is

$$G(s) = F(s)e^{(-d/v)s} \tag{2.95}$$

For $s = j\omega$, the exponential gives a phase lag proportional to the frequency. If the delay is long enough, the excess phase shift can be enough to make a feedback system unstable. Pure time delay is not readily treated in a state variable model. Over a finite frequency range, the effect can be approximated by a cascade of all-pass networks. These have functions of the form

$$G(s) = F(s)\frac{(1 - \tau_1 s)\,(1 - \tau_2 s)\,\cdots}{(1 + \tau_1 s)\,(1 + \tau_2 s)\,\cdots} \tag{2.96}$$

in which as many pole-zero pairs as needed (as few as possible) are used to approximate the phase shift of the exponential. The transfer function of a Butterworth low-pass filter can also be used. You can choose the corner frequency and number of sections to get as good a match as desired. Either of these approximations can be transformed to a state-space representation by the methods of Chap. 10.

2.11   TRANSDUCERS, SENSORS, AND ACTUATORS

Generally, the word "transducer" refers to a device which changes power from one form to another. Loudspeakers and microphones convert between electrical and acoustical power. Motors and generators convert between electrical and mechanical power. These transducers can be used either as actuators (movers) or sensors. A sensor which takes a variable in one physical form and converts it to another is not necessarily a transducer. A potentiometer is a converter between a mechanical variable—shaft position—and an electrical variable—wiper voltage. It can be used as a sensor, but it is not a transducer because no mechanical energy is converted to electrical energy. The mechanical variable just controls the electrical one. In most transducers the power conversion can be made in either direction, although in many cases the design is optimized for only one, so the reverse conversion is at greatly reduced efficiency. Imagine trying to use a potentiometer as a motor!

A sensor is intended to be a device which gives a faithful measure of some variable in an easily processed form. To this end, it must not significantly load the system it is measuring. It should have a flat frequency response over a range at least an order of magnitude wider than the system it is sensing. This keeps it from degrading system performance and allows it to be modeled by a transfer function which is just a constant. It must be accurate and noiseless, because its errors and its noise are not corrected by feedback.

The main business of an actuator is to move the load, but it must be able to do so like the dc motor, over a wide range of speed and power. It need not be very accurate or extremely linear, because these failings will be corrected by feedback. The actuator's response speed is usually the limiting factor in the system's bandwidth. Whether choosing or designing an actuator, the cost will be an increasing function of both the maximum power and acceleration requirements, and this will be the major hardware cost in many control systems.

2.12   NONLINEARITY

Linearity is an approximation to reality. Luckily, the people of the eighteenth and nineteenth centuries invented a number of devices that are very useful and good approximations to linear. For the past several decades designers have worked to make all components of a system as linear as possible. This means doing things like reducing starting friction and gear lash, choosing linear magnetic material and keeping it below saturation, choosing materials whose properties compensate for each other, etc. As a control engineer, the main thing you must do about nonlinearity is guard against it in your choice of components. In some cases, the nature of the plant or process is nonlinear and must be dealt with. There are many methods for handling nonlinearities of various sorts which are left either to further courses or independent study. Mild nonlinearities can be approximated by considering that the system will operate in a steady state for a constant input level and finding the slope(s) for small variations about the operating point. An example of this perturbation technique was given in Sec. 2.7. Formally, this method requires expanding the nonlinear function in a multivariable Taylor series and keeping only the linear terms. If we have a function of n variables $f(x_1, x_2, \ldots, x_n)$ and the equilibrium point of the system is such that $x_1 = X_1, x_2 = X_2, \ldots, x_n = X_n$, let us define the deviation from the operating point as $\Delta x_i = x_i - X_i, i = 1$ to n. Then

$$\Delta f(x_1, x_2, \ldots, x_n) - f(X_1, X_2, \ldots, X_n) \cong$$

$$\left. \frac{\partial f}{\partial x_1} \right|_{\mathbf{x} = \mathbf{X}} \Delta x_1 + \left. \frac{\partial f}{\partial x_2} \right|_{\mathbf{x} = \mathbf{X}} \Delta x_2 + \cdots + \left. \frac{\partial f}{\partial x_n} \right|_{\mathbf{x} = \mathbf{X}} \Delta x_n \qquad (2.97)$$

$$\Delta f = (\nabla f) \left. \right|_{\mathbf{x} = \mathbf{X}} \Delta x$$

where

$$\mathbf{x} = (x_1, x_2, \ldots, x_n)$$
$$\mathbf{X} = (X_1, X_2, \ldots, X_n) \tag{2.98}$$

and

$$\Delta\mathbf{x} = \mathbf{x} - \mathbf{X} \tag{2.99}$$

This follows because the operating-point value of f is the first term of the series and all higher-order terms involve products of small differences which make them even smaller compared to the linear terms. The vector notation is compact and can be used directly in APL.

■ Example 2.4

Some spool valves have the following relation between flow, opening area, and pressure drop.

$$F = CA_o\sqrt{P} \tag{2.100}$$

in which F is the flow, C is a constant, A_o is the opening area, and P is the pressure drop across the valve. We really want the dependence on valve displacement, so we replace A_o with wx, where w is the port width and x is the linear displacement from closure. We wish to linearize this relation about the operating point $A_o = wx_o$ and $P = P_s$ the system pressure.

$$\Delta F = Cw\sqrt{P_s} \quad \Delta x + Cw\frac{x_o}{2}\frac{\Delta P}{\sqrt{P_s}} \tag{2.101}$$

The system pressure is not apt to change much during operation, but the valve opening certainly could. This means that the approximation is good only for small-signal analysis and is not likely to be of much use for large-signal performance.   ■

FURTHER READING

Cochin, Ira, *Analysis and Design of Dynamic Systems*, New York, Harper and Row, 1980.

Ellison, Gordon N., *Thermal Computations for Electronic Equipment*, New York, Van Nostrand Reinhold, 1984.

Gibson, John E., and Tuteur, Franz B., *Control System Components*, New York, McGraw-Hill, 1958.

PROBLEMS

Section 2.1

2.1 Represent the following equations by block diagrams. In each case, x_i are inputs and y_j are outputs.
(*a*) $y = a_1 x_1 + a_2 (dx_2/dt)$
(*b*) $y = a_1 x + a_2 \int_0^t x(\tau)\, d\tau$
(*c*) $y + b\dot{y} = ax + c$
(*d*) $y_1 = a_1 x_1 + a_3 (dx_2/dt)$ $\qquad y_2 = a_2 x_2 + a_3 (dx_1/dt)$

2.2 Draw frequency domain block diagrams for the equations in Prob. 2.1.

2.3 Draw frequency domain signal flow graphs for the equations in Prob. 2.1.

2.4 In the equation $y + \dot{y} = x$ one can write $y = x - \dot{y}$ or one can write $\dot{y} = x - y$ and $y = \int \dot{y}\, dt$. Each form has its own diagram. The first has only differentiators, and the second has only integrators. Draw block diagrams both ways for the following equations:
(*a*) $6\ddot{y} + 3\dot{y} + 20y = x$
(*b*) $32\dot{y}_1 + y_2 = 8x_1 - 4x_2$
$\quad 16\dot{y}_2 - 4y_1 = 24x_1 + 48x_2$
(*c*) $5\dot{y} + 4y = 3x + 2\dot{x}$

2.5 Reformulate Prob. 2.4 in the frequency domain. Draw signal flow graphs without $1/s^k$ terms for the differentiator-only versions, and without s^k terms for the integrator-only cases.

Section 2.2

2.6 Convert the block diagrams shown in Fig. P2.6 to single loops.

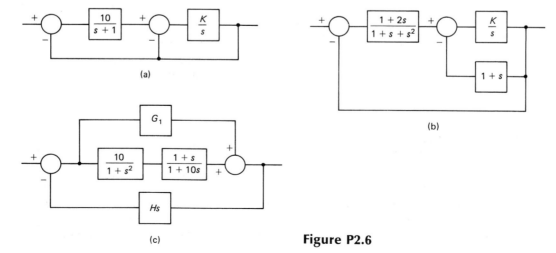

(a)

(b)

(c)

Figure P2.6

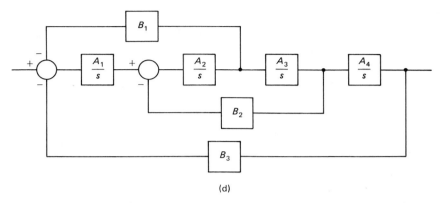

(d)

Figure P2.6   (*continued*)

Section 2.3

2.7   Convert the block diagrams in Fig. P2.6 to signal flow graphs. Find the input-output transfer function using Mason's gain rule.

2.8   Use Mason's gain rule to find the overall transfers in Fig. P2.8. For Fig. P2.8c is there a finite, nonzero value of a or b that gives zero transmission? Again for Fig. P2.8c, if $b = -1$, what values of a give infinite transmission?

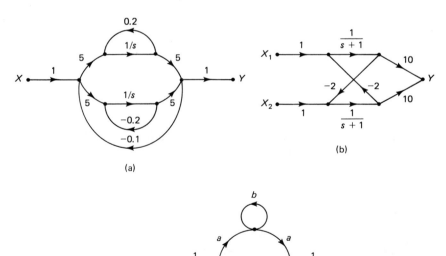

Figure P2.8

Section 2.4

2.9 Define and express the following system descriptions in vector-matrix state equations:

(a) $\dfrac{d^3y}{dt^3} + 3\dfrac{d^2y}{dt^2} + 6\dfrac{dy}{dt} + 3y = u(t)$

(b) $2\dfrac{d^4y}{dt^4} + 5\dfrac{d^2y}{dt^2} + 7y = 4u(t)$

(c) $3\dfrac{d^3y}{dt^3} + 18\dfrac{dy}{dt} = 51u(t)$

2.10 Find the Laplace transforms of the starting and final forms of the equations in Prob. 2.9. Arrange the initial conditions of the state variables in a separate vector.

2.11 The following systems have more than one output, so the state equation form will have an output vector. Transform the following descriptions into standard state equation form.

(a) $\begin{aligned} 16u(t) &= 10y_1 + 3\dfrac{dy_1}{dt} - 6y_2 - \dfrac{dy_2}{dt} \\ 0 &= -8y_1 + y_2 + 5\dfrac{dy_2}{dt} \end{aligned}$

(b) $\begin{aligned} 22u_1(t) &= 3\dot{y}_1 + 11y_1 - 9y_2 \\ 15u_2(t) &= -35y_1 + 140y_2 + 7\dot{y}_2 \end{aligned}$

(c) $\begin{aligned} \ddot{y}_2 + q(\dot{y}_2 - \dot{y}_1) + r(y_2 - y_1) &= 0 \\ p\ddot{y}_1 + q(\dot{y}_1 - \dot{y}_2) + r(y_1 - y_2) &= u(t) \end{aligned}$

2.12 Take the Laplace transforms of the given descriptions and your translations of Prob. 2.11. Put the state initial conditions in a separate vector.

Section 2.5

2.13 Define the energy storage variables in the circuits in Fig. P2.13 and use them to write state equations for each circuit.

2.14 Write the transfer functions for the circuits in Fig. P2.13. Check the dc and high-frequency values by inspecting the circuits with the capacitors blocking and shorting, respectively.

2.15 The circuit in Fig. P2.15 is designed to be an all-pass network. That is, its transfer function has constant magnitude at all frequencies, while the phase

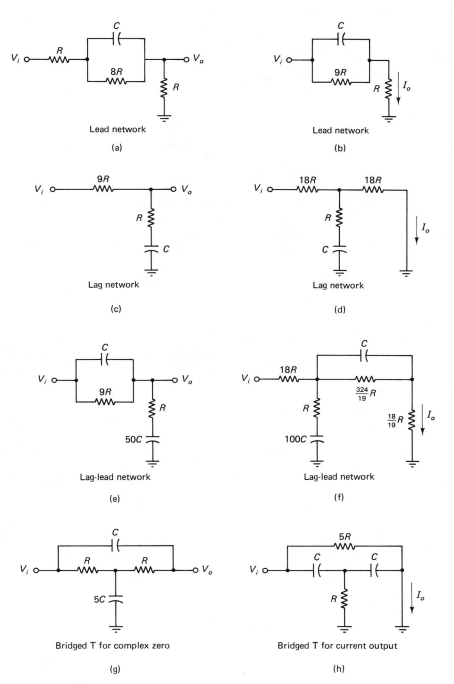

Lead network

(a)

Lead network

(b)

Lag network

(c)

Lag network

(d)

Lag-lead network

(e)

Lag-lead network

(f)

Bridged T for complex zero

(g)

Bridged T for current output

(h)

Figure P2.13 A group of circuits useful in compensator design.

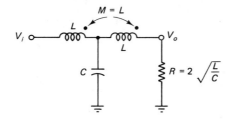

Figure P2.15 An all-pass network.

lag increases with frequency. Such networks are useful for approximating time delay or for phase correction over limited frequency bands.

(*a*) Write a set of state equations.

(*b*) Find the transfer function. Check your result by inspecting the circuit at high and low frequencies. Is the magnitude constant?

(*c*) Propose a network using Rs, Cs, and op amps to achieve an all-pass function.

Section 2.6

2.16 (*a*) Write state equations for the networks in Fig. P2.16. (*b*) Draw schematics and find the transfer functions. Check the low- and high-frequency values by physical inspection.

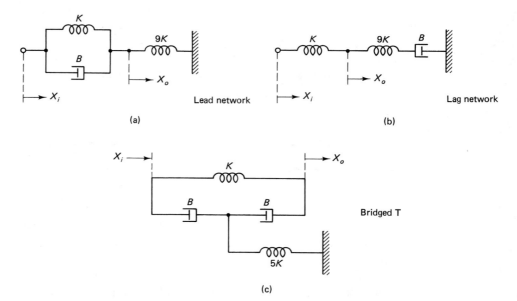

Figure P2.16 Some mechanical networks useful for compensator design.

2.17 Design a mechanical network which has the same form for its transfer function as the lag-lead electrical circuit in Fig. P2.13e.

2.18 For Fig. 2.14e, $M = 2Mg$. Find K and B so that the real part of the quadratic in $X_2(s)/X_1(s)$ is zero at $s = j2\pi/5$ and $|X_2/X_1| = 1/\sqrt{2}$ at this frequency.

2.19 Find the transfer function of the network in Fig. P2.19. Verify by physical inspection the values at $s = 0$ and infinity.

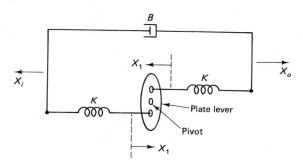

Figure P2.19   A mechanical all-pass network.

2.20 Find $\theta_i(s)/T_i(s)$ and $\theta_o(s)/T_i(s)$ for the system in Fig. P2.20. If $J_L = 10J_4$, $B_L = 0$, $N_1 = N_3$, $N_2 = N_4$, and the gears are all the same thickness, what value of N_1/N_2 will make the reflected load and gear inertias equal?

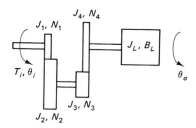

Figure P2.20   A two-stage gear train with load.

Section 2.8

2.21 Suppose that the motor in Example 2.3 is loaded with $B_L = 0$ and $J_L = 1.045$ mN-m-s^2. Find the factors of (2.76).

2.22 Suppose a dc servomotor is driven by a current source. Find $V_a(s)/I_a(s)$ and $\theta_m(s)/I_a(s)$ for $Y_L = J_L s^2 + B_L s$.

Section 2.9.1

2.23 A power transistor can be used as a current-controlled heater by operating it from a constant-voltage supply. Such a system is illustrated in Fig. P2.23. The main heat source is the collector current–supply voltage product, developed mostly across the collector-base junction. The heat Q_J causes the junction temperature T_J to rise so that heat flows through the thermal resistance R_J to the plate at temperature T_P. The heat spreads quickly throughout the plate, raising its temperature by way of thermal capacitance C_P, which causes some heat loss through the surface thermal resistance R_A to the air at temperature T_A. Draw time and frequency domain block diagrams. Write a state equation for the system. Find the transfer functions from I_b to T_J and T_P, and from T_A to T_J and T_P.

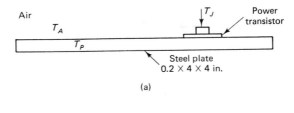

(a)

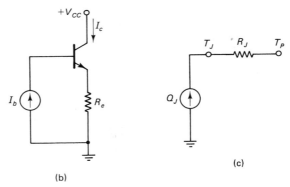

(b)

(c)

Figure P2.23   (a) Plate with heater. (b) Current-controlled heater schematic. (c) Thermal circuit for the transistor.

Section 2.10

2.24 Suppose that a system has a time delay of 0.1 s. Model the time delay with an all-pass stage having the transfer function

$$A(s) = \frac{1 - s/a}{1 + s/a}$$

If the frequency range of interest is 1 to 10 rad/s, what is the best choice for
a? Choose one of the following error criteria:

(*a*) Minimum average error magnitude

$$\min_{a} \ \text{avg}_{\omega} \ | \ \omega\tau \ + \ \angle A(j\omega) \ |$$

(*b*) Minimum integral squared error

$$\min_{a} \int_{1}^{10} [\omega\tau \ + \ \angle A(j\omega)]^2 \ d\omega$$

(*c*) Minimum maximum error

$$\min_{a} \ \max_{\omega} \ | \ \omega\tau \ + \ \angle A(j\omega) \ |$$

You may find it useful to write a computer program to evaluate your criterion
as a function of *a*.

2.25 If the delay in a system is 1 s and the frequency range of interest is 1 to 10
rad/s, what is the minimum number of all-pass sections needed to approximate the delay?

Section 2.12

2.26 A problem to challenge your ideas on basic mechanics involves the cable
and reel illustrated in Fig. P2.26. An applied force on the cable causes it to
feed off the reel at a speed $v(t)$. As the cable feeds off, the reel loses mass
as shown, and consequently the inertia and radius decrease. Assume a viscous drag also acts on the reel. It is the only linear quantity in this system.
Write the equation(s) of motion for this system, treating force as the input
and velocity as the output. Linearize the equations about an operating point
and obtain a transfer function from force to velocity. As a start, you might
take the view that torque is equal to the rate of change in angular momentum.

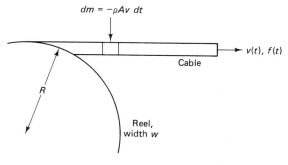

Figure P2.26 Cable and reel problem.

2.27 A portion of a flyball governor system is shown in Fig. P2.27. It is built
around two collars, the lower one of which is pinned to the rotating engine
shaft. The flyballs are attached to arms with a 90° bend hinged to the lower
collar. These arms are linked to the upper collar by short, pivoting arms.
As the shaft spins, the flyballs swing out, lifting the upper collar. The inputs
to this system are the shaft speed and a load force F_L on the upper collar,
and the output is the displacement of the upper collar x. Ignore the flyball
dynamics, assuming that they are in equilibrium between the centrifugal
force, gravity acting on the balls, and the load force. Write the equations
that describe this equilibrium. Linearize them to get an input-output relationship. If the speed range is such that the arm angle varies between 10 and
60°, $l_1/l_2 = 6$, and $a = l_2$, what is the ratio of the transfer coefficients over
this range?

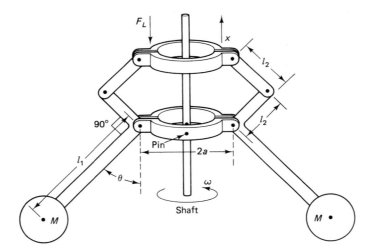

Figure P2.27   Portion of a flyball governor system.

Chapter 3

FREQUENCY RESPONSE ANALYSIS

3.1   OPEN- AND CLOSED-LOOP POLES AND ZEROS

In general, a *transfer function* is a ratio of polynomials in the complex frequency variable s. Furthermore, all the input signals of interest can be expressed also as ratios of polynomials in s. In both cases, these polynomials can be expressed as products of factors. I will use these facts to present a summary of the effects of factors that you've seen before in circuits courses. Suppose we have a system whose closed-loop transfer function is

$$H(s) = K \frac{\prod\limits_{i=1}^{m} (s - z_i)}{\prod\limits_{i=1}^{n} (s - p_i)} \tag{3.1}$$

with an input function

$$R(s) = R_0 \frac{\prod\limits_{k=1}^{q} (s - w_k)}{\prod\limits_{k=1}^{u} (s - v_k)} \tag{3.2}$$

The numerator factors are called *zeros* because they force the function to be zero when $s = z_i$ or w_k. The value of s which causes the function to be zero is also called a zero. The denominator factors are called *poles* because they force the function to infinity when s has a value that makes one of these factors zero. The corresponding value of s is also called a pole. The system output is

$$C(s) = H(s)R(s)$$

$$= KR_0 \frac{\prod_{i,k} (s - z_i)(s - w_k)}{\prod_{i,k} (s - p_i)(s - v_k)} \tag{3.3}$$

To find to the time response, one converts from a product form to a sum form using partial-fraction expansion. I have ignored repeated roots for simplicity. This gives

$$C(s) = \sum_{j,r} \frac{A_j}{s - p_j} + \frac{B_r}{s - v_r} \tag{3.4}$$

where

$$A_j = KR_0 \left. \frac{\prod_{i,k} (s - z_i)(s - w_k)}{\prod_{i \neq j,k} (s - p_i)(s - v_k)} \right|_{s = p_j} \tag{3.5}$$

and

$$B_r = KR_0 \left. \frac{\prod_{i,k} (s - z_i)(s - w_k)}{\prod_{i,k \neq r} (s - p_i)(s - v_k)} \right|_{s = v_r} \tag{3.6}$$

The output is now a sum of pole terms, some of which come from the transfer function and some of which come from the input function. Notice that each transfer function pole amplitude contains the input function evaluated at that pole, and that each input function pole amplitude has the transfer function evaluated at that pole. This is expressed as

$$A_j = K \left. \frac{\prod_{i} (s - z_i)}{\prod_{i \neq j} (s - p_i)} \right|_{s = p_j} R(p_j) \tag{3.7}$$

$$B_r = H(v_r)R_0 \left. \frac{\prod_{k} (s - w_k)}{\prod_{k \neq r} (s - v_k)} \right|_{s = v_k} \tag{3.8}$$

With inverse transforming, the time response is

$$c(t) = \sum_j A_j e^{p_j t} + \sum_r B_r e^{v_r t} \tag{3.9}$$

If I had included repeated roots, there would be a general sum of powers of t multiplying the exponential terms.

The last three equations are generalizations of circuit properties. The time response has a "natural" component due to the poles of the transfer function and a "forced" response due to the poles of the input function. These words are in quotes because both responses are excited by the input function, as I have assumed zero initial conditions. A system is stable and useful if the output is approximately a copy of the input. This can happen only if the frequency content of the output contains, after some initial time period, only those frequencies that are in the input signal. This requires that the exponentials in the natural response decay. For this to happen, p_j, which are either real or complex conjugates, must have negative real parts. This is the basis of all stability tests given in this book.

The amplitude of each term in the natural response depends on where the other poles and zeroes of the transfer function are with respect to the pole whose response is the term in question, and the amount of input signal at the pole frequency. Thus, if a zero is close to the pole, the amplitude will be small, while if another pole is close, the amplitude will be large. The same sort of thing is true for the forced response; the amplitude depends on the input amplitude at that frequency, and the transfer function (frequency response) value at that pole frequency. In studying stability and degree of stability, we are interested in the natural response. In studying performance, we are interested in the forced response as well.

The discussion in the preceding paragraph concerning closeness of zeros and poles leads to the idea of a pictorial display of these values in the s plane. The function

$$H(s) = 10 \frac{1 + s/10}{[(1 + s/(1 + j)][1 + s/(1 - j)]} \tag{3.10}$$

has a zero at -10, a pole at $-1 - j$, and a pole at $-1 + j$. This is displayed in Fig. 3.1. By consensus, a zero is always shown by a circle and a pole by an x. In this case, you can see that the poles are much closer to each other than the zero is to either of them. In fact, for any value of s, a factor $s - a$ has the magnitude of the line between s and a. (Check this by drawing phasors for s and a if you haven't seen that done before.) Since the poles must have negative real parts for stability, the $j\omega$ axis is an important boundary in the s plane.

From Chaps. 1 and 2, you have seen that it is easier to get an open-loop description of a system than a closed-loop one. Also, for design we work with the open-loop description because added elements directly affect this description.

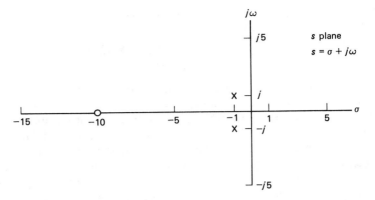

Figure 3.1   Pole-zero map.

What are the relations between open- and closed-loop transfer functions? They are not obvious. Consider first the very simple system shown in Fig. 3.2. The open-loop transfer function is

$$\frac{C(s)}{E(s)} = \frac{K}{1 + s} \tag{3.11}$$

The closed-loop transfer function is

$$\frac{C(s)}{R(s)} = \frac{K}{K + 1 + s} = \frac{K}{K + 1}\frac{1}{1 + s/(K + 1)} \tag{3.12}$$

The pole for the open-loop function is fixed at -1, but the pole for the closed-loop function depends on the gain and is at $-(K + 1)$. The maps are shown in Fig. 3.3. Next, look at the system in Fig. 3.4. It has an integrator and a real zero in the forward path. The open-loop transfer function is

$$\frac{C(s)}{E(s)} = \frac{1 + s/a}{s} \tag{3.13}$$

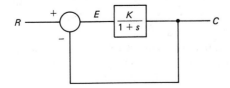

Figure 3.2   A simple feedback for Section 3.1.

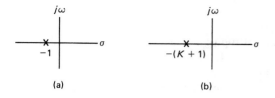

(a)   (b)

Figure 3.3 Pole-zero maps for the system of figure 3.2. K > 0. (a) Open loop. (b) Closed-loop.

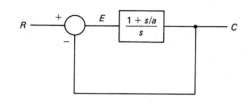

Figure 3.4 Another simple system.

and the closed-loop transfer function is

$$\frac{C(s)}{R(s)} = \frac{1 + s/a}{1 + s/a + s} = \frac{1 + s/a}{1 + ((a + 1)/a)s} \tag{3.14}$$

The maps are shown in Fig. 3.5. This time I've fixed the gain and made the open-loop zero variable. The zero is reproduced in the closed-loop function, and the pole ranges between 0 and −1 for a > 0. What's the range of the pole value in Fig. 3.3b? These two examples show you, I hope, both the complexity of the relation between open- and closed-loop transfer functions and that it is possible to go from one to the other. For systems of moderate complexity, the root locus method of the next chapter allows mapping from the s-plane zeros and poles of the open-loop function to the s-plane poles of the closed-loop function. For systems of immoderate complexity, the methods of this chapter allow us to deduce which half of the s plane the closed-loop poles are in and something about their magnitude and their effects on transient response.

(a)   (b)

Figure 3.5 Pole-zero maps for Figure 3.4, a > 1 (a) Open loop. (b) Closed loop.

3.2.1 Frequency Response Data

As mentioned in Chap. 1, the designers of high-gain feedback amplifiers found a way to determine stability from open-loop frequency response measurements. To use the method, they displayed the measurements in a special way. The measured frequency response is the value of the transfer function for $s = j$, and it is a complex number which can be expressed either as magnitude and phase or as real and imaginary parts. Ordinarily, one would display the data as a function of frequency, in two plots. However, for the Nyquist plot one uses the complex function plane and displays each data point without explicit reference to its frequency. For example, if $K = 10$ in (3.11), then a measurement of frequency response from 0.1 to 10 rad/s will give the data in Table 3.1.

Table 3.1 Frequency response for $10/(1 + s)$

Radians/second	Real Part	Imaginary Part
0.1	9.9	−0.99
0.2	9.6	−1.92
0.5	8	−4
1	5	−5
2	2	−4
5	0.38	−1.92
10	0.099	−0.99

The plot in the function plane is shown in Fig. 3.6. Although you can and should produce this plot with a software package, you should also check those values that are easy to do by hand. These are usually at zero and infinite frequency and at each frequency where a factor has equal real and imaginary parts. In this case, the values at zero and infinity are readily seen to be 10 and 0. At $s = j$, the denominator has the polar value $\sqrt{2}\angle 45°$, so the function has the polar value $5\sqrt{2}\angle -45° = 5 - j5$. These are exact values, and more complicated functions can be approximated fairly readily by choosing frequencies in the same way.

Systems with integrators in the forward path are of special interest because an integrator has infinite gain at zero frequency, which drives the static error to zero. It is useful to know some special approximations for the low-frequency behavior of the Nyquist plots for such systems. Consider the system shown in Fig. 3.7. The loop gain function is

$$G(s)H(s) = \frac{K(1 + s/b_1)(1 + s/b_2)}{s(1 + s/a_1)(1 + s/a_2)} \tag{3.15}$$

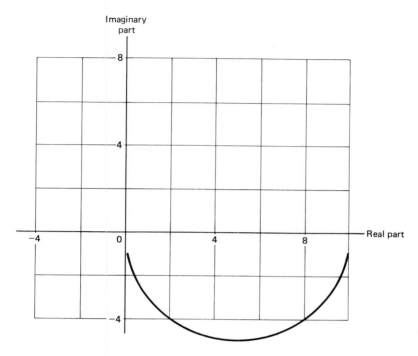

Figure 3.6 Nyquist plot for $10/(1 + s)$.

For very low frequencies, the denominator terms can be moved to the numerator by using the approximation

$$\frac{1}{1 + j\omega/p} \approx 1 - \frac{j\omega}{p} \qquad |\omega| \ll |p| \tag{3.16}$$

$$GH(j\omega) \approx \frac{K}{j\omega}\left(1 + \frac{j\omega}{b_1}\right)\left(1 + \frac{j\omega}{b_2}\right)\left(1 - \frac{j\omega}{a_1}\right)\left(1 - \frac{j\omega}{a_2}\right) \tag{3.17}$$

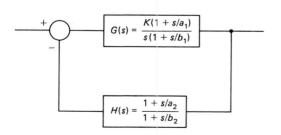

Figure 3.7 A single-integrator system.

Each factor is now 1 plus a small number. When they are multiplied together, they will give 1 plus linear terms plus even smaller product terms, from which only the 1 and the linear terms need be kept. This yields

$$GH(j\omega) \approx \frac{K}{j\omega} \left[1 + j\omega \left(\frac{1}{b_1} + \frac{1}{b_2} - \frac{1}{a_1} - \frac{1}{a_2} \right) \right]$$

$$\approx \frac{K}{j\omega} + K \left(\frac{1}{b_1} + \frac{1}{b_2} - \frac{1}{a_1} - \frac{1}{a_2} \right) \qquad |\omega| \ll |a_i|, |b_j| \qquad (3.18)$$

We now have an imaginary part which is very large, which you probably expected for small ω, and a constant real part. This constant real part is an asymptote for low frequency, and a boundary, if all the factors are real zeros and poles, for the plot for all $|\omega| > 0$.

The result generalizes as follows. For a loop transfer function given by

$$GH(s) = \frac{K \prod\limits_{i} (1 + s/b_i)}{s \prod\limits_{k} (1 + s/a_k)} \qquad (3.19)$$

the denominator terms are brought to the numerator to form

$$GH(j\omega) \approx \frac{K}{j\omega} \prod\limits_{i,k} \left(1 + \frac{j\omega}{b_i} \right) \left(1 - \frac{j\omega}{a_k} \right) \qquad (3.20)$$

This multiplies out to

$$GH(j\omega) \approx \frac{K}{j\omega} + K \left(\sum\limits_{i} \frac{1}{b_i} - \sum\limits_{k} \frac{1}{a_k} \right) \qquad |\omega| \ll |a_k|, |b_i| \qquad (3.21)$$

Some examples are given in Figs. 3.8–3.14.

■ Example 3.1

For Fig. 3.8, the function has its low-frequency asymptote at $10(-1/10) = -1$, and at infinite frequency the function goes to zero at an angle of $-180°$. At $s = j10$, the function is $10/[j10(1 + j)] = 0.707\angle -135° = -0.5 - j0.5$. You can see these three conditions on the plot. ■

■ Example 3.2

In Fig. 3.9, a zero factor at -2 has been added to the system. This shifts the asymptote to $10(0.5 - 0.1) = 4$. At infinity, the function still goes to zero because there are more poles than zeros, but the phase becomes $-90°$ because

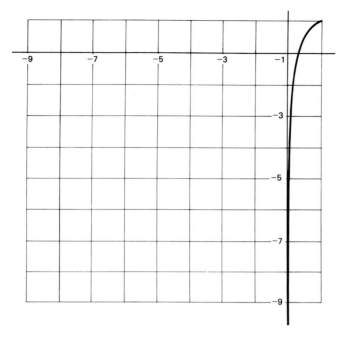

Figure 3.8 Nyquist plot for $10/[s(1 + s/10)]$ from 1 to 100 rad/s.

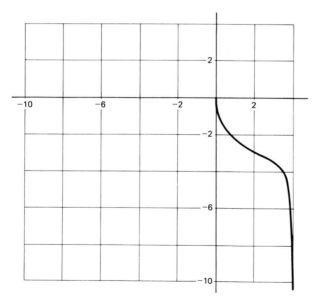

Figure 3.9 Nyquist plot for $10(1 + s/2)/[s(1 + s/10)]$ from 1 to 100 rad/s.

the net number of poles is 1. To get easy estimates at the intermediate frequencies, I'm going to use some rather crude approximations. At $s = j2$, I will treat the pole as about equal to 1. Then the function is $10(1 + j)/j2 = 7.07\angle -45° = 5 - j5$. While this is not close enough for design, it's close enough for a rough check. At $s = j10$, I will treat the zero as approximately its larger part, $j5$. The function is approximately $10j5/[j10(1 + j)] = 3.54\angle -45° = 2.5 - j2.5$. Again, close enough for a rough check, and I didn't even need a calculator.   ■

■ Example 3.3

The system in Fig. 3.10 is like that in Fig. 3.8 with an added resonant pole pair. To figure the low-frequency asymptote, one can ignore the s^2 in the quadratic, since it produces only the same kind of second-order terms we threw out in deriving the asymptote. Thus, the asymptote is at $10(-0.8 - 0.1) = -9$. At infinite frequency, the function goes to zero with a phase of $-360°$, since there is a net of four poles. The interesting thing about the quadratic is that its small value at resonance causes an increase in magnitude that overcomes the general decrease in magnitude with increasing frequency. Real factors, whether poles or zeros, don't do this. The resonant frequency is at $s = j/\sqrt{0.8}$. At this frequency, the function is about $10/[j1.1j0.9) = -10$. Right on! At $s = j10$, the quadratic will

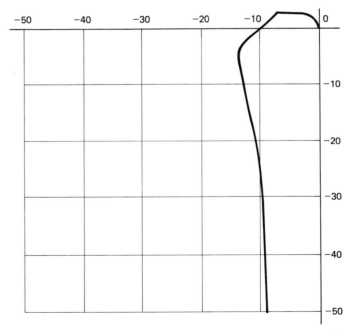

Figure 3.10   Nyquist plot for $10/[s(1 + 0.8s + 0.8s2)(1 + s/10)]$ from 0.2 to 100 rad/s.

force too small a magnitude to be seen on the scale of this plot. It is $10/[j10(-80)(1 + j)] = 0.009\angle -315°$. ∎

■ **Example 3.4**

Figure 3.11 shows the effects of adding a zero at -2 to the system in Fig. 3.10. Since the zero is so close to the resonance, its phase lead shifts the curve down into the third quadrant. When the curve crosses the negative real axis, it does so at a higher frequency and with a consequently smaller magnitude. Try putting the zero at -1.1 in your software package. ∎

■ **Example 3.5**

Figures 3.12–3.14 concern a system with two real poles in addition to the integrator. For Fig. 3.12, the asymptote is $10(-1 - 0.1) = -11$. Since there are no zeros, as the frequency goes to infinity, the magnitude goes to zero at an angle of $-270°$. The scale of the plot does not allow the angle to be seen. At $s = j$, the function is approximately $10/[j(1 + j)] = 7.07\angle -135° = -5 - j5$. Pretty close. At $s = j10$, the function is about $10/[j10j10(1 + j)] = 0.0707\angle -225° = -0.05 + j0.05$. This is too small for this scale. To get a better look at the

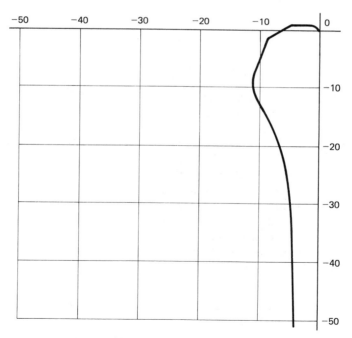

Figure 3.11 Nyquist plot for $10(1 + s/2)/[s(1 + 0.8s + 0.8s2)(1 + s/10)]$ from 0.2 to 100 rad/s.

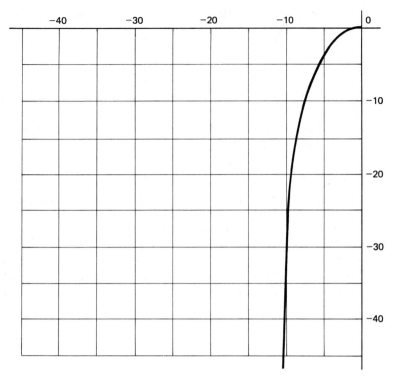

Figure 3.12 Nyquist plot for $10/[s(1 + s)(1 + s/10)]$ from 0.2 to 100 rad/s.

function's behavior around the origin, one must restrict the low-frequency start of the plot. Figure 3.13 does this, and you can see that it does go into the second quadrant and that the last point calculated is also quite close to the curve. In Fig. 3.14, a zero is put in at -3, close to the geometric mean of the nonzero poles. The asymptote becomes $10(0.33 - 1 - 0.1) = -7.7$. The phase at high frequency goes to $-180°$ now. At $s = j$, the approximate value doesn't change because the zero, being at a higher frequency, is treated as 1. It misses the curve by quite a bit more though, because 1 is a crude approximation for $1 + j/3$. At $s = j3$, the function is about $10(1 + j)/(j3j3) = 1.56\angle -135° = -1.1 - j1.1$. This is quite close to the curve, the symmetrical choice of frequency producing symmetrical errors. Again, the overall effect of the zero is to raise the frequency at which a given phase is reached. This effect allows the poles to reduce the magnitude at that phase angle. ■

Suppose we are dealing with a two-integrator system now. A general expression for such a system is

$$GH(s) = \frac{K \prod_i (1 + s/a_i)}{s^2 \prod_k (1 + s/b_k)} \tag{3.22}$$

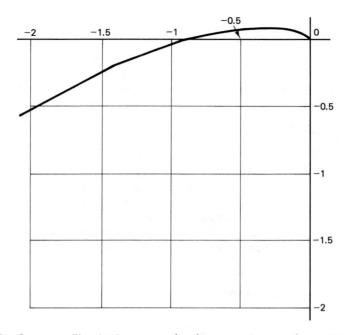

Figure 3.13 Same as Fig. 3.12 except the frequencies are from 2 to 200 rad/s.

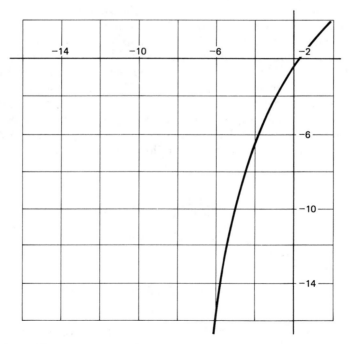

Figure 3.14 Nyquist plot for $10(1 + s/3)/[s(1 + s)(1 + s/10)]$ from 0.5 to 50 rad/s.

Immediately you can see that the phase will start at $-180°$ at low frequency and that the magnitude will drop even faster than for a corresponding single-integrator system. A low-frequency approximation can be found in the same way as (3.21):

$$GH(j\omega) \approx -\frac{K}{\omega^2} - \frac{jK}{\omega}\left[\sum_i \frac{1}{a_i} - \sum_k \frac{1}{b_k}\right] \qquad |\omega| \ll |a_i|, |b_k| \quad (3.23)$$

Now both the real and imaginary parts are inverse powers of frequency, with the real part being negative and much larger than the imaginary part. The bracketed value is mainly important because it determines the sign of the imaginary part and thereby which quadrant the curve lies in. If this value is positive, the plot will lie in the third quadrant, and if it's negative, the plot will lie in the second quadrant. Try adding another origin pole to the systems plotted in Figs. 3.8–3.14 in your software package.

3.2.2 The Conceptual Part

In order to detect poles in the right half s plane of the closed-loop transfer function, Nyquist showed that one must plot the open-loop transfer function for a set of s values that enclose the RHP. In effect, one must treat the transfer function as a mapping or transformation of a contour from the s plane to the function plane. The standard contour in the s plane is shown in Fig. 3.15. It is made up of several parts. We have dealt with only one, the measurable part, in the last subsection.

The measurable part is the $+j\omega$ axis. If there are poles on this axis, they must be excluded by taking semicircular detours around them, as shown for the pole at the origin in the figure. The values of the transfer function for s on the $-j\omega$ axis are necessarily conjugates of the measurable values. Therefore the plot for these two parts of the contour has even symmetry about the real axis. For all transfer functions representing real systems, which must be low pass, the large semicircle maps into the origin. When there are poles at the origin, requiring the tiny detour, their mapping is as follows. For a function expressed as

$$GH(s) = \frac{K \prod_i (1 + s/a_i)}{s^P \prod_k (1 + s/b_k)} \qquad (3.24)$$

it will become

$$GH(\rho e^{j\theta}) \rightarrow \frac{K}{\rho^P} e^{-jp\theta} \qquad (3.25)$$

when s is on the detour. The tiny radius is converted by the poles into a very large one, and the traverse of the small semicircle is mapped into a traverse of p

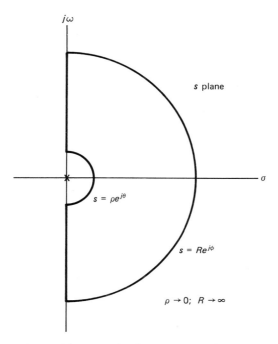

Figure 3.15   The standard Nyquist s-plane contour.

semicircles in the function plane. Finally, note the sign change. This causes a change in the direction of travel in mapping from the s plane to the function plane. If one starts at $\theta = -90°$, the start in the function plane will be at an angle $+p90°$. As θ progresses to 0 through decreasing negative values, the function angle progresses through decreasing positive values, also to 0. The rest of the semicircle is the conjugate of this quarter-circle, so the function plane plot will again have even symmetry about the real axis. Figure 3.16 shows the mapping for $p = 1, 2$, and 3.

3.2.3   Putting the Parts Together

Scale is a major problem with the complete Nyquist plot. In practice, as you will learn in the next section, we are interested in numerical detail only when the function is in the neighborhood of -1. Otherwise, only the general shape is important. A reasonable procedure is to make two plots. The first one is for the measurable frequencies to find both numerical detail and the shape over the range of frequencies covering the nonzero poles and zeros. The second is a sketch using the shape of the first one and the appropriate low-frequency closure. I will work through a few cases for you.

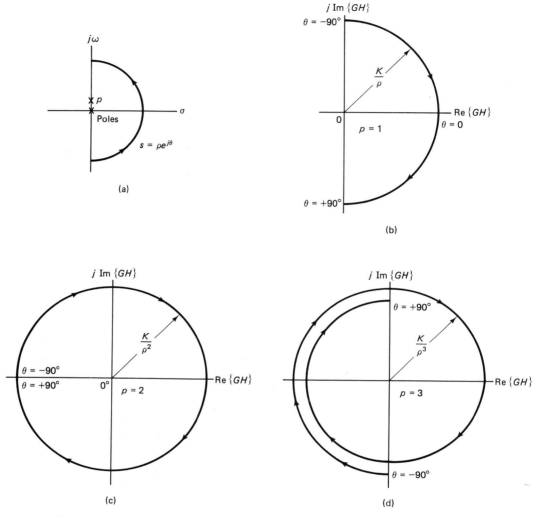

Figure 3.16 Mappings of the detour around poles at the *s*-plane origin. (a) The detour. (b) One pole. (c) Two poles. (d) Three poles.

■ Example 3.6

The transfer function whose data are shown in Fig. 3.6 has no origin pole, so the closure is simply a point. The complete plot is the full circle formed by conjugating the measurable data. This is shown in Fig. 3.17a. ■

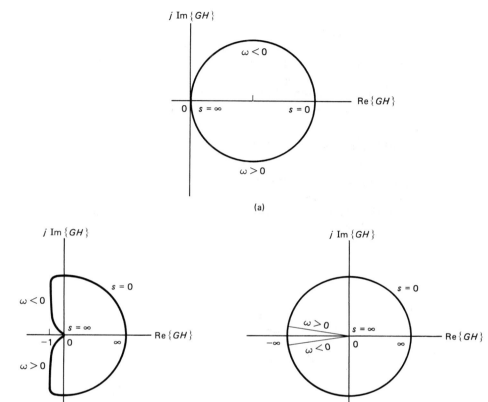

Figure 3.17 Nyquist sketches. (a) A real pole, plus one integrator (b), plus another integrator (c).

■ Example 3.7

The measurable data for the same system with an added integrator are shown in Fig. 3.8. Again the plot is conjugated to get the mapping from the $-j\omega$ axis, and the appropriate closure is an infinite semicircle from Fig. 3.17b. The result is sketched in Fig. 3.18b.   ■

■ Example 3.8

Adding a second integrator to the system will give a measurable plot in the second quadrant. I can say this without actually doing the calculation, because I can see that the imaginary part of the low-frequency approximation (3.23) is positive. From symmetry, the $-j\omega$ axis will map into the third quadrant, and the low-frequency closure is a full circle as in Fig. 3.16c. The complete sketch is shown in Fig. 3.17c. Notice that, in this case, the circle is slightly more than full—it crosses itself. This is because the closure always starts at the low-frequency end of the $-j\omega$ axis mapping, goes around, and finishes at the low-frequency end of the $+j\omega$ axis mapping. Since the start is in the third quadrant and the traverse is clockwise, it must cross itself at the end to finish in the second quadrant. ■

3.3   THE NYQUIST STABILITY CRITERION

Suppose we have a function which is a simple zero, $f(s) = s + z$. How does it map a contour from the s plane to the f plane? Evidently, it will just be pushed over by the distance z. The important feature of the mapping is not so obvious though, and that is the matter of how the angle of $f(s)$ changes for a complete

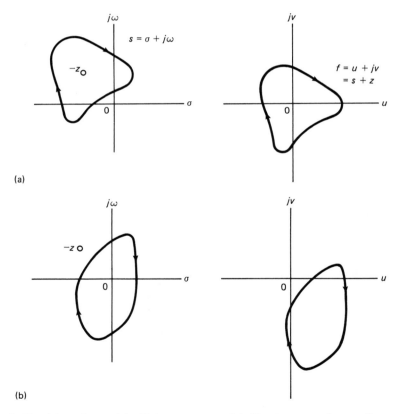

Figure 3.18 Mapping with $f(s) = s + z$. (a) Contour encloses the zero. (b) Contour excludes the zero.

trip around the contour. There are two possibilities, either the s-plane contour encloses $-z$ or it doesn't. Two such contours and their mappings into the f plane are shown in Fig. 3.18. In Fig. 3.18a, the s-plane contour encloses the zero. The function is the phasor from the zero to the contour, and the phasor makes a complete 360° rotation in one trip around the path. Therefore the origin is enclosed in the function plane. In Fig. 3.18b, on the other hand, the zero is outside the contour, so the maximum angle change is limited to less than 360°, and the origin in the function plane is not enclosed. Note also that the direction of travel is the same in the two planes.

Next, consider a simple pole, $f(s) = 1/(s + p)$ as shown in Fig. 3.19. This time the mapping is the reciprocal of the phasor in the s plane from the pole to the contour. Thus small magnitudes map into large ones, and angles have their signs changed. In Fig. 3.19a, the pole is enclosed, so the phasor makes a full rotation and so does the function's angle, but it does so in the opposite direction. Again, if the pole is not enclosed, the origin in the function plane is not enclosed.

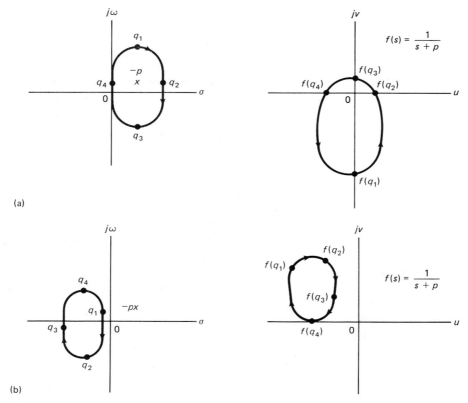

Figure 3.19 Mapping with a pole. (a) Contour encloses pole. (b) Contour excludes pole.

We can safely generalize these observations for polynomial ratio functions, since they are made up of products of pole and zero functions of the types we have just looked at. The angle of such a product of functions is the sum of their individual angles, and any contour either includes or excludes the zero or pole of each factor. One can tell the difference between the number of zeros and the number of poles inside an s-plane contour by counting the number and direction of enclosures of the origin of the mapping of that contour into the function plane. It appears that we might have a tool for detecting s-plane poles of the closed-loop transfer function. To use it, we need to look again at the relations between the open- and closed-loop functions.

Let the forward-path transfer function be

$$G(s) = \frac{N_1(s)}{D_1(s)} \tag{3.26}$$

and the feedback transfer function be

$$H(s) = \frac{N_2(s)}{D_2(s)} \tag{3.27}$$

where the N_i are the numerator polynomials and the D_i are the denominator polynomials. The closed-loop transfer function can be written as

$$T(s) = \frac{G}{1 + GH} = \frac{N_1 D_2}{D_1 D_2 + N_1 N_2} \tag{3.28}$$

We can see that the closed-loop zeros are the forward-path zeros and the feedback path poles, so if we know these functions, we know the closed-loop zeros. The poles of $T(s)$ are the zeros of $1 + GH$, which we don't know and are trying to find out about. From the preceding, if we use an s-plane contour that encloses the RHP, and if we know how many poles of $1 + GH$ are in the RHP, we can use $1 + GH$ as a mapping function to determine how many zeros of $1 + GH$, poles of T, lie in the RHP. But the poles of $1 + GH$ are the poles of GH. If we have a system which is measurable, it must be open-loop-stable, which means there are no poles of the open-loop transfer function in the RHP. If we have GH analytically from models, we usually have D_i in factored form so we know where the poles of GH are. Of course, what we want is *no* poles of $T(s)$ in the RHP— not even close (in some sense to be defined shortly) to it. There's one more elegant observation to make our life simpler: A zero of $1 + GH$ means that $GH = -1$. Therefore, if we use GH as the mapping function, enclosure of -1 becomes the critical question, and -1 is called the critical point.

If we travel the s-plane Nyquist contour in the clockwise direction, denote the number of encirclements of -1 in the GH plane by N_e ($+$ if clockwise, $-$ if counterclockwise), the number of RHP poles of $GH(s)$ by N_o, and the number of RHP poles of $T(s)$ by N_c, then

$$N_c = N_e + N_o \tag{3.29}$$

This is the Nyquist stability criterion.

The test is quite easy to use. For example, all the open-loop transfer functions whose Nyquist diagrams are sketched in Fig. 3.17 have their poles either in the LHP or at the origin. The poles at the origin are excluded by the small detour, so there are no poles inside the s-plane contour. $N_o = 0$. For Fig. 3.17a and b, there are no encirclements; $N_e = 0$, so $N_c = 0$ also. These sketches represent systems that will be stable when the loops are closed. Figure 3.17c shows two clockwise encirclements. You may wonder about the count when the contour crosses itself, but if you look at the number of times it passes the critical point on the right, it becomes clear. $N_c = N_e = 2$, so there will be a pair of RHP poles if the feedback loop is closed on this system.

3.4   STABILITY MARGINS

If the system is stable, the next question is, How stable? Since the boundary is an open-loop gain of 1 at a phase angle of $-180°$, we measure the degree of stability, the distance from having poles in the RHP, by the changes that can be made in gain and phase to bring the transfer function to the stability boundary. In particular, the *gain margin* (GM) is the number by which the gain at $-180°$ can be multiplied to make it 1. This definition points to its historical origin; early designers knew that a feedback amplifier could be stabilized by turning the gain down. Formally, this can be expressed as

$$GM = \frac{1}{|GH(j\omega_p)|} \qquad \text{for } \angle GH(j\omega_p) = -180° \tag{3.30}$$

A second measure of closeness is the difference between the phase at unity gain and $-180°$. This is called the *phase margin* (PM):

$$PM = 180° + \angle GH(j\omega_c) \qquad \text{for } |GH(j\omega_c)| = 1 \tag{3.31}$$

Figure 3.20 illustrates these definitions.

■ Example 3.9

In order to apply these definitions, one must have a plot with sufficient resolution to see the contour behavior near -1. Figure 3.11 is an example of this. In this figure, it appears to me that the phase margin is only a couple of degrees.

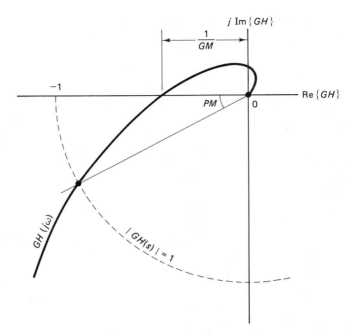

Figure 3.20   Illustration of gain and phase margins.

The negative real axis crossing is at about -0.8, so the gain margin is about 1.25. This represents a system for which a small change in some part value could cause oscillation. ∎

3.5   THE BODE MAGNITUDE PLOT

The design objective for a feedback system is to have sufficiently high loop gain so that the closed-loop performance is mainly controlled by precisely made passive elements in the feedback path. This often turns out to mean that the designer is not permitted to tinker with the low-frequency response of the system to produce a better stability margin. The designer may *not* turn the gain down. Therefore, one must use networks that reshape the frequency response in the neighborhood of unity gain to achieve good margins, and also good transient response, as you will see in later chapters. A weak point of the Nyquist diagram as a tool is that frequency is not explicitly displayed on the plot, and it is not easy to tell what will happen when the frequency response is altered by adding or changing networks. Once the given condition of the system is established by an initial Nyquist diagram, one would like a better tool for reshaping the frequency response.

Bode went back to the idea of plotting magnitude and phase separately against frequency. He found that if he plotted log (magnitude) against log (frequency), he could make some simple approximations that allow a quite accurate sketch of the frequency response for an analytic expression. This method provided the tool needed to easily see what modifications could be made and how to make them to improve a control system's performance.

To develop this idea, consider a simple pole, $G(s) = 10/(1 + s)$. This could be the transfer function of a low-pass filter. On inspection, you can see that the response is about constant for frequencies below 1 and decreases as the inverse of frequency for values much above 1 rad/s. A plot of the magnitude of G for frequencies from 0 to 20 rad/s is shown in Fig. 3.21. This plot shows the behavior in the stop band ($\omega > 1$ rad/s) quite well. It is a curve asymptotically approaching the horizontal. It hardly shows the pass band (0–1 rad/s) at all. If one wished to display the pass band in more detail, one would have to choose a scale which ends closer to 1 and thereby cut out the stop band behavior. If one wanted to do the plot by hand, one would have to calculate several points in the stop band to define the curve. As an alternative, consider the log-log plot in Fig. 3.22. The logarithmic frequency scale gives as much space to low frequencies as it does to high. Also, the plot can be quite well approximated by two straight-line segments—a horizontal one for the pass band and one with a -1 slope in the stop band. Lay a straightedge along the sloping portion and you will see that it intersects

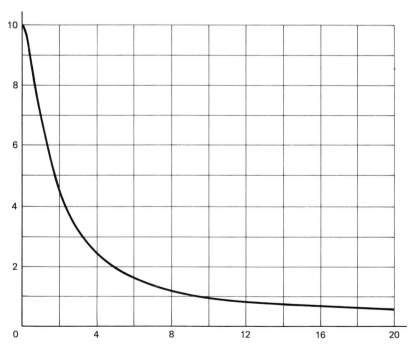

Figure 3.21 Magnitude versus ω for $G(j\omega) = 10/(1 + j\omega)$.

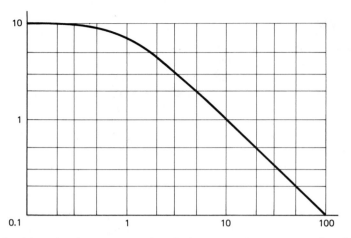

Figure 3.22 Magnitude versus ω for $G(j\omega) = 10/(1 + j\omega)$ on log-log scales.

the point (10,1) which is the extension of the pass band response to the corner (boundary) frequency between them. These two features, equal display space for high and low frequencies and good approximation by straight-line segments, are general properties of the log-log plots of transfer function magnitudes. It turns out that we can do a great deal of design using the straight-line version of the plots.

Consider a general pole of the form

$$G(s) = \frac{K}{(1 + s/\omega_c)^p} \tag{3.32}$$

$$|G(j\omega)| = \frac{K}{[1 + (\omega/\omega_c)^2]^{p/2}} \tag{3.33}$$

$$\log|G(j\omega)| = \log K - \frac{p}{2}\log\left[1 + \left(\frac{\omega}{\omega_c}\right)^2\right] \tag{3.34}$$

$$\approx \log K \qquad \omega \ll \omega_c \tag{3.35}$$

$$\approx \log K - p\log\omega + p\log\omega_c \qquad \omega \gg \omega_c \tag{3.36}$$

Let $Y = \log|G|$, $X = \log\omega$, and $C = \log K + p\log\omega_c$. Then (3.36) becomes

$$Y = C - pX \tag{3.37}$$

For a real pole, the plot is a horizontal line below the corner frequency and a line of slope $-p$ above. The result has the same form for a real zero, except that the slope is positive. What is the constant? What is the true value at the corner?

Next let's look at a function with two different corner frequencies:

$$G(s) = \frac{K}{(1 + s/a)(1 + s/b)} \quad a < b \quad (3.38)$$

For $\omega < a$, both factors are treated as 1. So

$$\log |G(j\omega)| = \log K \quad \omega < a \quad (3.39)$$

For $a < \omega < b$, the first factor is treated as ω/a, while the second is still treated as 1. Thus

$$\log |G| = \log K + \log a - \log \omega \quad a < \omega < b \quad (3.40)$$

For $b < \omega$ both factors are treated as being proportional to frequency:

$$\log |G| = \log K + \log a + \log b - 2 \log \omega \quad b < \omega \quad (3.41)$$

These results show that each factor affects the plot above its corner frequency, and the effect is to change the slope. This leads to the simple sketching method presented in the next two paragraphs.

If you are going to do design work without a computer software package to draw the plots, you should use commercial log-log paper for the sketching. If you are going to do design work using computer-generated plots and you want a sketch to check the plots as you go along, then you can make sketches on scratch paper or engineering paper by making your own log scales. I will describe the latter approach, but you should use commercial paper a few times for the experience. Log scales are set up in decades. That is, a ratio of 10 : 1 is represented by a fixed distance on the paper. Before you set up the scales, you should estimate how many decades you will need for each axis. Usually, you will want to bracket fairly closely the corner frequencies in the transfer function, so the minimum number of decades that will do this establishes the frequency axis. Estimating the amplitude range is a bit like knowing the answer ahead of time. For many open-loop transfer functions, an initial guess of two decades vertical for each decade horizontal is a good place to start. Of course, if you are trying to verify a computer plot, you will use the scales generated by the computer. I use equal-size decades for the amplitude and frequency scales so that the actual line slopes follow the applicable approximation formula, and so a given slope looks the same every time I do another plot. Since the slope is intimately related to the phase shift of the transfer function, a matter discussed in the next two sections, experience with magnitude plots having true slopes sometimes allows one to dispense with the phase plot and also provides another check on numerical results and machine plots. Once the decade size has been established by the number of decades and the dimensions of your paper, lay out the axes and mark and label the decade points. Next, divide each decade into thirds. The values here are about 2 and 5 times the decade starting value. Thus, in the decade from 0.1 to 1, these points are 0.2 and 0.5. Then the middle of the interval can be marked. Its exact value is $\sqrt{10}$, but 3 is good enough most of the time. Since a given length represents a fixed value ratio, the value at the middle of any line segment is the geometric mean of the values at its ends. (Prove it.) So, for instance, 0.7 is about halfway between 0.5 and 1. Using this fact, you can lay in approximations for other values, but don't clutter up the scales by doing this except for corner frequencies and amplitudes.

After you set up the axes and scales, start the plot at the magnitude of the function below its lowest corner frequency. Move along horizontally until you reach this frequency. At this point, change the slope to the value of the degree of the factor and make it positive for a zero or negative for a pole. Thus, if the lowest corner frequency belongs to a zero which is cubed, the slope increases by $+3$. Proceed to draw on this slope to the next corner frequency and again change the slope according to the nature and degree of the factor. If the next corner frequency belongs to a pole and the factor is squared, decrease the slope by 2, to $+1$. Carry on this process through all the corner frequencies in the range you have chosen. Now it's time for a few examples.

■ Example 3.10

$$G(s) = \frac{10(1 + s/20)}{(1 + s/10)(1 + s/100)} \tag{3.42}$$

Since all the corner frequencies are in the range 10–100 rad/s, only one frequency decade is required, although one may wish to show the behavior on either side of this range so that three decades might also be used. Since the function goes pole-zero-pole, the slope is never steeper than -1, so two decades, 0.1 to 10, might work for the amplitude. Figure 3.23a shows the scales and the first plot segment.

The level of the constant is 10, so the first segment is at 10 and has zero slope. The first corner frequency is 10 rad/s and belongs to a first-degree pole, so the slope will change to -1 starting at 10 rad/s. The next corner frequency is at 20 rad/s. Since the ratio of the two corner frequencies is 2, the ratio of the amplitudes on this -1 slope segment must also be 2, implying that the magnitude at 20 rad/s is 5. The sketch to this point is shown in Fig. 3.23b. Since the 20 rad/s belongs to a first-degree zero, the slope changes, by $+1$, back to zero, and the plot continues horizontally over to 100 rad/s. This step is shown in Fig. 3.23c. Since 100 rad/s belongs to another first-degree pole, the final slope is -1. At 1000 rad/s, 10:1 further on, the level must be down by 10, so it is 0.5 at the end of the plot. The finished straight-line plot is shown in Fig. 3.23d, and a machine-generated plot is shown in Fig. 3.24 with the straight-line approximation drawn on for easier comparison. ■

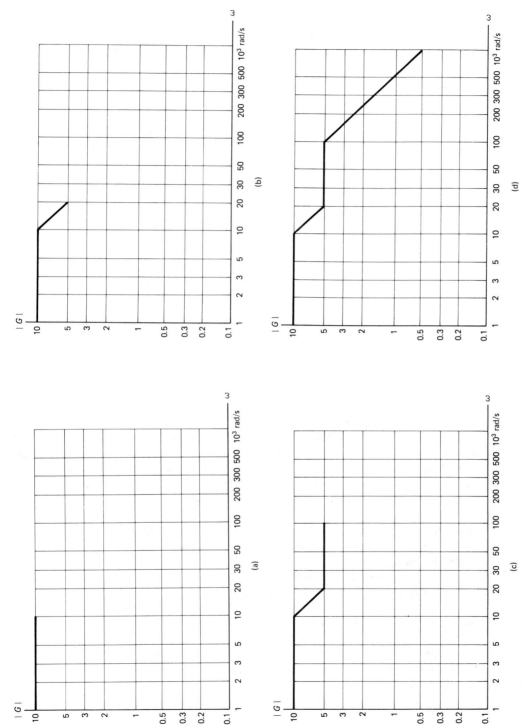

Figure 3.23 Stages in a straight-line Bode plot, Example 3.10.

93

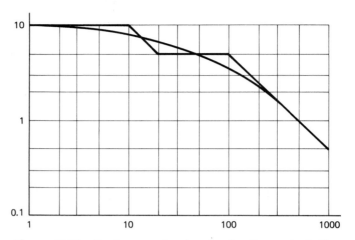

Figure 3.24   Bode magnitude plot for Example 3.10.

■ Example 3.11

$$G(s) = \frac{10(1 + s/10)}{(1 + s/2)^2(1 + s/50)^2} \tag{3.43}$$

In this case, all the action takes place (corner frequencies lie) in the two decades from 1 to 100 rad/s. Looking at the sequence of poles and zeros, you can see that the slope is always negative after the first corner. You might choose a vertical scale of four decades topped by 10 on this basis. After drawing and labeling the scales, the sketch begins with a zero slope segment from 1 to 2 rad/s at a magnitude of 10. At 2 rad/s, G has a second-degree pole which means a change to a -2 slope. This slope segment will stop at 10 rad/s, the next corner, but you can establish the slope by using the fact that a one-decade increase in frequency will give a two-decade drop in amplitude on a -2 slope. Since this segment starts at (2, 10), its extension would pass through (20, 0.1). The two corners have a 5:1 frequency ratio, so the amplitude at 10 rad/s will be $10/5^2 = 0.4$. The corner at 10 rad/s is due to a zero of first degree, so the slope increases by $+1$, to -1. Again, the slope can be established by marking the point (100, 0.04) and drawing toward it, stopping at 50 rad/s. Since the end frequencies for this segment are again 5:1 and the slope is -1, the magnitude at 50 rad/s is $0.4/5 = 0.08$. Finally, the last corner is due to a double pole, so the slope will change by -2, to -3. The level at the end of the sketch, 100 rad/s, is $0.08/2^3 = 0.01$. The sketch is shown in Fig. 3.25 drawn on a machine-generated plot. The mark at (10, 0.4) was made by using the fact that the distance from 0.2 to 0.4 is the same as the distance from 0.1 to 0.2. ■

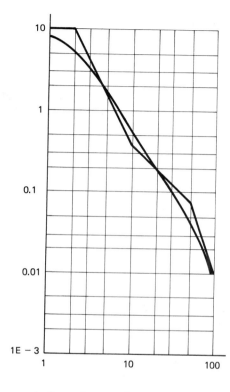

Figure 3.25 Bode magnitude plot for Example 3.11.

A few observations on these two examples are in order now. First, the transfer function factors are all in a special form, $1 + s/a$. If one starts with factors in the form $s + a$, one should convert to the former by writing $s + a = a(1 + s/a)$ and absorbing the constants into a new value. For example,

$$\frac{K(s + b)}{s + a} = \frac{(Kb/a)(1 + s/b)}{1 + s/a} = \frac{K'(1 + s/b)}{1 + s/a}$$

On any line segment of slope n, two amplitudes $A(\omega_1)$ and $A(\omega_2)$ have the ratio

$$\frac{A(\omega_2)}{A(\omega_1)} = \left(\frac{\omega_2}{\omega_1}\right)^n \tag{3.44}$$

Remember that n is positive for an upward slope and negative for a downward slope. This equation is useful in finding compensator frequencies, as well as in doing sketches.

By now, you might have said to yourself, These are all real poles and zeros. What about poles or zeros at the origin, and complex pairs (quadratic factors)? In the examples, I started the sketches with a line of zero slope. If there is a pole at the origin (an s in the denominator of G), the initial slope of the sketch is above the corner frequency, therefore it is -1. Since zero can't be chosen as a frequency on a log scale, one must set the scales on the basis of the nonzero corner frequencies. This also means that one needs to establish a starting amplitude. If the gain constant is 3 and the lowest frequency is 0.1 rad/s, the starting amplitude for one pole at the origin is $3/0.1 = 30$. Likewise, for two poles at the origin the starting level is $3/0.1^2 = 300$ and the starting slope is -2. In general, if the transfer function is

$$G(s) = \frac{Ks^n(\ldots)}{(\ldots)} \tag{3.45}$$

and the plot starts at w rad/s, where w is lower than all the nonzero corner frequencies, the starting amplitude is

$$A(w) = Kw^n \tag{3.46}$$

and the initial slope is n.

Quadratic factors are not so easily handled. To be specific, let's discuss the standard form

$$F(s) = 1 + 2\zeta\,\frac{s}{\omega_0} + \frac{s^2}{\omega_0^2} \tag{3.47}$$

$$F(j\omega) = 1 - \left(\frac{\omega}{\omega_0}\right)^2 + j2\zeta\,\frac{\omega}{\omega_0} \tag{3.48}$$

The ζ in the center term is called the *damping ratio*. If $\zeta > 1$, the quadratic is the product of two distinct real-rooted factors. At $\zeta = 1$ the roots become the same, and for $\zeta < 1$ they are complex, which is the case we are interested in here. For $\zeta < 1$ the quadratic is said to be *underdamped*, and those with values of less than about 0.3 are often called *lightly damped*. Clearly, for small ω/ω_0 the magnitude is close to 1, and for large ω/ω_0 the magnitude increases as the square of the frequency. A straight-line approximation with zero slope to the left of ω_0 and a $+2$ slope to the right represent this approximation. However, if the quadratic is lightly damped, the error in the neighborhood of the corner frequency is not acceptable. The previous examples showed that if one were to take the straight-line segment plot as a guide, staying inside the corners, one could make a pretty fair freehand sketch of the actual Bode plot. Figure 3.26 shows that this is not the case for a lightly damped quadratic. The value at resonance on the straight-line approximation is 1, while the actual value is 2ζ, which can be a very small number and far away on a logarithmic scale. The solution to this problem is to use the resonant value to add a point to the straight-line sketch to allow an ap-

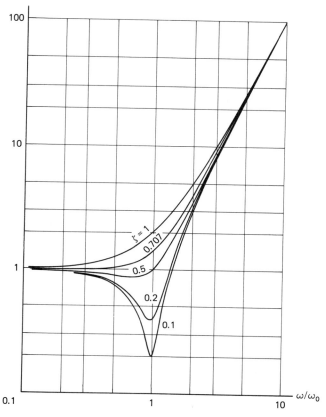

Figure 3.26 Magnitude of a quadratic factor for damping factors of 1, 0.707, 0.5, 0.2, and 0.1.

proximation to the resonance. Since this is only a factor in the total transfer function, the level at resonance of the diagram is multiplied by 2ζ if the factor is in the numerator or divided by 2ζ if it is in the denominator.

■ Example 3.12

$$H(s) = \frac{50(1 + s/3.16)}{s^2(1 + s/79)} \qquad (3.49)$$

This example shows how to handle origin poles. Since $n = -2$ and the two decades from 1 to 100 rad/s span the nonzero corner frequencies, 1 rad/s is the start and 50 is the level at that point. The first corner frequency is 3.16, which belongs to a zero. The level at this frequency is $50/3.16^2 = 5$. The slope changes by $+1$, to -1. The ratio of the two corner frequencies is $79/3.16 = 25$, so the level at 79

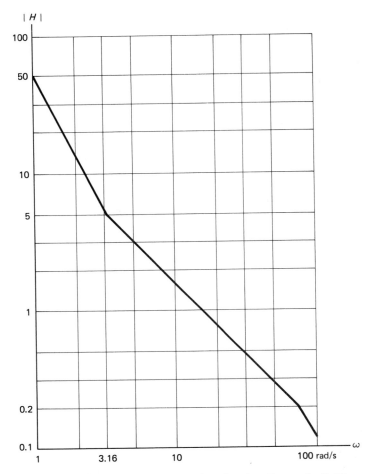

Figure 3.27 Bode magnitude plot for Example 3.12.

rad/s is $5/25 = 0.2$. The slope changes back to -2 at this corner. Although it isn't necessary to do the calculation, the level at the edge of the plot is $0.2[(100/79)^{-2}]$ $= 0.125$. To draw the sketch, one needs a minimum of two decades of frequency and three decades of amplitude, 0.1 to 100. It is shown in Fig. 3.27. ∎

∎ Example 3.13

$$H(s) = \frac{10}{(1 + s/0.1)(1 + 0.2s + s^2)} \qquad (3.50)$$

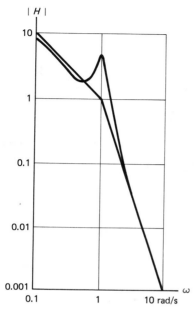

Figure 3.28 Bode magnitude sketch for Example 3.13.

In this example I have placed a lightly damped pole pair at 1 rad/s and another pole with a corner at 0.1 rad/s. Since the level is constant to the left of 0.1 rad/s, I start the sketch at this frequency with a -1 slope from level 10. At 1 rad/s, a 10:1 ratio, the amplitude is 1. The slope changes to -3 so the level plummets to 0.001 at 10 rad/s. At resonance, the quadratic has the magnitude 0.2, so the line segment level is multiplied by 5 to give 5 as the approximate value. These considerations give four decades as the amplitude range. The sketch is shown in Fig. 3.28. From Fig. 3.26, I saw that the quadratic factor for $\zeta = 0.1$ is fairly close to the straight-line values at a frequency ratio of 2 either side of resonance. Therefore I brought the freehand sketch close to the segments near these values.   ■

3.6   THE BODE PHASE PLOT

Phase doesn't do well on a log-log plot. For real poles and zeros, phase is the arctangent of a frequency ratio, which implies that it would look good plotted as linear phase against logarithmic frequency. Figure 3.29 shows both magnitude and phase plotted for a simple pole, $1/(1 + s)$. The phase scale is linear, and the magnitude and frequency scales are logarithmic. You can see that the phase has odd symmetry about the corner frequency and that it starts and finishes changing

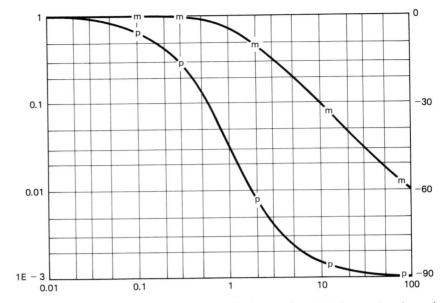

Figure 3.29 Bode magnitude (m) and phase plots (p) for a simple pole.

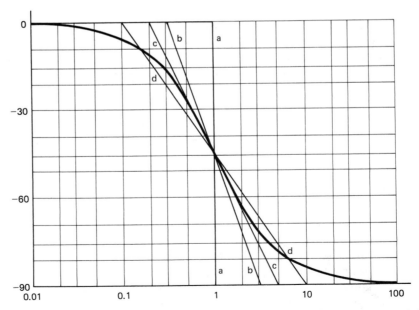

Figure 3.30 Some straight-line approximations to the phase curve. (a) Step-wise. (b) 90°/decade. (c) Matched slope. (d) 45°/decade.

at frequencies much further from the corner frequency than do the magnitude curve slopes. Any straight-line approximation to the phase curve should have horizontal segments for the initial and final values and should pass through $-45°$ at the corner frequency. After that, the choice is up to you.

Figure 3.30 shows four different straight-line approximations that are sometimes used in other texts or in practice. In general, the range is from easiest to use and least accurate to most accurate. If one is going to draw a freehand curve using straight-line segments as a guide, the stepwise approximation is not as bad as it looks. One merely uses the steps as an indicator of the trend of the curve, without making sharp changes. For a check on what a computer has produced, this is often good enough. The 90°/decade approach is more accurate and requires the slope change to happen a half-decade before the corner frequency. This makes the sketching easier in a short frequency range than the following methods. By making the slope change at a frequency one-fifth of the corner frequency and finishing the segment at 5 times the corner frequency, one obtains a good match to the slope of the phase curve at its midpoint and very good accuracy for a 4:1 frequency range. The 45°/decade line is the most accurate in the sense that it has the smallest maximum error, but it means using a frequency scale at least two decades longer than the span of the corner frequencies. The following two examples illustrate how to use these approximations.

■ Example 3.14

Consider again the transfer function in (3.42). Its magnitude plot was prepared for Example 3.10 in Figs. 3.23 and 3.24. Since this is a pole-zero-pole sequence, the phase ranges between 0 and $-90°$. I choose a frequency scale that runs from a decade below the lowest corner to a decade above the highest, 1–1000 rad/s. For the stepwise approach, I simply draw a horizontal line at 0° from 1 to 10 rad/s. Since the first corner is a pole, I step down to $-90°$ and continue horizontally until the next corner, at 20 rad/s. This corner belongs to a zero, so I step up 90°, back to 0. I carry this value along to 100 rad/s, the last corner. Again, since this is a pole, I drop to $-90°$. A sketch, with a freehand interpolation, is shown in Fig. 3.31a. Figure 3.31b shows the 45°/decade version. To use any of the sloping approximations, you should go along the frequency axis and mark the frequencies for the start and finish of each component slope. For this one, the corner at 10 needs marks at 1 and 100, the corner at 20 needs marks at 2 and 200, and the corner at 100 needs marks at 10 and 1000 rad/s. I start at 1 rad/s with a down slope of 45°/decade, going along until I read 2 rad/s. At this point, the zero contributes an up slope, so the net slope is zero. The constant level is maintained until 10 rad/s, where the second pole contributes a down slope of 45°/decade again. This is followed until I reach 100 rad/s, where the effect of the first pole stops. This leaves a net zero slope until 200 rad/s, where the effect of the zero terminates. From this point to 1000 rad/s, I draw the final slope of $-45°$/decade. This must necessarily finish at $-90°$.   ■

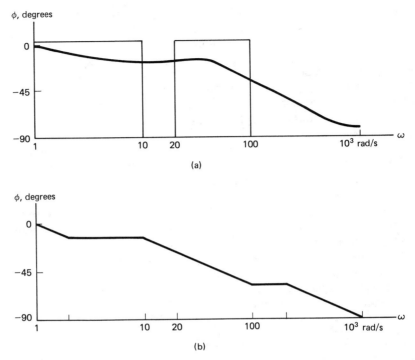

Figure 3.31 Phase plots for Example 3.14. (a) Stepwise. (b) 45°/decade.

■ Example 3.15

For this example, I use the matching slope method on the transfer function in (3.43). Figure 3.32 shows the work. The start and finish frequencies are 5:1 away from the corners, so they are 0.4, 2, 10, 50, and 250 rad/s. The frequency scale must go from 0.1 to 1000 rad/s, and the phase range is 0 to $-270°$ because there is a net of three poles at high frequencies. I start the sketch with a horizontal line at 0° over to 0.4 rad/s. The first corner belongs to a double pole, so I need a down slope that aims at $-90°$ at 2 rad/s. At 2, the zero takes effect, so the slope decreases to $-45°$ (5:1 ratio), aiming at $-180°$ at 50 rad/s. At 10 rad/s, the poles at 2 stop their effect, but the poles at 50 start, so there is no net change and the slope continues until the zero's effect runs out at 50 rad/s. At this point, the final slope is due to the two poles at 50, so it is a $-90°$ drop to $-270°$ at 250 rad/s. A line at $-270°$ to 1000 rad/s completes the plot. ■

Poles or zeros at the origin simply shift the starting level of the phase plot by $90n°$. Quadratics have a phase transition from 0 to 180° whose sharpness depends on the damping factor. Phase plots for the standard quadratic, (3.47), are shown in Fig. 3.33.

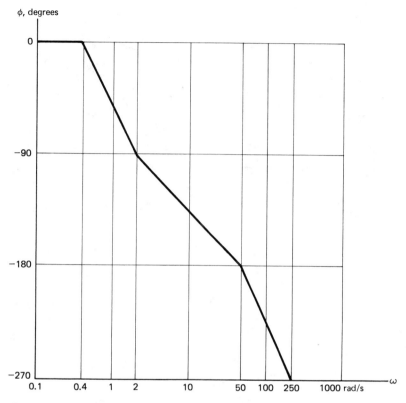

Figure 3.32 Bode phase plot for Example 3.15, matching slope method.

3.7   STABILITY MARGINS ON THE BODE PLOT

Gain and phase margins are quite easy to find on a Bode plot, especially if gain and phase curves are plotted over the same frequency axis. Such is the case in Fig. 3.34 for $G(s) = 3/(1 + s)(1 + s + s^2/2)$. To find the phase margin, locate the point where the gain is equal to 1, place a straightedge vertically through this point, and mark its intersection with the phase curve. The phase reading (which is usually negative) added to 180° is the phase margin or the vertical distance between the last point found and the $-180°$ line. To find the gain margin, first locate the point where the phase curve crosses the $-180°$ line. The point on the gain curve directly above this is $1/GM$. For this figure, the phase margin is about 75°, and the gain margin is about 4.5.

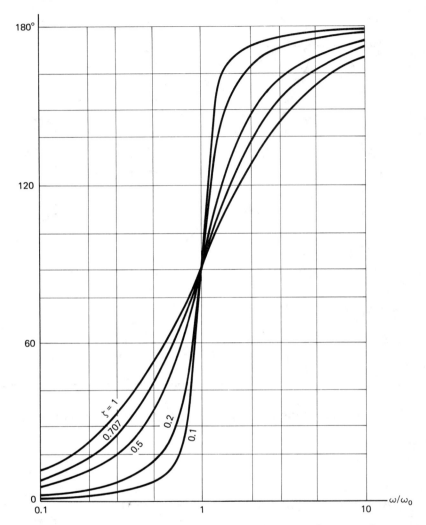

Figure 3.33 Phase plots for a quadratic with damping factors of 1, 0.707, 0.5, 0.2, and 0.1.

3.8   TIME DELAY EFFECTS

From Chap. 2, we know that a delay in time translates to an exponential function in the frequency domain. The effect on frequency response is to add phase lag without changing the magnitude of a function. An example of this effect is shown in the Nyquist plot in Fig. 3.35. A system consisting of an integrator and a 1s time delay is tested over a decade and a half. If the delay were not present, the

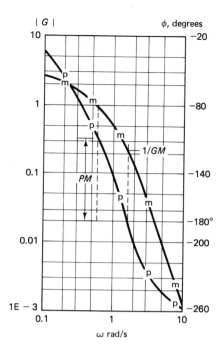

Figure 3.34 Gain (m) and phase (p) margins on a Bode plot.

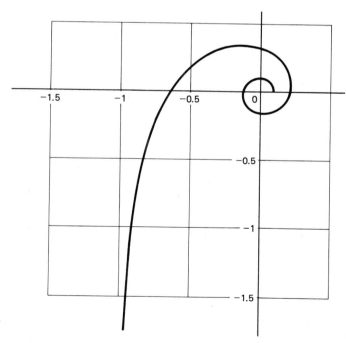

Figure 3.35 Nyquist plot for $G(s) = (e^{-s})/s$ with $0.5 \leq \omega \leq 10.77$.

plot would just lie on the imaginary axis. The added phase shift moves it left at low frequencies and then causes it to spiral into the origin at high frequencies. You can see that if the gain were raised, this would be an unstable system. If the phase of the exponential were plotted on linear coordinates, it would be a straight line with a negative slope. Plotted on linear-log coordinates like a Bode phase plot, it is a curving, diving line, as shown in Fig. 3.36. Generally speaking, it is very hard to compensate a system whose gain is above unity in the frequency range for which a time delay causes 90° or more of the phase lag. If the time delay can't be removed or reduced, the designer has to reduce the high-frequency gain, which is equivalent to slowing the system down so the time delay is not important.

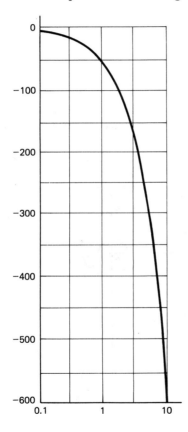

Figure 3.36 Phase plot for a 1 s time delay.

3.9   RELATIONS BETWEEN MAGNITUDE AND PHASE

You have no doubt noticed that both magnitude and phase are determined by the corner frequencies in a transfer function. There appears to be a close connection between the slope of the magnitude plot and the value tended toward on the phase plot. For a long -1 slope the phase tends toward $-90°$, for a long 0 slope the phase tends toward 0, and so on. Except for the transform of time delay, all the factors and functions we have studied so far are a type called *minimum phase*. That is, they have the minimum phase lag consistent with their effect on the magnitude. The delay exponential is not minimum phase, because it has no effect on amplitude (which ordinarily means zero phase lag) but produces phase lag. Another type of factor which is not minimum phase is a right-half-plane zero, $1 - s/a$. This zero gives a phase lag instead of a phase lead as does a left-half-plane (LHP) zero. Its Bode plot will have a $+1$ slope and a $-90°$ phase shift at high frequencies. If we know that a system is minimum phase, then we can look at its magnitude plot and determine its phase plot. Of course, if we have only the magnitude plot, we will have to fit it with appropriate straight-line segments to make a guess as to the transfer function it represents. Sometimes this is a useful step before the design begins, for example, if we have a plant for which no analytical model is available and we can make frequency response measurements. In any case, Bode found that one can take the magnitude data directly and calculate the phase. Since, as pointed out above, the real relation is between the magnitude slopes and phase, and the phase at a given frequency is influenced by slope at all frequencies, the relation is a weighted integral (limit of a weighted sum) of the slope.

Let $F(s)$ be a minimum-phase (no RHP zeros or poles) transfer function. Let

$$u = \ln\left(\frac{\omega}{\omega_0}\right) \tag{3.51}$$

and

$$L(u) = \ln|F(j\omega)| \tag{3.52}$$

L is the actual vertical distance on the magnitude plot, and u is a scale of actual horizontal distance whose zero is set at ω_0. The slope is

$$n = \frac{dL}{du} \tag{3.53}$$

Bode's weighting function is

$$W = \ln\left(\coth\left|\frac{u}{2}\right|\right) \tag{3.54}$$

The phase at ω_0 is

$$\phi(\omega_0) = \frac{1}{\pi}\int_{-\infty}^{\infty} nW\,du \text{ radians} \tag{3.55}$$

$W(u)$ weights the slope at ω_0 very heavily and can be approximated by an impulse function.

The Bode phase integral is mainly important because it gives theoretical proof to our experience in practice and sets the conditions for which our experience is valid. As with many general propositions, some people will find many uses and some will never use a particular one. In my case, I used the phase integral once to find the phase for a magnitude function which had been measured with a spectrum analyzer. It was very valuable that I remembered having seen the integral 20 years earlier.

Symmetry and approximate symmetry are helpful to notice in any analysis problem. For example, a magnitude plot with a slope sequence of -2, -1, -2 has odd symmetry, and the corresponding phase plot has even symmetry. If a and b are the two corner frequencies, the frequency for symmetry is the geometric mean, $c = \sqrt{ab}$. Because of the odd symmetry, the actual magnitude plot crosses the straight-line plot at c. The phase curve starts and finishes at $-180°$ and peaks at c at a value depending on the length of the -1 slope (the ratio b/a). A magnitude plot with a slope sequence of -1, -2, -3 doesn't appear to have any symmetry, but its phase companion does. The phase sequence is $-90°$, $-180°$, $-270°$. Again, the geometric mean of the corner frequencies is an exact value, this time on the phase plot where the phase is exactly $-180°$. Two long slopes, say 100:1 each in frequency ratio, will have a phase at their corner very close to the average of their slopes times $90°$. These are some of the more frequently occurring cases. Be alert!

3.10   CLOSED-LOOP FREQUENCY RESPONSE

Generally speaking, the design objectives for a control system require that the transfer function in the feedback path be constant, at least for all frequencies of interest. A system of this sort is shown in Fig. 3.37. Presumably, there is some frequency range over which it is desired that $T(s)$ be approximately equal to $1/B$, which implies that $A(s)B \gg 1$. The behavior of the system gets interesting when $A(j\omega)B$ is in the neighborhood of 1. The behavior of the open-loop gain in this frequency range determines whether the closed-loop frequency response will

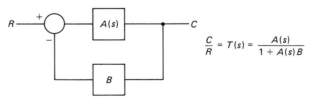

$$\frac{C}{R} = T(s) = \frac{A(s)}{1 + A(s)B}$$

Figure 3.37   General control system with constant feedback.

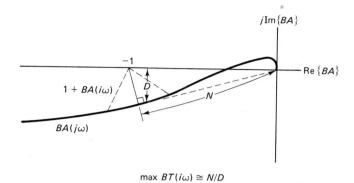

$$\max BT(i\omega) \cong N/D$$

Figure 3.38 Nyquist plot for a system with resonant closed-loop frequency response.

roll off smoothly, have a resonant peak, or do something in between. For $A(j\omega)B \ll 1$, $T(j\omega) = A(j\omega)$, so the ultimate roll-off will be that of the forward-path portion of the system. If there is to be a peaking effect, the denominator must become less than 1, that is, $BA(j\omega)$ must get close to -1 for some frequency range. This is illustrated in Fig. 3.38. The length of the line between the Nyquist plot and the -1 point is the magnitude of the denominator $T(j\omega)$, and the plot itself is proportional to the numerator $A(j\omega) = BA(j\omega)/B$. For the case shown, you can see that the line from the -1 point to the plot varies faster in the neighborhood of its minimum length than does the magnitude of the plot itself, so that the peak is proportional to the ratio of the two lines from the minimum distance point. For a small enough phase margin, one can derive an expression (requested in Prob. 3.15) for the peak value in terms of the phase margin and B.

Specifications for control systems may be given either in terms of time domain performance, frequency domain performance, or both. Time domain performance is discussed in Chap. 5, and some relations between time and frequency domain performance for the canonic second-order system are given in Chap. 6. Figure 3.39 illustrates generally used frequency domain terms for specifications. Note that the frequencies are given as cycle frequency hertz rather than radians/second and that only the magnitude curve is pertinent.

FURTHER READING

The topics treated in this chapter are part of every text in control systems published since 1950. Books published in the 1950s and 1960s have more detailed expositions because they have fewer topics to cover than more recent works. Two good books from different periods are

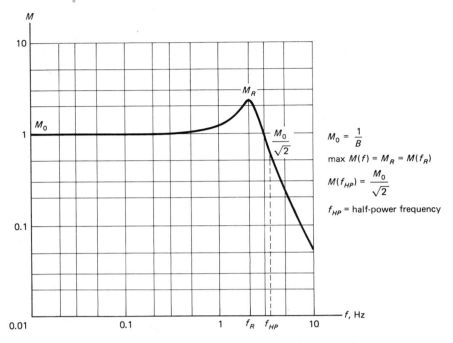

Figure 3.39 Closed-loop frequency response definitions.

Bower, John L., and Schultheiss, Peter M., *Introduction to the Design of Ser-vomechanisms*, New York, John Wiley and Sons, 1958.

Kuo, Benjamin C., *Automatic Control Systems*, 4th ed., Englewood Cliffs, N.J., Prentice-Hall, 1982.

PROBLEMS

Section 3.1

3.1 A system has the transfer function

$$T(s) = \frac{25}{s^2 + 5s + 25}$$

(*a*) Make a map showing the closed-loop poles.

(*b*) If the system is a unity-feedback loop, map the open-loop poles.

(*c*) If the system is driven by $r(t) = 3u(t) \sin (2t)$, where $u(t)$ is the unit step function, find the natural and forced time responses.

3.2 Repeat Prob. 3.1 for $T(s) = 36/(s^2 + 36)$.

3.3 Repeat Prob. 3.1 with $T(s) = 1/(1 + s/2)$ and $r(t) = u(t) = $ impulse (t).

3.4 Repeat Prob. 3.1 with $T(s) = (1 + s)/(2 + s + s^2)$ and $r(t) = tu(t)$.

Sections 3.2 and 3.3

3.5 Sketch by hand the Nyquist plots for $s = j\omega$, $\omega > 0$, for the following functions:

(a) $3/(1 + s/3)$ (b) $3(1 + s)/(1 + s/5)$ (c) $21(1 + s)/(7 + s)$
(d) $21(1 + s)/(7 + 6s + s^2)$ (e) $100(1 + s)/[s(100 + s) + s^2]$
(f) $81(1 + s/3)/[s^2(81 + 18s + s^2)]$ (g) $81(1 - s/3)/[s^2(81 + 12s + s^2)]$
(h) $5/[s(1 - s/5)(1 + s/20)]$
(i) $5(1 + s/5)/[s(1 - s/5)(1 + s/20)]$
(j) $5(1 - s/5)/[s(1 + s/5)(1 + s/20)]$

3.6 Make a complete Nyquist sketch for each of the functions in Prob. 3.5. For each case, give the number of encirclements of the critical point and the number of RHP poles of the corresponding closed-loop function.

Section 3.4

3.7 A system has $H(s) = 1$ in the feedback path and $G(s) = 3/(s + s^2/3)$ in the forward path. What are the gain and phase margins?

3.8 For Prob. 3.7, if the gain is raised from 3 to 6 (doubled), what are the GM and PM?

3.9 Repeat Prob. 3.7 with $G(s) = 50(1 + s)/[s^2(1 + s/50)]$.

3.10 Repeat Prob. 3.7 for $G(s) = K/[s(1 + s/2)(1 + s/20)]$ with K at
(a) 1,   (b) 2,   (c) 6.32.

Sections 3.5–3.7

3.11 Sketch straight-line Bode magnitude and phase plots for the functions in Prob. 3.5. Calculate the level at each corner frequency and the frequencies for unity gain and $-180°$ phase on the straight-line plots for each case. Estimate the gain and phase margins where appropriate.

Section 3.8

3.12 In Fig. 3.36 it appears that there is a low-frequency asymptote at -1. Prove that this is indeed the case.

Section 3.9

3.13 Which of the functions in Prob. 3.5 are nonminimum phase?

3.14 Write a computer program (subroutine, procedure, APL function) to calculate the Bode phase integral for a finite set of amplitude and frequency data. Determine the effects of truncation by testing a known transfer function. Add extrapolation to your program and test for a suitable frequency range extension.

Section 3.10

3.15 For a small phase margin, the shortest line representing $1 + BA(j\omega)$ is approximately $1 + BA(j\omega_c)$, where ω_c is the unity-loop-gain frequency. Use this to find an approximation for M_p. Test your result with $A(s) = K/s(1 + s/10)$, $B = 0.1$, for $K = 100, 316, 1000$.

Chapter 4

THE ROOT LOCUS METHOD

4.1 INTRODUCTION

The first section of the last chapter is also an introduction to this one. That section showed you that the response of a linear feedback system to an input is determined by the zeros and poles of its closed-loop transfer function, and that the poles of the closed-loop transfer function determine the system's stability. In principle, the Nyquist criterion allows us to deal with systems that are open-loop-unstable, but in practice, the Nyquist plot is not a good tool for design, as pointed out in the last chapter. A right-half-plane pole makes phase lead on a Bode plot, giving the unwary designer a false impression of good phase margin. W. R. Evans had to deal with aircraft subsystems which regularly had RHP open-loop poles. He felt he needed a clear picture of how the closed-loop poles were behaving as a function of the gain constant, something like the pole-zero plots in Sec. 3.1. Consider again the system in Fig. 3.2, whose closed-loop pole is shown in Fig. 3.3b. The pole is located at $-(K + 1)$ so that a plot showing the pole location for each $K > 0$ will be a straight line heading left from -1. This is shown in Fig. 4.1. Such a set of points is called a locus, and in this case it is a root locus because it shows the location of the root of the denominator of the closed-loop transfer function.

A second-order system is shown in Fig. 4.2. Its closed-loop transfer function is

$$\frac{C}{R} = \frac{K}{s^2 + s + K} \tag{4.1}$$

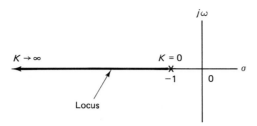

Figure 4.1 Root locus for the system of Fig. 3.2.

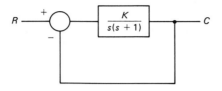

Figure 4.2 A second-order system.

and the roots (poles) are

$$p = -\tfrac{1}{2} \pm \sqrt{\tfrac{1}{4} - K} \tag{4.2}$$

Since there are two roots and I want to talk about them individually, let me give them separate names.

$$p_1 = -\tfrac{1}{2} + \sqrt{\tfrac{1}{4} - K}$$
$$p_2 = -\tfrac{1}{2} - \sqrt{\tfrac{1}{4} - K} \tag{4.3}$$

At $K = 0$, they are both real and equal to the open-loop poles, as in the previous case. As K increases, p_1 moves left, toward $-\tfrac{1}{2}$, and p_2 moves right. At $K = \tfrac{1}{4}$, the radical is zero and $p_1 = p_2 = -\tfrac{1}{2}$. For larger K, the radical becomes imaginary, so that both roots have a constant real part of $-\tfrac{1}{2}$ and an increasing imaginary part. This is shown in Fig. 4.3.

Suppose the system has a zero added, as in Fig. 4.4. The closed-loop transfer function becomes

$$\frac{C}{R} = \frac{K(1 + s/2)}{s^2 + s(1 + K/2) + K} \tag{4.4}$$

with denominator roots at

$$p_1 = -\left(\frac{1}{2} + \frac{K}{4}\right) + \left(\frac{1}{4} - \frac{3}{4}K + \frac{K^2}{16}\right)^{1/2}$$

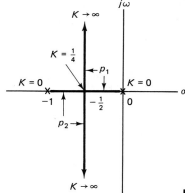

Figure 4.3 Root locus for the system in Fig. 4.2.

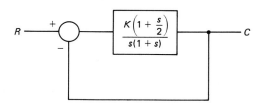

Figure 4.4 A zero added to the system in Fig. 4.2.

$$p_2 = -\left(\frac{1}{2} + \frac{K}{4}\right) - \left(\frac{1}{4} - \frac{3}{4}K + \frac{K^2}{16}\right)^{1/2} \tag{4.5}$$

This time the radical contains a quadratic in K, which means there are two values of K for which it is zero and the roots are equal. Setting this quadratic to zero,

$$K^2 - 12K + 4 = 0 \tag{4.6}$$

The values of K are

$$K_1 = 6 - \sqrt{36 - 4} = 6 - 4\sqrt{2} \approx 0.3$$
$$K_2 = 6 + 4\sqrt{2} \approx 11.7 \tag{4.7}$$

The corresponding root values are

$$p(K_1) = -(\tfrac{1}{2} + \tfrac{6}{4} - \sqrt{2}) \approx -0.6$$
$$p(K_2) = -(\tfrac{1}{2} + \tfrac{6}{4} + \sqrt{2}) \approx -3.4 \tag{4.8}$$

For $K < K_1$ the roots are real, and for $K > K_2$ they are also real. In between, the locus must have an imaginary part. One can obtain the shape of the s-plane curve by eliminating K from the expressions for σ and ω.

$$\sigma = - \left(\frac{1}{2} + \frac{K}{4} \right) \Rightarrow K = -2 - 4\sigma \tag{4.9}$$

$$\omega^2 = K - \sigma^2 = -2 - 4\sigma - \sigma^2 \tag{4.10}$$

$$(\sigma + 2)^2 + \omega^2 = 2 \tag{4.11}$$

It turns out the shape is a circle centered at -2 with a radius of $\sqrt{2}$. For large K,

$$p_2 \approx -\frac{K}{2} \rightarrow -\infty \tag{4.12}$$

$$p_1 = - \left(\frac{1}{2} + \frac{K}{4} \right) + \frac{K}{4} \left(1 - \frac{12}{K} + \frac{4}{K^2} \right)^{1/2}$$

$$\approx -\frac{1}{2} - \frac{K}{4} + \frac{K}{4} \left(1 - \frac{6}{K} + \frac{2}{K^2} \right) \rightarrow -2 \tag{4.13}$$

Figure 4.5 shows the locus.

The three cases I have shown illustrate a number of properties which will be derived and discussed in the next section. First, the locus is always symmetrical about the real axis. This is so because the roots must always be complex conjugates. Second, the locus is made of as many tracks as there are poles in the

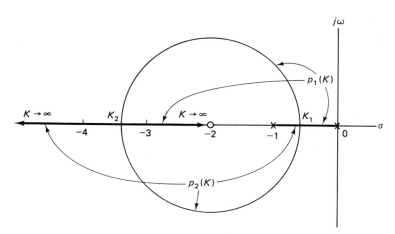

Figure 4.5   Root locus for the system in Fig. 4.4.

open-loop system, and each track starts (thinking of $K = 0$ as a start) on one of the open-loop poles. The tracks either go off to infinity or end on an open-loop zero. To unify this property the excess of open-loop poles over open-loop zeros are called and counted as zeros at infinity. You will see later that these are not always on the negative real axis as in these examples.

The last case involved a modest amount of analysis. The effort rapidly grows with the order of the system. The sketching rules given in the next section mainly have the purpose of avoiding direct solution for the roots. As you might imagine, repeated formation of the denominator polynomial and factoring by hand calculator is a tedious job. Programs are now available for both large and small computers to generate the plots, some of them based on repeated factoring and some that mechanize the sketching rules. Whatever the package available to you does, it will be useful to you to know its basis and its limitations. The hand-sketching process both trains your intuition and helps you to avoid buying computer-generated trash results.

4.2   THE LOCUS EQUATIONS

The general form of the system I will discuss is given in Fig. 4.6. The denominator of the closed-loop transfer function is

$$\Delta = 1 + KG(s)H(s) \tag{4.14}$$

Since we are not concerned with $G(s)$ and $H(s)$ as separate entities, I will use the common shorthand $GH(s)$ for the open-loop transfer function:

$$G(s)H(s) = GH(s) = \frac{N(s)}{D(s)} \tag{4.15}$$

$N(s)$ and $D(s)$ are polynomials. One can express the condition that $\Delta = 0$ either as

$$1 + KGH(s) = 0 \tag{4.16}$$

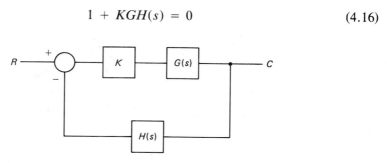

Figure 4.6   System with gain block shown explicitly.

or as

$$D(s) + KN(s) = 0 \qquad (4.17)$$

The latter form is the conventional polynomial form and it will be used to derive some of the sketching rules for special cases. The form in (4.16) gives direct conditions for a point to be on the locus and for the gain at that point. For a value $s = p(K)$ to be on the locus, $GH(p)$ must be real and negative.

$$\angle GH(p) = \pi + 2i\pi \qquad i \text{ an integer} \qquad (4.18)$$

Once the phase condition is satisfied, the point is on the locus because one can always find a gain which satisfies (4.16).

$$K = \frac{1}{|GH(p)|} \qquad (4.19)$$

The phase condition is very important for several of the sketching rules and is sometimes called the *locus equation*. It has a direct geometrical meaning as follows. If

$$GH(s) = \frac{\displaystyle\prod_{i=1}^{m} (s - z_i)}{\displaystyle\prod_{k=1}^{n} (s - d_k)} \qquad (4.20)$$

then the angle at a particular value of s is the sum of the angles of the zero factors minus the sum of the angles of the pole factors. Each of these factors can be represented by a line in the s plane from the zero or pole in question to the

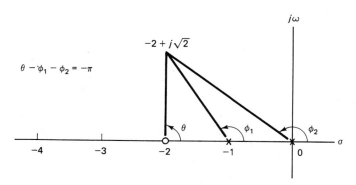

Figure 4.7   Example of the locus angle condition.

particular point. For example, the point at the top of the circle in Fig. 4.5 is $-2 + j\sqrt{2}$. Figure 4.7 shows this point, with lines drawn to it from each of the poles and the zero. From right-triangle geometry, $\theta = \pi/2$, $\phi_1 = \pi - \tan^{-1}\sqrt{2}$, $\phi_2 = \pi - \tan^{-1}\dfrac{\sqrt{2}}{2} = \pi - \tan^{-1}(1/\sqrt{2}) = \pi/2 + \tan^{-1}\sqrt{2}$.

$$\angle GH(-2 + j\sqrt{2}) = \theta - (\phi_1 + \phi_2) = \frac{\pi}{2} - \left(\pi + \frac{\pi}{2}\right) = -\pi$$

What is the value of K for this locus point? Note that if you use the line segment lengths in the figure, your answer will be wrong. Why?

4.3   PROPERTIES AND SKETCHING RULES

Equation (4.17) shows why the tracks start at the open-loop poles. When $K = 0$, only these poles remain in the polynomial. Likewise, as K becomes very large, the only finite values of s which can drive (4.17) to zero are near the zeros of $N(s)$, so some tracks must terminate on these zeros. Remember that real systems are low pass, so that the degree of D is always higher than that of N. Those tracks that do not terminate on the finite zeros must go to infinity in a symmetrical way. The details of this behavior will be given a little later. I will present the sketching rules in the order in which one actually uses them, which also tends to be the order of increasing complexity.

Rule 1: Real-Axis Segments

In the cases we have discussed so far, the real axis has had major portions of the locus. The angle condition gives us an easy tool to find the general rule. Look at the cases in Fig. 4.8. If the point in question is to the left of a real pole or zero, the angle contribution is π, while if it is to the right, the contribution is zero, as in Fig. 4.8a. The net contribution when p is to the left of two such points is a multiple of 2π, as in Fig. 4.8b. Complex factors make a net contribution of zero, as shown in Fig. 4.8c. The general rule is thus

Points on the real axis which are to the left of an odd number of real poles or zeros are locus points and not otherwise.

For cases with a number of real-axis poles and zeros, start with the rightmost one, draw to the next one at the left, stop drawing, start drawing at the next one at the left (third one now), stop at the next (fourth one), and so on. If the total number of real poles and zeros is odd, the last segment will go off to infinity at the left. Look back at the cases in Sec. 4.1. For a double singularity, such as a double pole at the origin, simply think of them as two separate roots with a vanishing space between them.

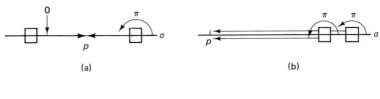

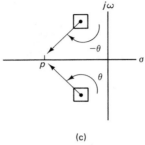

(c)

Figure 4.8 Angles to a real-axis point. The squares may be either poles or zeros.

Rule 2: Asymptotes and the Zeros at Infinity

To deal with the zeros at infinity we would like a simple approximation that will tell us how the tracks behave as they go off. From the cases so far, it appears that the tracks go off in straight lines, and this is generally true, sometimes taking an initial curve to get headed in the right direction. The most direct way to go is to use (4.17) with GH formulated as in (4.20). This defines N and D as

$$N(s) = \prod_{i=1}^{m} (s - z_1) = \sum_{i=0}^{m} b_i s^i \qquad b_m = 1$$

$$D(s) = \prod_{k=1}^{n} (s - d_k) = \sum_{k=0}^{n} q_k s^k \qquad a_n = 1 \qquad (4.21)$$

This allows (4.17) to be reformulated as

$$\frac{D(s)}{N(s)} = -K = Ke^{j(\pi + 2i\pi)} \qquad i \text{ an integer}$$

$$\frac{s^n + a_{n-1}s^{n-1} + \cdots}{s^m + b_{m-1}s^{m-1} + \cdots} = Ke^{j(\pi + 2i\pi)} \qquad (4.22)$$

Since we are looking at large s, the crudest approximation is to take the highest power in each polynomial and omit the rest:

$$s^{n-m} \cong Ke^{j\pi} \qquad n > m$$

$$s = K^{1/(n-m)} \exp j \frac{\pi + 2i\pi}{n - m} \qquad 1 \le i \le n - m \tag{4.23}$$

This result says that, there are approximately $n - m$ straight lines coming out from the origin and equally spaced, in angle, that the locus tends toward. A look at the second-order case in Sec. 4.1 will tell you that the directions check, but the lines the locus actually follows don't start at the origin. As is so often the case, an adequate approximation requires keeping the first two terms in the series or sequence. The first two terms of synthetic division of the polynomials give

$$s^{n-m} + (a_{n-1} - b_{m-1})s^{n-n-1} \cong Ke^{j(\pi + 2i\pi)} \tag{4.24}$$

Now what we want is the same result, but with the center of the lines shifted. This is described by the form

$$(s + q)^t = s^t + tqs^{t-1} + \cdots \tag{4.25}$$

Comparing forms,

$$s^{n-m} + (a_{n-1} - b_{m-1})s^{n-m-1} \approx \left(s + \frac{a_{n-1} - b_{m-1}}{n - m} \right)^{n-m} \tag{4.26}$$

which leads to

$$s = -\frac{a_{n-1} - b_{m-1}}{n - m} + K \exp j \frac{\pi + 2i\pi}{n - m} \tag{4.27}$$

$$a_{n-1} = -\sum_{k=1}^{n} d_k \qquad b_{m-1} = -\sum_{i=1}^{m} z_i \tag{4.28}$$

The excess poles tend to go to infinity along a set of straight lines centered at

$$\sigma_A = \frac{1}{n - m} \left(\sum_{k=1}^{n} d_k - \sum_{i=1}^{m} z_i \right) \tag{4.29}$$

with angles of

$$\theta_A = \frac{\pi}{n - m}(1 + 2i) \qquad 0 \le i \le n - m - 1 \tag{4.30}$$

Rule 3: Breakpoints

In the plots in Fig. 4.3 and 4.5, you saw that for low K the tracks went along the real axis toward each other, met, and then left the real axis in opposite directions. The point at which tracks meet, which is a multiple-root point, is called a *breakpoint*. I have been talking about a point on the locus as a function of gain $p(K)$. Now think of the relationship in its inverse, the gain as a function of position on the locus $K(p)$. From this point of view, as one approaches a breakpoint the gain increases, hits a maximum, and then decreases if we pass through the breakpoint in the same direction. Then $dK/dp = 0$ at the breakpoint. Reformulating (4.17) once again,

$$K(p) = -\frac{D(p)}{N(p)} \tag{4.31}$$

from which

$$\frac{dK}{dp} = -\frac{D'N - DN'}{N^2} = 0$$

$$N\frac{dD}{dp} - D\frac{dN}{dp} = 0 \tag{4.32}$$

It is important to realize that if p is a breakpoint, then the derivative is zero, but not necessarily the converse. In most cases, (4.32) has more roots than there are breakpoints, so you should know that you need to look for breakpoints before using this equation. You know you have to look for breakpoints when you have a real axis segment bounded by either two poles or two zeros. In the case of two poles, two tracks are heading toward each other, which will force a breakpoint, and in the case of two zeros, two tracks are moving away from each other, which means they must have a common start on the segment.

If a breakpoint value is required, it will be a root of equation (4.32).

For the system in Fig. 4.4,

$$GH(s) = \frac{K(1 + s/2)}{s(1 + s)} = \frac{K'(s + 2)}{s(s + 1)} \qquad K' = \frac{K}{2} \tag{4.33}$$

and

$$N(s) = s + 2 \qquad N' = 1$$
$$D(s) = s^2 + s \qquad D' = 2s + 1 \tag{4.34}$$

so that

$$ND' - DN' = 2s^2 + 5s + 2 - s^2 - s = s^2 + 4s + 2 = 0 \qquad (4.35)$$

and the breakpoints are

$$\sigma_B = -2 \pm \sqrt{2} \qquad (4.36)$$

as before.

Rule 4: Departure and Arrival Angles

Now we venture out into the finite complex plane. For poles and zeros on the real axis, it is usually easy to tell how the locus tracks depart and arrive, but off-axis poles and zeros need some calculation. Suppose we are at a point on the locus which is very near (compared to the other poles and zeros) to a complex pole d_s as shown in Fig. 4.9. Since p is on the locus, it must satisfy the angle criterion:

$$\theta - \phi - \theta_d = \pi(1 + 2i)$$

so

$$\theta_d = \theta - \phi - \pi(1 + 2i) = \theta - \phi - \pi \qquad (4.37)$$

Since p is very close to d_x, the angles from the other poles and zeros are very close to the angles for lines drawn to the pole itself. In the limit then, θ is the angle of the line from the zero to d_x, and ϕ is the angle of the line from the pole

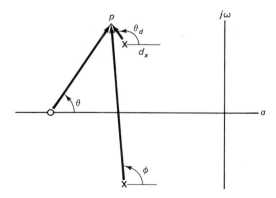

Figure 4.9 Departure angle for a track leaving a complex pole.

to d_x. If d_x is a multiple pole of power m, then $m\theta_d$ must satisfy the angle criterion and

$$\theta_d = \frac{1}{m} [\theta - \phi - \pi(1 + 2i)] \tag{4.38}$$

and there will be m distinct values of i for m tracks leaving the multiple pole. Finally, although only one zero and one pole are shown for clarity, θ is replaced by the sum of the angles from all the zeros to d_x, and ϕ is replaced by the sum of the angles of the lines from all the other poles to d_x.

$$\theta_d = \frac{1}{m} \left[\sum \theta - \sum \phi - \pi(1 + 2i) \right] \tag{4.39}$$

Similarly, you can show that the arrival angle at a complex zero of multiplicity m is

$$\theta_a = \frac{1}{m} \left[\sum \phi - \sum \theta + \pi(1 + 2i) \right] \tag{4.40}$$

Remember that the root locus is symmetrical about the real axis, so that the departure angle from $d_x{}^*$ is the negative of that from d_x, and likewise for complete conjugate zeros.

Let p be the location of an off-axis pole or zero. A locus track departs from or arrives at p at an angle given by (4.39) for a pole, and (4.40) for a zero. m is the multiplicity of the pole or zero at p, $\sum \theta$ is the sum of the angles of lines drawn from these zeros to p, $\sum \phi$ is the sum of the angles of lines drawn from the poles to p, and i is an integer.

These are all the concrete aids to sketching I can offer you. You have probably noticed that these rules give you starts and finishes but tell you nothing about the great in between. This gap is filled by intuition gained from following examples and working out problems. Now I will present a few examples to show how the work progresses.

■ Example 4.1

$$GH(s) = \frac{K}{s(1 + s + s^2/2)} = \frac{K'}{s(s^2 + 2s + 2)} \qquad K' = 2K$$

$$= \frac{K'}{s(s + 1 + j)(s + 1 - j)} \tag{4.41}$$

This transfer function has one real pole at the origin and a complex conjugate pair. I did the sketch for the root locus with pencil and ruler on engineering paper, on a scale of four squares/unit frequency. After I drew the scales, the first step was to find the real-axis segments that are part of the locus. Since there is only one pole and no zeros on the real axis, all of the axis to the left of the origin is on the locus. Since there are three poles and no zeros in the system, all three tracks must go to infinity. The track leaving the origin must simply go along the real axis. The other two stay out in the plane. The center of asymptotes is at

$$\sigma_A = \frac{-1 - j - 1 + j}{3} = -\frac{2}{3}$$

The asymptote angles are

$$\theta_A = \frac{\pi}{3} (1 + 2i) = \frac{\pi}{3}, \pi, \tfrac{5}{3}\pi$$

One asymptote has already been covered by the locus on the negative real axis, while the other two are symmetrical about the real axis, as they must be. There are no breakpoints to calculate, since no tracks meet on the real axis. The departure angle at the upper pole is

$$\theta_d = \pi - \left(\frac{\pi}{2} + \frac{3}{4} \pi \right) = -\frac{\pi}{4}$$

and that at the lower pole is its negative. I know the departure angle for the tracks and where they are headed, so I just draw a smooth transition to make the tracks. The steps and results are shown in Fig. 4.10.

The value of K was not specified. It may be our choice in a real situation. The maximum value for stability is that which corresponds to the point where the locus crosses the imaginary axis. This is a value similar in interest to the gain margin from a Bode plot. From the sketch, the crossover frequency (note the difference in meaning compared to the usage in Chap. 3) is about $\omega_c = 1.4$ rad/s. I measured the distance from each pole to this point to find $K_c' = K'(\omega_c)$.

$$K_c' = \text{product of pole distances} = 1.05 \times 1.5 \times 2.6 = 3.82$$

$$K_c = \frac{K_c'}{2} = 1.9$$

I could have used the Routh test from App. A to find K_c, but I feel that the objective of a sketching method is to minimize the amount of calculation. After all, I can (and did) use a software package to draw the plot and find values of K

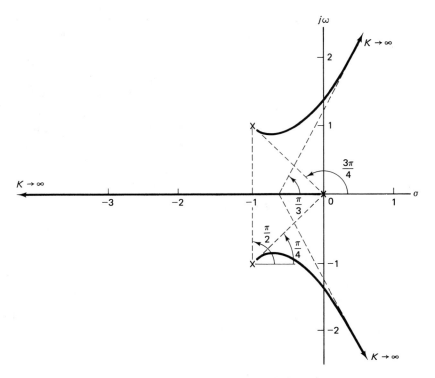

Figure 4.10 Root locus sketch for (4.41).

to any useful accuracy. Figure 4.11 shows the plot made using the Linear Systems Package (discussed in App. B and hereafter referred to as the LSP) on a personal computer (PC) dot-matrix printer. The program shows only the upper half, in order to make maximum use of the display area. The agreement turned out to be quite good, even before my sketch was improved by a graphic artist. The LSP gave $K_c = 1.96$, also a good agreement. ■

■ Example 4.2

Suppose that the performance requirements for the system in (4.41) could not be met with $K = 1.9$. From the sketching rules and the cases discussed in Sec. 4.1, we know that tracks go to zeros, so we might consider adding a zero to the system. A simple network for doing this is called a *lead network,* and it provides a low-frequency zero with a high-frequency pole. Suppose we put the zero at -1, near the system poles, and the pole far away from the action, at -10. This gives the new transfer function

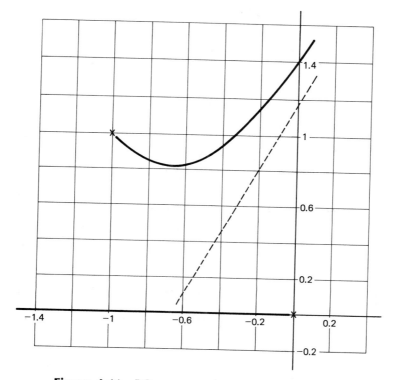

Figure 4.11   PC-generated root locus for (4.41).

$$GH(s) = \frac{K(1 + s)}{s(1 + s + s^2/2)(1 + s/10)}$$

$$= \frac{K'(s + 1)}{s(s^2 + 2s + 2)(s + 10)}, \qquad K' = 20K \quad (4.42)$$

Since the magnitude of the largest pole or zero is now 10, I chose a coarser scale, two squares/unit frequency, for the sketch. This time there are two real poles and a real zero, so the segments of the real axis that are on the locus are between the pole at 0 and the zero at -1, and to the left of the pole at -10. Again, there are three net poles (four poles minus one zero), so there are three asymptotes—at the same angles as before. The center of asymptotes is now

$$\sigma_A = \frac{-10 - 1 - j - 1 + j + 1}{4 - 1} = \frac{-11}{3} = -3.67$$

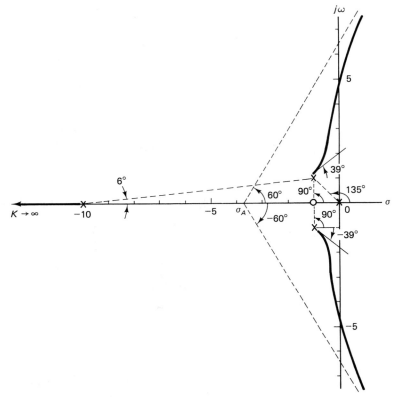

Figure 4.12 Locus construction for Example 4.2.

This time I used a protractor to read the angle from -10 to the upper pole as $6°$, so I calculated the departure angle in degrees:

$$\theta_d = 180 + 90 - (90 + 135 + 6) = 39°$$

This time the asymptote is above the upper pole, so the track departs on a positive bearing. Drawing in a freehand curve for each track completes the sketch (Fig. 4.12).

The imaginary axis crossover frequency is about 4.5 this time, and $\omega_c = 4.5$.

$$K_c' = 4.5 \times 3.6 \times 5.6 \times 10.9/4.7 = 210.4$$

$$K_c = \frac{K_c'}{20} = 10.5$$

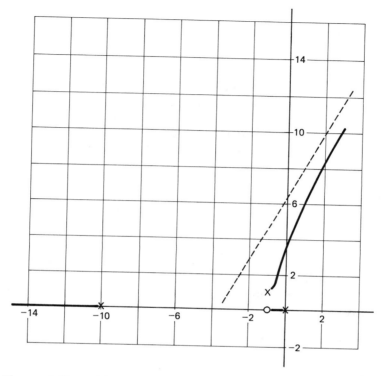

Figure 4.13   Adding a lead network to the locus in Fig. 4.11.

The LSP produced Fig. 4.13. You can see that the shape is right, but I approached the asymptotes too fast. This gave me too high a crossover frequency. Using the computer to check the gain and phase error, I found that crossover frequency is 3.4 rad/s and $K_c = 5.9$. This is an improvement, but not as much as the sketch indicates. ∎

∎ Example 4.3

$$GH(s) = \frac{K[1 + (s/3) + s^2/18]}{s(1 + s/10)[1 + (s/5) + s^2/10]} = \frac{K'(s^2 + 6s + 18)}{s(s + 10)(s^2 + 2s + 10)}$$

$$= \frac{K'(s + 3 + j3)(s + 3 - j3)}{s(s + 10)(s + 1 + j3)(s + 1 - j3)} \qquad K' = \frac{100K}{18} = 5.55K \quad (4.43)$$

This transfer function could have started with a system having the three poles nearest the origin. Its locus would be similar to that for Example 4.1, with instability occurring for some low value of K. To reduce the effect of the complex poles, a compensator with complex zeros and a real pole is proposed. The zeros are close to, but not on, the complex poles to reduce the sensitivity of the locus to component value changes from one unit to the next. Note that if the zeros covered the poles, the system would be of second order and have a locus like that in Fig. 4.3, with vertical tracks passing through -5. After reading through this example, you ought to run a series of root loci for values of the zeros approaching the poles to see how the track behavior changes.

Again, for the pencil-and-ruler sketch I used a scale of two squares/unit frequency to get the -10 pole on the paper. Marking in the poles and zeros and looking at the real axis, I saw that the segment between the two real poles must be on the locus. Since this segment is bounded by poles, it must have a breakpoint. Before finding it though, I found everything else. There are two excess poles, so there must be asymptotes at $\pm 90°$. Their center is at

$$\sigma_A = (-10 - 1 - 1 + 3 + 3)/(4 - 2) = -3$$

Notice that I didn't bother with the imaginary parts, since they always cancel. The departure angle for the upper pole is found by drawing lines and using a protractor to measure the angles that are not obvious from the other poles and zeros:

$$\theta_d = 180 + (0 + 72) - (18 + 90 + 108) = 36°$$

Likewise, the arrival angle at the upper zero is

$$\theta_a = -180 + (23 + 180 + 108 + 135) - 90 = 176°$$

The phase criterion requires 180 plus a multiple of 360°, so I hold off on the first value until I see how the pole and zero angles add up. All the simple things have been done, so next I have to find the real-axis breakpoint. Forming numerator and denominator polynomials and their derivatives gives

$$N = s^2 + 6s + 18 \quad N' = 2s + 6 \quad D = s^4 + 12s^3 + 30s^2 + 100s,$$
$$D' = 4s^3 + 36s^2 + 60s + 100$$
$$ND' = 4s^5 + 60s^4 + 348s^3 + 1108s^2 + 1680s + 1800$$
$$N'D = 2s^5 + 30s^4 + 132s^3 + 380s^2 + 600s + 0$$
$$ND' - N'D = 2s^5 + 30s^4 + 216s^3 + 728s^2 + 1080s + 1800$$

From POLYFAC in the LSP, the roots are $-0.39 \pm j1.91$, -5.077, $-4.57 \pm j5.076$. The only real root is the one I need for the breakpoint, $\sigma_B = -5.08$.

Since the tracks leave the complex poles heading right, and those coming into the zeros do so from the left, it is unlikely that they are the same tracks. I

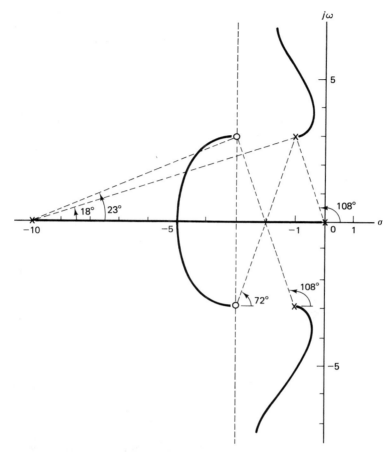

Figure 4.14   Real locus sketch for Example 4.3.

presumed that the zeros were approached from the breakpoint and that the tracks leaving the complex poles turned and headed for the asymptotes. I couldn't tell, without using the Routh test, if the latter crossed the imaginary axis. I guessed that they didn't (Fig. 4.14), and this was confirmed by the computer-generated plot in Fig. 4.15.

One can tell quite a bit about the change in the locus as the zeros move toward the complex poles by looking at the changes in the above calculations. The center of asymptotes moves left toward -5 as the zeros move right. The upper pole departure angle moves counterclockwise to a limit of 58°, while the upper zero arrival angle moves clockwise to a limit of 126°. This suggests that at

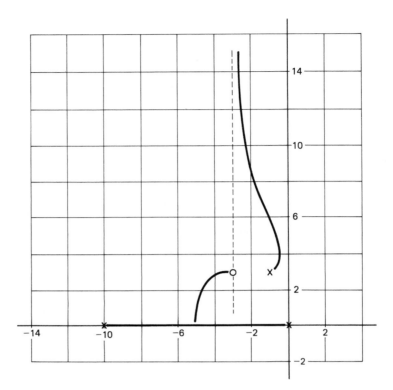

Figure 4.15   PC-generated root locus for Example 4.3.

some position of the zeros the tracks leaving the complex poles will loop toward them, while the tracks leaving the real axis will head for the aysmptotes.   ■

4.4   THE LOCUS FOR A PARAMETER OTHER THAN GAIN

One can use the form of (4.17) to study polynomial root behavior for problems other than control system gain. Any problem in which the root variation as a function of a parameter is sought can be dealt with using root locus rules under the following conditions: The parameter can be arranged as multiplying part of the polynomial and is present in only the first power, and the degree of s in $D(s)$ is not less than that in $N(s)$. In addition, negative values of the parameter can be studied by changing the phase condition (4.18) and the rules that depend on it. I will change the parameter symbol to P, to avoid the association with gain, and reformulate the problem as follows:

$$D(s) + PN(s) = 0 \qquad (4.44)$$

is equivalent to

$$1 + P\frac{N(s)}{D(s)} = 0 \tag{4.45}$$

For $P > 0$, $N/D = -1/P$ to satisfy (4.45) which says

$$\angle\left(\frac{N(s)}{D(s)}\right) = \pi + 2n\pi \qquad n \text{ an integer} \tag{4.46}$$

$$P = \left|\frac{D(s)}{N(s)}\right| \tag{4.47}$$

For $P < 0$, $N/D = 1/P$, so that

$$\angle\left(\frac{N(s)}{D(s)}\right) = 2n\pi \qquad n \text{ an integer} \tag{4.48}$$

$$P = -\left|\frac{D(s)}{N(s)}\right| \tag{4.49}$$

The change in the sign of P essentially removes a π from the asymptote angle rule and the angle of departure and arrival rules. It also alters the real-axis rule to say that segments to the *right* of an odd number of zeros and poles are on the locus. The following examples will not treat the negative parameter case.

■ Example 4.4

Suppose an open-loop transfer function has the form

$$GH(s) = \frac{10}{s(1 + bs + s^2/100)} \tag{4.50}$$

and we want to study the system's closed-loop poles as a function of b. The denominator of the closed-loop transfer function is $1 + GH(s)$ and the equation defining its poles is

$$1 + GH(s) = 0 = \frac{s^3}{100} + bs^2 + s + 10 \tag{4.51}$$

Rearranging this polynomial into a part multiplied by b and the rest of the terms gives

$$\frac{s^3}{100} + s + 10 + bs^2 = 0 \tag{4.52}$$

The polynomial which will be the denominator of the new locus function is not in either of the two standard forms because neither its first nor last coefficient is 1. The two standard forms are produced by multiplying through by 100:

$$s^3 + 100s + 1000 + (100b)s^2 = 0 \qquad \text{let } b' = 100b \qquad (4.53)$$

and by dividing through by 10,

$$\frac{s^3}{1000} + \frac{s}{10} + 1 + \frac{b}{10}s^2 = 0 \qquad \text{let } b'' = \frac{b}{10} \qquad (4.54)$$

For sketching, (4.53) is the more useful form, while the LSP uses (4.54). For (4.53),

$$N(s) = s^2 \qquad D(s) = s^3 + 100s + 1000 \qquad (4.55)$$

Factoring,

$$D(s) = (s + 6.82)(s - 3.41 - j11.6)(s - 3.41 + j11.6) \qquad (4.56)$$

We have two zeros at the origin, one real pole in the LHP, and a complex pair of poles in the RHP. Although I ran this locus on the LSP, it is always good to make simple checks to verify the result. From the sketching rules, the axis to the left of -6.82 is on the locus. Since there are two complex poles left and two zeros, their tracks must go to these zeros. The departure angle for the upper track is

$$\theta_d = 180 + (2 \times 73.6) - (90 + 48.6) = 188.6°$$

This track arrives at

$$\theta_a = (180 + 0 + 106.4 - 106.4)/2 = 90°$$

The plot in Fig. 4.16 agrees with these results. This agreement provides confidence in both the calculations and the plot. We see that the track crosses the imaginary axis at 10, and the LSP tells me that $b'' = 0.01$ at this point on the locus. The minimum b for the stability is therefore 0.1. You can find b' by measuring the phasor values on the plot to check this. The damping of the dominant roots is measured by the angle of a line from the origin to the upper root. The larger this angle, the more damped, less oscillatory, the system's step response will be. It appears that the maximum in this case is around $s = -0.8 + j5$. Again, the LSP tells me the gain at this point is $b'' = 0.041$, which means that $b = 0.41$. For this value of b, the roots of the quadratic in the open-loop transfer function are -2.6 and -38.4. The quadratic thus corresponds to a decaying response with no ripple,

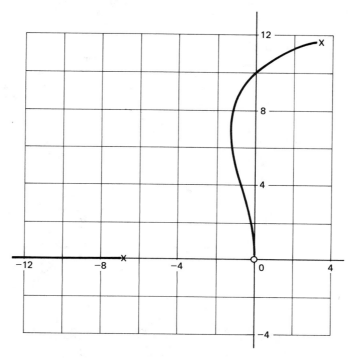

Figure 4.16   Root locus for Example 4.4.

while the closed-loop system is similar in response, as you will see in later chapters, to a rather underdamped second-order system. ■

■ Example 4.5

I have considered a position control system in this example. Suppose that the motor in Table 2.1 is used to move a drilling table in a computer-controlled drill press system. The motor is connected to a threaded shaft which passes through a threaded hole in the table to convert the motor rotation to linear motion at the rate of 40 turns/inch. This is called a screw actuator. A linear position sensor provides 5 V/in. for position feedback to a differential amplifier–power amplifier whose other input is the reference or control voltage and which drives the motor. The motor has a tachometer, so it can be used for rate feedback to provide a reduced driving voltage during transients when the position error is large and the motor would otherwise tend to overshoot as a result of excessive speed. That's a long way of saying the rate feedback provides damping. A block diagram representing the system is shown in Fig. 4.17a, and a reduced version is shown in Fig. 4.17b. The load is represented as a friction constant, reflected back

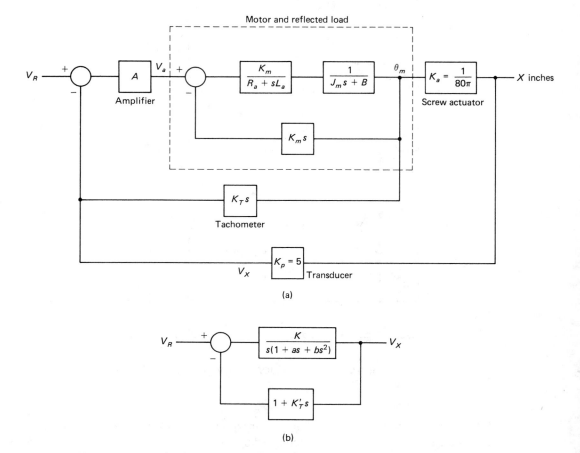

Figure 4.17 Block diagrams for Example 4.5. A linear position control system with tachometer feedback. (a) Blocks for system elements. (b) Reduced, with transducer voltage as output.

and added to the motor friction. The value chosen for this problem is $B_L = 0.2$ N $-$ m/krpm $= 6/\pi$ mN $-$ m $-$ s.

Equation (2.78) gives the transfer from applied voltage to shaft angle. Let

$$Q_m = BR_a + K_m^2 \tag{4.57}$$

Then

$$a = \frac{BL_a + JR_a}{Q_m} \tag{4.58}$$

$$b = \frac{JL_a}{Q_m} \quad (4.59)$$

Let

$$c = K_a K_p \quad (4.60)$$

$$K = \frac{AcK_m}{Q_m} \quad (4.61)$$

$$K_T' = \frac{K_T}{c} \quad (4.62)$$

From Table 2.1, with $J = J_m$ and $B = B_L + B_m$,

$$a = 0.0079 \quad b = 19.922 \times 10^{-6} \quad c = 0.0199 \quad \frac{cK_m}{Q_m} = 0.11296$$

The poles of the quadratic are $-198.3 \pm j104.3$.

There are two parameters which can be adjusted to set the system performance, the amplifier gain and the amount of rate feedback. These are contained in K and K_T' respectively. First, I looked at the root locus with no tachometer feedback, which is shown in Fig. 4.18. The j-axis crossing frequency is about 224

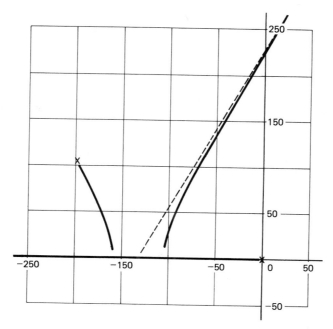

Figure 4.18   Root locus for Example 4.5 with no tachometer damping.

rad/s, and the LSP gives $K = 396.4$ for this point. Next, considering K fixed, I set up the locus equation for K_T' as the parameter:

$$\frac{b}{K} s^3 + \frac{a}{K} s^2 + \frac{s}{K} + 1 + K_T's = 0 \tag{4.63}$$

With $K = 396.4$, the poles of the cubic are -396.5 and $-0.01773 \pm j224$. The root locus for K_T' is shown in Fig. 4.19. The effect of increasing the tachometer constant is to move the complex poles toward the asymptote at -198.2, which is generally good, but a real pole moves toward the origin. I chose to look for a value that gives maximum damping of the complex poles. The track tangent which passes through the origin makes contact at about $-127 + j350$. The parameter at this point has a value of 0.00639. It also has this value on the real axis at about -143.5. In the sense that the poles are roughly all the same distance from the imaginary axis, this seems like a pretty good choice for K_T'. The frequency of the feedback-path zero is $-1/K_T' = -159.5$ rad/s. Since this value is near the magnitude of the forward-path poles, it seems reasonable.

The response to a unit step input (step response) is shown in Fig. 4.20. It appears to be produced by a real pole (an exponential approach to the final value) with some ripple from the complex poles. I moved the zero to the forward path and ran another step response for comparison, which is in Fig. 4.21. Here you

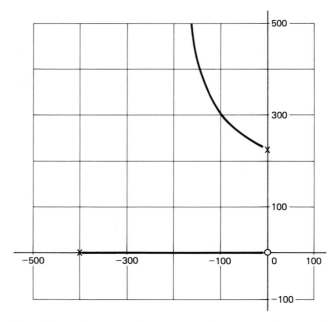

Figure 4.19   Root locus for Example 4.5, as a function of speed feedback, with maximum gain.

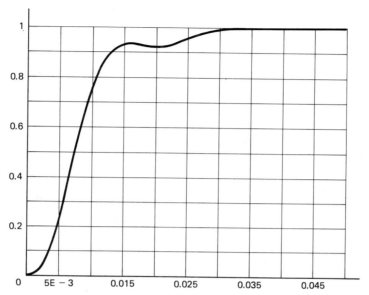

Figure 4.20 Step response for Example 4.5. The loop gain constant is 396.4; the tach. feedback is set to give a zero at −159.5.

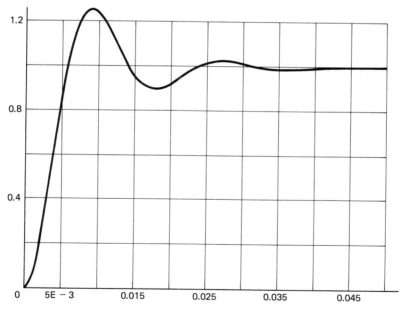

Figure 4.21 Same system as for Fig. 4.20, except that the zero has been moved to the forward path.

see behavior that is dominated by the complex poles. Why the difference? Write the closed-loop transfer functions for the two cases and examine the poles and *zeros* for each.

If the performance is satisfactory, the designer needs to know the amplifier gain and tachometer constant required. $A = 396.4/0.11296 = 3509$, and $K_T = 0.00627 \times 0.0199 = 0.0001248$ V-s. The tachometer supplied with the motor has an output of 0.0668 V-s, so an attenuator of gain $0.0001248/0.0668 = 0.001868$ would have to be designed.

The main purpose of this example is to show the use of the root locus to study the effect of the amount of rate feedback on the system, not to arrive at a final design for the control system. To accomplish the latter, one needs to have a set of performance specifications to work toward and a philosophy of design. Even as it stands, you might explore the limits of performance possible with this system for various combinations of amplifier gain and rate feedback. What combination of values will give a dominant complex pole pattern with good damping? Let's say "dominant" means that the real pole is at least five times the real part of the complex poles, and "good damping" means that the angle on the complex pole is at least 135°. The complex poles should have as large a magnitude as possible to get the quickest system response. There are other performance measures which will be discussed in the design section, and this sytem will be revisited there. Familiarity with it from the root locus viewpoint and also the frequency response viewpoint will help you then. ∎

4.5   THE ALGEBRAIC APPROACH

In Sec. 4.1, I showed that the off-axis locus for the system in Fig. 4.4 is a circle. I did this by taking the equations for the real and imaginary parts of the roots and eliminating K to find an equation relating σ and ω. The idea of finding equations that give the off-axis tracks is the basis of the algebraic approach. Such equations allow one to fill in the in-between parts that the sketching rules only point to. We cannot solve explicitly for the roots for higher-order systems, but we can nevertheless get the track equations by splitting (4.17) into real and imaginary parts. Let

$$N(\sigma + j\omega) = N_r + jN_i \quad \text{and} \quad D(\sigma + j\omega) = D_r + jD_i \quad (4.64)$$

Since K is real,

$$D_r + KN_r = 0$$
$$D_i + KN_i = 0 \quad (4.65)$$

Eliminating K from the two equations gives

$$D_i N_r - D_r N_i = 0 \qquad (4.66)$$

Since the only restriction on K is that it be real, this result gives the closed-loop poles for all values of the parameter from minus to plus infinity. Also, $\omega = 0$, the entire real axis, is always a solution of (4.66). One can separate the portions for positive K with the help of the sketching rules. A couple of simple examples show the application and some possibilities of this approach.

■ Example 4.6

Let's revisit the system in Fig. 4.4.

$$D(s) = s^2 + s \qquad N(s) = 1 + \frac{s}{2} \qquad (4.67)$$

$$D(\sigma + j\omega) = \sigma^2 + j2\omega - \omega^2 + \sigma + j\omega$$
$$D_r = \sigma^2 + \sigma - \omega^2 \qquad D_i = (2\sigma + 1)\omega \qquad (4.68)$$

$$N(\sigma + j\omega) = 1 + \frac{\sigma}{2} + j\frac{\omega}{2}$$

$$N_r = 1 + \frac{\sigma}{2} \qquad N_i = \frac{\omega}{2}$$

$$D_r N_i = \frac{\omega}{2}(\sigma^2 + \sigma - \omega^2) \qquad (4.69)$$

$$D_i N_r = \omega(2\sigma + 1)\left(1 + \frac{\sigma}{2}\right)$$

$$= \frac{\omega}{2}(2\sigma^2 + 5\sigma + 2)$$

$$D_r N_i - D_i N_r = -\frac{\omega}{2}(\sigma^2 + 4\sigma + 2 + \omega^2) = 0 \qquad (4.70)$$

implies $\omega = 0$ as one solution, and

$$(\sigma + 2)^2 + \omega^2 = 2 \qquad (4.71)$$

as the off-axis equation, which is the same as (4.11).

This system is always stable, but its closed-loop poles do have a minimum damping angle for some K, which should correspond to minimum phase margin. A line passing through the origin is described by

$$\omega = -b\sigma \qquad (4.72)$$

for $b > 0$ and a negative slope. The intersection with the circle is given by substituting for ω in (4.71):

$$(1 + b^2)\sigma^2 + 4\sigma + 2 = 0$$

$$\sigma = \frac{-2}{1 + b^2} \pm \left[\frac{4}{(1 + b^2)^2} - \frac{2}{1 + b^2} \right]^{1/2} \tag{4.73}$$

The tangent is the solution for b which makes the radical zero:

$$\frac{4}{1 + b^2} - 2 = 0$$

$$1 + b^2 = 2 \qquad b = 1 \tag{4.74}$$

$$\sigma_T = -1 \qquad \omega_T = 1 \tag{4.75}$$

$$K_T = -\frac{D_r}{N_r} = -\frac{-1}{\frac{1}{2}} = 2 \tag{4.76}$$

The middle of the -2 slope on a Bode plot is at $\omega = \sqrt{2}$.

$$G(j\sqrt{2}) = \frac{2(1 + j/\sqrt{2})}{j\sqrt{2}(1 + j\sqrt{2})} = \frac{\sqrt{2} + j}{j(1 + j\sqrt{2})}$$

$$= 1\angle - 2\tan^{-1}\sqrt{2} \tag{4.77}$$

which is unity gain and the maximum phase point. ∎

■ Example 4.7

Consider next the three-pole system described by (4.41), whose locus is given in Fig. 4.11. For this system,

$$D(s) = s^3 + 2s^2 + 2s \qquad N(s) = 1 \tag{4.78}$$

$$D(\sigma + j\omega) = \sigma^3 + 3j\sigma^2\omega - 3\sigma\omega^2 - j\omega^3$$
$$+ 2\sigma^2 + 4j\sigma\omega - 2\omega^2 + 2\sigma + 2j\omega \tag{4.79}$$

$$D_r = \sigma^3 + 2\sigma^2 + 2\sigma - (3\sigma + 2)\omega^2$$

$$D_i = \omega(3\sigma^2 + 4\sigma + 2 - \omega^2) \tag{4.80}$$

$$N_r = 1 \qquad N_i = 0$$

$$D_i N_r = 0 = \omega[3(\sigma + \tfrac{2}{3})^2 + \tfrac{2}{3} - \omega^2] \tag{4.81}$$

$$\left(\sigma + \frac{2}{3} \right)^2 - \frac{\omega^2}{3} = -\frac{2}{9} \tag{4.82}$$

Again, $\omega = 0$ is a solution. Equation (4.8s) describes a hyperbola, opening upward and downward. There is no real solution for when $\omega = 0$. If you happen to know how to construct a hyperbola knowing its equation, you can do so, otherwise, you can always put in a few values of σ and solve for ω. Note that for large values, (4.82) is satisfied by two straight lines passing through $-\frac{2}{3}$, 0 with slopes $\pm 1/\sqrt{3}$, which is to say at angles of $\pm 60°$ with respect to the real axis. At $\sigma = -\frac{2}{3}$, $\omega = \sqrt{2/3} = 0.8165$. This is the lowest point of the hyperbola. The imaginary axis crossing is at $\sigma = 0$, $\omega_c = \sqrt{2}$. Suppose one wishes to have the complex poles at an angle of $120°$ from the real axis. The upper one must lie on a line whose slope is $-\sqrt{3}$:

$$\omega = -\sqrt{3}\sigma \tag{4.83}$$

Substituting into (4.82), $\frac{4}{3}\sigma + \frac{4}{9} = -\frac{2}{9}$

$$\sigma_p = -\frac{1}{2} \qquad \omega_p = \frac{\sqrt{3}}{2}$$

$$K_p' = -D_r\left(-\frac{1}{2} + j\frac{\sqrt{3}}{2}\right) = 1 \tag{4.84}$$

For comparison, the gain at the imaginary axis crossover is

$$K_c' = -D_r(0 + j\sqrt{2}) = 4 \tag{4.85}$$

In this case, we can also find the value of the real pole for $K' = 1$ by solving

$$D_r(\sigma + j0) + K'N_r(\sigma + j0) = 0 \tag{4.86}$$

$$\sigma^3 + 2\sigma^2 + 2\sigma + 1 = 0 \tag{4.87}$$

to obtain

$$\sigma_r = -1 \tag{4.88}$$

∎

■ Example 4.8

For a slightly more complex case, I've taken the system in Example 4.3, with the open-loop transfer function in (4.43). This one has four poles and two zeros.

$$D(s) = s(s + 10)(s^2 + 2s + 10) = s^4 + 12s^3 + 30s^2 + 100s$$

$$N(s) = s^2 + 6s + 18$$

$$D(\sigma + j\omega) = \sigma^4 + j4\sigma^3\omega - 6\sigma^2\omega^2 - j4\sigma\omega^3 + \omega^4 + 12\sigma^3$$
$$+ j36\sigma^2\omega - 36\sigma\omega^2 - j12\omega^3 + 30\sigma^2 + j60\sigma\omega$$
$$- 30\omega^2 + 100\sigma + j100\omega$$

$$D_r = \sigma(\sigma^3 + 12\sigma^2 + 30\sigma + 100) - 6(\sigma^2 + 6\sigma + 5)\omega^2 + \omega^4$$

$$D_i = \omega[4(\sigma^3 + 9\sigma^2 + 15\sigma + 25) - 4(\sigma + 3)\omega^2] \qquad (4.89)$$

$$N(\sigma + j\omega) = \sigma^2 + j2\sigma\omega - \omega^2 + 6\sigma + j6\omega + 18$$

$$N_r = \sigma^2 + 6\sigma + 18 - \omega^2$$

$$N_i = 2\omega(\sigma + 3) \qquad (4.90)$$

$$D_r N_i = 2\omega(\sigma + 3)[\sigma(\sigma^3 + 12\sigma^2 + 30\sigma + 100)$$
$$- 6(\sigma^2 + 6\sigma + 5)\omega^2 + \omega^4]$$

$$D_i N_r = 4\omega(\sigma^2 + 6\sigma + 18 - \omega^2)$$
$$\times [\sigma^3 + 9\sigma^2 + 15\sigma + 25 - (\sigma + 3)\omega^2]$$

$$D_i N_r - D_r N_i = 0 = (\sigma + 3)\omega^4 - (-2\sigma^3 - 18\sigma^2 - 36\sigma + 68)\omega^2$$
$$+ (\sigma^5 + 15\sigma^4 + 108\sigma^3 + 364\sigma^2 + 540\sigma + 900) \qquad (4.91)$$

For $\omega = 0$,

$$\sigma_B = -5.0775 \qquad (4.92)$$

Notice that for each value of σ, the equation is a quadratic in ω^2. This makes stepping along in σ and solving for ω a lot faster than stepping in K' and factoring (4.17), which is fourth-order for this system. The locus is shown in Fig. 4.15. Its off-axis parts consist of two tracks, one below and the other above $j3$. One might suspect that a well-damped system could be obtained if the gain were set so that the lower track pole is on a line of slope $-\frac{1}{2}$ from the origin. This should put the upper pole at a value where the frequency response is down enough to make its effect small. I chose this slope because a slope with a convenient value a little higher would miss the lower track. In fact, you can see that there is a range of slopes for lines passing through the origin so that they do not hit either track. For $\omega = -\sigma/2$, I had to go to (4.91) and substitute, removing ω. It helps to arrange each term in a row with descending powers of σ and the coefficients of equal powers lined up in columns:

$$\frac{1}{16} \times \quad \sigma^5 + \quad 3\sigma^4$$
$$8\sigma^5 + \quad 72\sigma^4 + \quad\quad 144\sigma^3 - \quad 272\sigma^2$$
$$16\sigma^5 + 240\sigma^4 + \quad 1728\sigma^3 + 5824\sigma^2 + 8640\sigma + \quad 14,400$$
$$25\sigma^5 + 315\sigma^4 + \quad 1872\sigma^3 + 5552\sigma^2 + 8640\sigma + \quad 14,400 = 0$$

The real root is a $\sigma_p = -4.54428$, so $\omega_p = 2.27214$.

$$K_p' = \frac{-D_i}{N_i} = 2\omega_p^2 - 2\,\frac{\sigma_p^3 + 9\sigma_p^2 + 15\sigma_p + 25}{\sigma_p + 3}$$

$$= 73.57 \qquad K_p = 0.18K_p' = 13.25. \tag{4.93}$$

On the other track, $K = K_p$ at about $-1.6 + j7$. This is unfortunate because $7/2.27 = 3.08$, which means the very lightly damped poles are too close to the chosen ones and will have a strong effect on the system response. Figure 4.22 shows the unit step response. It has a lot of ripple, with a period of a little under 1 s, or a ripple frequency of a bit over 1 Hz = 6.28 rad/s. The response can be smoothed either by reducing the gain or by moving the zeros closer to the poles. ∎

These examples illustrate some general properties of the algebraic approach. When (4.66) is arranged as a polynomial in ω^2 with coefficients which are functions of σ, the pole determination involves factoring a polynomial whose order is equal to or less than half that of the system polynomial. Since we know the open-loop poles and zeros, we know the portion of the s plane in which we want to see the locus, while we have little or no idea ahead of time about the corresponding gain

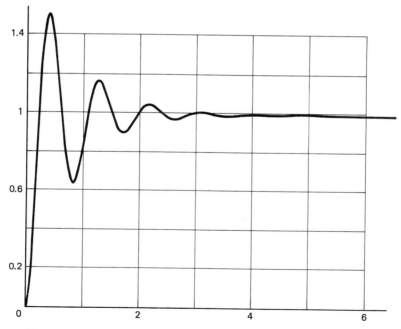

Figure 4.22   Step response for Example 4.8, $K = 13.25$.

range. Arithmetic solutions for pole locations with a specified angle, for the imaginary axis crossing, and for the gain at a given root value can be found with any desired precision—much more precision than that to which one usually knows the system parameters. There are only two significant disadvantages to this method of finding the root locus: In my view the amount of algebra necessary to find the coefficient equations for (4.66) poses a high danger of mistakes, and the method is less useful for a system with time delay because the equations for ω become transcendental. The time delay problem is treated in the next section.

In 1972, Harley and Power published a paper on the analytical (algebraic) approach in which they included a table for the off-axis locus equations in terms of the system polynomial coefficients for systems up to five poles. Although this table is not exhaustive, it is a useful place to start and reduces the chance of error through algebra. One could program a programmable calculator or personal computer with these equations and generate loci for the systems covered. The sketching rules help to separate the roots for positive parameter values from those for negative parameter values (positive gain from negative gain). On a more general level, one can write a program to generate the coefficients for (4.66) from the open-loop poles and zeros and a given value of σ. I did this as an APL function by treating each factor as a real polynomial and an imaginary polynomial and using another function, PMUL, which takes the coefficients of two polynomials and produces the coefficients of their product. In this way, I looped through the input polynomial, $D(s)$ or $N(s)$, to produce real and imaginary polynomials to pass to another function which uses (4.66) and LROOTS (Laguerre root finder) to find the corresponding ω. The development was quite fast, and the result runs fast enough to be useful even though it's run under an interpreter, whatever the level of your computational ability and resources, I recommend the algebraic approach as a useful tool for the root locus analysis and design of a system.

4.6 LOCI FOR SYSTEMS WITH TIME DELAY

If a system has significant time delay, one must introduce an exponential function of s into its transfer function, as discussed in Chap. 2. The general case is shown in Fig. 4.23, with time delays in both the forward and feedback paths. The closed-loop transfer function is

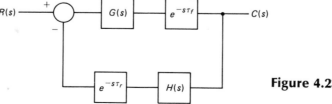

Figure 4.23 General system with time delays.

$$\frac{C}{R} = \frac{G(s)e^{-s\tau_f}}{1 + G(s)H(s)e^{-s(\tau_f + \tau_r)}} \tag{4.94}$$

Its poles are given by

$$1 + GH(s)e^{-s\tau} = 0 \qquad \tau = \tau_f + \tau_r \tag{4.95}$$

Using polynomials, this is

$$1 + \frac{N(s)}{D(s)} Ke^{-s\tau} = 0 \tag{4.96}$$

and replacing s by $\sigma + j\omega$,

$$1 + \frac{N}{D} Ke^{-\sigma\tau}e^{-j\omega\tau} = 0 \tag{4.97}$$

For $K > 0$, the angle condition now becomes

$$\angle \frac{N}{D} = \pi + \omega\tau + 2n\pi \qquad n \text{ an integer} \tag{4.98}$$

and the magnitude equation is

$$K = e^{\sigma\tau} \left| \frac{D}{N} \right| \tag{4.99}$$

The general effect of the angle condition is to force locus points to the right of where they would be otherwise, so that N/D will have more lead (or less lag) than would be the case without the time delay. Since, by definition, $\sigma < 0$ in the LHP, the magnitude condition says that the gain on a locus point will be smaller the further it is into the LHP. The net effect is a curving of the locus toward and into the RHP, as shown in Fig. 4.24. Another interesting result is that there is an infinity of tracks going left to right in the s plane. This is caused by the angle condition in which the $\omega\tau$ term cycles through 2π periods as ω increases. Only the first period is shown in Fig. 4.24. Some of the sketching rules are still useful or can be modified. This is left for you to do in the problem set.

Recasting (4.96) into the form of (4.17) gives us

$$D(s) + N(s)Ke^{-s\tau} = 0 \tag{4.100}$$

Evidently factoring polynomials is out as a way of finding the roots. We can, however, follow the method of the algebraic approach and find something useful. By writing the polynomials and the exponential in real and imaginary parts,

$$D_r + jD_i + Ke^{-\sigma\tau}(N_r + jN_i)[\cos(\omega\tau) - j\sin(\omega\tau)] = 0 \tag{4.101}$$

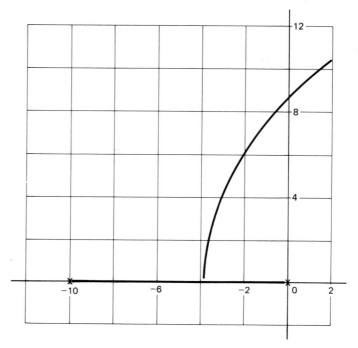

Figure 4.24 Root locus for a two-pole system with a 0.1 s time delay.

and then separating the real and imaginary parts of the equation

$$Dr + Ke^{-\sigma\tau}[N_r \cos(\omega\tau) + N_i \sin(\omega\tau)] = 0 \qquad (4.102)$$

$$D_i + Ke^{-\sigma\tau}[N_i \cos(\omega\tau) - N_r \sin(\omega\tau)] = 0 \qquad (4.103)$$

Again, one can eliminate the gain and the real exponential:

$$Ke^{-\sigma\tau} = \frac{-D_r}{N_r \cos(\omega\tau) + N_i \sin(\omega\tau)} \qquad (4.104)$$

to give

$$(D_iN_r - D_rN_i)\cos(\omega\tau) + (D_iN_i + D_rN_r)\sin(\omega\tau) = 0 \qquad (4.105)$$

We can't choose σ and solve for ω because of the trigonometric functions, but we can choose ω and solve for σ by way of polynomial formation and factoring. Drawing on the examples of the last section, these will be polynomials of degree at least equal to that of the system polynomial without the time delay, but this procedure is the only "direct" method I know. The alternative is to use a search method which looks for the point which satisfies the angle condition. The LSP uses the search method, employing the modified sketching rules for starting points. In general, one cannot solve (4.105) for poles of a specified angle or for the imaginary axis crossing because of the transcendental functions. One is reduced to estimating from a plot and checking by calculating the transfer function at the estimated point.

FURTHER READING

Power, Henry M., and Simpson, Robert J., *Introduction to Dynamics and Control*, New York, McGraw-Hill, 1978.

PROBLEMS

Section 4.1

4.1   Derive the locus equations for

$$GH(s) = \frac{K'(s + z)}{s(s + p)}$$

That is, find the function $\omega(\sigma)$ by factoring the polynomial for the closed-loop poles and eliminating K' from the results.

4.2   Write the open-loop transfer function whose root locus is shown in Fig. P4.2. What is the root value for $\sigma = -\omega$?

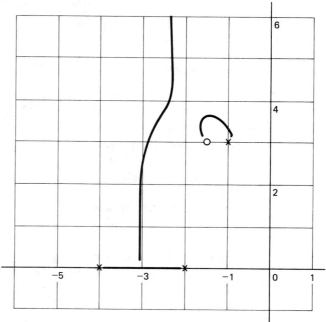

Figure P4.2.   Upper half of a root locus.

Section 4.2

4.3 Calculate $\angle GH(p_i)$ for the points p_i shown in Fig. P4.3. State whether each is a locus point.

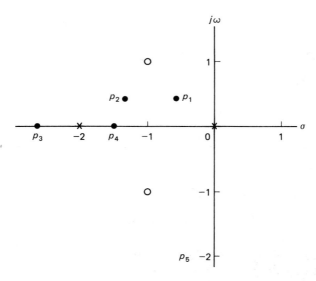

Figure P4.3.

4.4 What is (4.17) if

$$GH(s) = \frac{A(s + 1 + j5)(s + 1 - j5)}{s(s + 2)(s + 6)(s + 30)} ?$$

Section 4.3

4.5 Verify (4.40).

4.6

$$GH(s) = \frac{K'(s^2 + 2qs + q^2 + 9)}{(s + 2)(s + 4)(s^2 + 2s + 10)}$$

For $q = 0.2, 0.5, 1, 2$, find the
(a) Real axis locus segments
(b) Center of asymptotes
(c) Departure angle
(d) Arrival angle.
Organize the results in a table.

4.7 Use a computer program to generate the loci for Prob. 4.6. In each case do the following:
(a) Find the root for which $\sigma = -\omega$.
(b) Find the roots on the other tracks which have the same gain as that for the root in (a).
(c) Arrange the results in a table.

Section 4.4

4.8 For $P < 0$ derive:
   (a) The real-axis rule
   (b) The asymptote rules
   (c) The angle of departure and arrival rules.

4.9

$$GH(s) = \frac{K[1 + (q/5)s + s^2/10]}{(1 + s/2)(1 + s/4)[1 + (s/5) + s^2/10]}$$

   (a) Plot the root loci for q, with $K = 1, 2,$ and 5.
   (b) In each case, what is the maximum value obtainable for the angle of the least-damped root? Make a table showing q and the pole values for this condition.

Section 4.5

4.10 Use the algebraic approach on (4.42) and verify the results in Example 4.2.

Section 4.6

4.11 Find the modifications caused by time delay to the following rules:
   (a) Real-axis rule
   (b) Center of asymptotes
   (c) Arrival and departure angles
   (d) Breakpoints.

4.12

$$GH(s) = \frac{Ke^{-0.1s}}{s(1 + s/10)}$$

Use (4.105) to form the locus equation as a polynomial in σ whose coefficients are functions of ω. Find and plot σ for at least 10 values of ω, $0 \le \omega \le 10$.

Chapter 5

TIME RESPONSE

5.1   STEADY-STATE ERROR AND SYSTEM TAPE

In Chap. 1 it was observed that the most important advantage of feedback in a control system is error reduction. In Chap. 2 we examined mathematical models for components and systems, and in Chaps. 3 and 4 we studied the basis of frequency domain analysis with, you may have noticed, emphasis on the main problem with feedback, stability. Now it's time to take an analytic look at the advantage of feedback, partly to remind you of the motivation for all the past and current effort put into this area and partly to lay the foundation for measures of quality of control systems.

As with electric circuits, the time response of control systems is roughly divided into two regimes, the transient response and the steady-state response. Unlike electric circuits, the test signals of interest are not usually sine waves. Instead, the signals used to test and specify control system response are the impulse (Dirac delta) function and its integrals, the step, the ramp, and occasionally the parabola functions of time. I list these functions and their Laplace transforms in (5.1) to (5.4) for reference, with the assumption that you have seen them in other courses, possibly with different symbols.

Impulse:
$$\text{imp}(t) = 0 \qquad \text{for all } t <> 0 \tag{5.1}$$
$$\int_{-e}^{+e} \text{imp}\,(t)\,dt = 1 \qquad \text{for any } e;\ \text{Imp}(s) = 1$$

Unit step:
$$u(t) = 0 \qquad \text{for } t < 0 \tag{5.2}$$
$$= 1 \qquad \text{for } t > 0 \qquad U(s) = 1/s$$

Ramp: $\text{ramp}(t) = tu(t)$ $\text{Ramp}(s) = 1/s^2$ (5.3)

Parabola:
$$p(t) = \frac{t\,\text{ramp}(t)}{2}$$

$$= \frac{t^2 u(t)}{2} \qquad P(s) = \frac{1}{s^3} \qquad (5.4)$$

The function most commonly used to specify steady-state response is the step, followed by the ramp. Because position control systems are so common, the step input is often called a *constant-position* or simply *position input,* and the ramp input is likewise often called a *constant-velocity* or simply *velocity input.* In this vein, a parabola input is a constant-acceleration input. Why isn't the impulse function used to specify steady-state response?

■ Example 5.1

An operational amplifier circuit with a desired net gain of -10 is shown in Fig. 5.1. What is its steady-state error in response to a unit step input?

Figure 5.1b shows a block diagram for the circuit model, which is based on the node equations for the labeled voltages. For a stable system, the steady-state value of the output can be found by applying the Laplace transform final value theorem to the product of the transfer function and the transform of the input function. Since I am interested in the error, I define it as the difference between the desired response and the actual response. The closed-loop transfer function is

$$\frac{V_o}{V_i} = \frac{G_1}{\displaystyle\sum_i} \frac{-A(s)\left(G_0 / \displaystyle\sum_o\right)}{1 + \left[A(s)G_0 G_2 \Big/ \displaystyle\sum_i \sum_o\right]} \qquad (5.5)$$

$$= \frac{-A_0\left(G_1 G_0 \Big/ \displaystyle\sum_i \sum_o\right)}{s + 1 + A_0\left(G_2 G_0 \Big/ \displaystyle\sum_i \sum_o\right)}$$

$$\text{where } \sum_i = G_1 + G_2 + G_i, \quad \sum_o = G_2 + G_0 + G_L \qquad (5.6)$$

For a unit step input, $V_i(s) = 1/s$, the desired output is $-10/s$. Therefore the error is $-(10/s) - V_o(s)$:

$$\text{Error} = -\frac{10}{s} + \frac{\left(G_1 G_0 \Big/ \displaystyle\sum_i \sum_0\right) A_0}{s\left[s + 1 + \left(G_2 G_0 \Big/ \displaystyle\sum_i \sum_o\right) A_0\right]} \qquad (5.7)$$

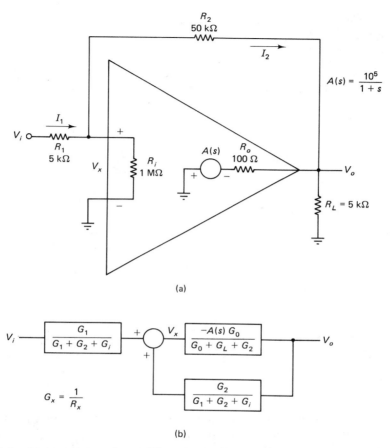

Figure 5.1 (a)

(b)

Figure 5.1   An operational amplifier circuit for a gain of -10. (a) Circuit model. (b) Block diagram.

Since the system is obviously stable,

$$\text{Steady-state error} = \lim_{s \to 0} s \text{ Error } (s) = -10 + \frac{G_1}{G_2} \frac{G_2 G_0 A_0 \Big/ \sum_i \sum_o}{1 + G_2 G_0 A_0 \Big/ \sum_i \sum_o} \qquad (5.8)$$

As you will see shortly, for the typical numbers chosen in this case the second term in (5.8) is very close to $G_1/G_2 = R_2/R_1$. To the usual first approximation then, the steady-state error is the amount by which this ratio differs from 10. From the point of view of many writers in control theory, this ratio *is* the desired value. R_1 and R_2 are the precision passive parts the designer choses to set the closed-loop gain, so it is up to the rest of the system to make the output match their ratio. If $G_1/G_2 = 10$,

$$E_{ss} = -\frac{G_1}{G_2} + \frac{G_1}{G_2}\frac{p-1}{p} = \frac{G_1}{G_2}\frac{-1}{p} \tag{5.9}$$

where

$$p = 1 + \frac{G_2 G_0}{\displaystyle\sum_i \sum_o} A_0 \tag{5.10}$$

is the closed-loop pole.

$$\frac{G_2}{\displaystyle\sum_i} = \frac{2 \times 10^{-5}}{20 \times 10^{-5} + 2 \times 10^{-5} + 10^{-6}} = \frac{2}{22.1} = 0.0905$$

$$\frac{G_0}{\displaystyle\sum_o} = \frac{0.01}{0.01 + 2 \times 2 \times 10^{-5}} = \frac{1}{1.004} = 0.996$$

$$p = 1 + 0.0905 \times 0.996 \times 10^5 = 9015 \text{ rad/s}$$

$$\frac{G_1}{G_2 p} = 0.00111$$

The purpose of doing the arithmetic separately for each factor is to show the effects of the amplifier input and output conductances. If the feedback resistor is chosen to be in the same order of magnitude as the input resistance, or if the load is of the same order as the output resistance, the loop gain will decrease significantly and the error will consequently increase. As things are, favorable numbers have been used so that the main contribution to error is the finite gain of the amplifier.   ■

The block diagram in Fig. 5.1b is a special case of the general form shown in Fig. 5.2. For this single-pole system,

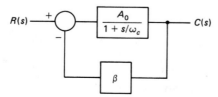

Figure 5.2   Single-pole system.

$$\frac{C}{R} = \frac{A_0}{1 + (s/\omega_c) + A_0\beta} = \frac{A_0/(1 + A_0\beta)}{1 + s/(\omega_c(1 + A_0\beta))} \qquad (5.11)$$

$$\text{Ideal } \frac{C}{R} = \frac{1}{\beta} \qquad (5.12)$$

$$\text{Error}(s) = \left[\frac{1}{\beta} - \frac{A_0}{1 + A_0\beta} \frac{1}{1 + s/(\omega_c(1 + A_0\beta))} \right] R(s) \qquad (5.13)$$

For $R(s) = 1/s$,

$$E_{ss} = \frac{1}{\beta}\left(1 - \frac{A_0\beta}{1 + A_0\beta}\right) = \frac{1}{\beta}\frac{1}{1 + A_0\beta} \qquad (5.14)$$

A position error constant is usually defined as

$$K_p = A_0\beta \qquad (5.15)$$

In the last example, the steady-state error was about 0.1%. In some applications, it might be desirable to have an accuracy of 0.01%. It is possible to make film resistors to this tolerance, but the op amp gain may vary over an order of magnitude from one unit to the next. Since an integrator has infinite gain at zero frequency, one might use an integrator in the forward path in addition to the amplifier. Figure 5.3 shows a general system with one pole at the origin and other finite poles and zeros—some in the forward path and some in the feedback path. The constant H_0 sets the scale factor between the dc values of R and C, and the constant G_0 is the low-frequency gain constant. Let's condense the notation by defining the forward gain block as $G(s)/s$, where

$$G(s) = G_0 \frac{\prod_j (1 + s/q_j)}{\prod_k (1 + s/p_k)} \qquad (5.16)$$

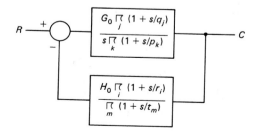

Figure 5.3 General single-integrator system.

and let the feedback block be $H(s)$. Then the transfer function from input to error is

$$\frac{\text{Error}(s)}{R(s)} = \frac{1}{H_0} - \frac{C(s)}{R(s)} = T_E(s) \tag{5.17}$$

$$= \frac{1}{H_0} - \frac{G(s)/s}{1 + G(s)H(s)/s}$$

$$T_E(s) = \frac{1}{H_0} - \frac{G(s)}{s + G(s)H(s)} \tag{5.18}$$

For small s, the pole and zero factors not at the origin go to 1, $G(s) \rightarrow G_0$ and $H(s) \rightarrow H_0$. So

$$T_E(s) \rightarrow \frac{1}{H_0} - \frac{G_0}{s + G_0 H_0} = \frac{s}{H_0(s + G_0 H_0)} \tag{5.19}$$

If $R(s) = 1/s$, a unit step in time, the steady-state error is

$$E_{ss} = \lim_{s \to 0} s T_E(s) \frac{1}{s} = 0 \tag{5.20}$$

and if $R(s) = 1/s^2$, a unit ramp in time,

$$E_{ss} = \lim_{s \to 0} s T_E(s) \frac{1}{s^2} = \frac{1}{H_0 G_0 H_0} = \frac{1}{H_0 K_v} \tag{5.21}$$

The introduction of the integrator does indeed drive the step input error to zero, and it also makes the response to a ramp input a ramp with a constant lag in the steady state. $1/H_0$ is the output-input scale factor, and the open-loop gain constant product $G_0 H_0$ is called the velocity error constant K_v. Suppose now a unit parabola in time is used to test the system. $R(s) = 1/s^3$ and the steady-state error goes to infinity.

$$E_{ss} = \lim_{s \to 0} s T_E(s) \frac{1}{s^3} = \lim_{s \to 0} \frac{1/s}{H_0(s + K_v)} \rightarrow \infty \tag{5.22}$$

Evidently, it takes a double-integrator system to follow a parabola with finite error.

Now let's go for the general case. Let $G(s)$ and $H(s)$ be defined as above, and let there be n integrators in the forward path. Then, for small s,

$$T_E(s) = \frac{1}{H_0} - \frac{G_0}{s^n + G_0 H_0} = \frac{s^n}{H_0(G_0 H_0)} \qquad n \geq 1 \tag{5.23}$$

For an input whose transform is $R(s) = 1/s^u$, the steady-state error is

$$E_{ss} = \lim_{s \to 0} s \frac{s^n}{H_0 G_0 H_0} \frac{1}{s^u} = \lim_{s \to 0} \frac{s^{n+1-u}}{H_0 G_0 H_0} \quad\quad (5.24)$$

$$= 0 \quad\quad \text{if } n + 1 > u$$

$$= \frac{1}{H_0 G_0 H_0} \quad\quad \text{if } n + 1 = u$$

$$\to \infty \quad\quad \text{if } n + 1 < u \quad\quad (5.25)$$

At this point, I want to pick up a few loose ends left on the way to the grand result. First, you may have noticed that the summing junction output signal, labeled E in previous chapters, isn't the error signal. In fact, there is no error signal in the general case, since the error is the difference between a nonphysical (desired) output and the actual output. Taking the definition a step closer to an all-physical form by saying that the desired output is $1/H_0$ times the input makes E a scaled version of the error.

$$\text{Error}(s) = \frac{E(s)}{H_0}. \quad\quad (5.26)$$

If the feedback is unity, only then is $E(s)$ the error signal. Otherwise, it is sometimes called the return difference in the literature. This is a good name because it is the difference between the input and an (often) scaled-down copy of the output returned to the input.

Second, the product of the gain constants has a name which depends on how many integrators are used. This is an unfortunate bit of historical jargon which persists in current texts. You have already seen K_p for no integrators and K_v for one integrator. It's K_a (guess why) for two integrators and nameless for more. Another practice is to classify systems by the number of integrators: Type 0 means no integrator, type 1 means one integrator, and so on. This is more useful and certainly less confusing. Table 5.1 lists the steady-state errors for the cases of practical importance.

Table 5.1   Steady-state error $H_0 E_{ss}$ as a function of test signal and number of integrators n

	Input Signal		
n	$u(t)$	$ramp(t)$	$p(t)$
0	$1/(1 + G_0 H_0)$	∞	∞
1	0	$1/(G_0 H_0)$	∞
2	0	0	$1/(G_0 H_0)$
3	0	0	0

Finally, what is the steady-state output? Since the steady-state error is the difference between the desired and actual output values, the actual steady-state output must be the difference between the desired value and the error. This fact allows us to use the finite entries in Table 5.1 directly to find the steady-state output. For example, for a type-1 system with a step input $10u(t)$ the steady-state output must be $10/H_0$ because the error has gone to zero. For the same system with an input ramp $10ramp(t)$ the output will be $(10t/H_0) - (10/G_0H_0^2)$ in the steady state. Note the amplitude scaling in the error term, which is allowed because the system is linear.

5.2   TRANSIENT RESPONSE TO A STEP INPUT

The transient response is the initial part of the system's output from $t = 0$ to a defined closeness to the steady-state value. The time it takes to come close to the steady-state value is called the *settling time*. The purpose of this section is to illustrate these and some other definitions in connection with the system's response to a unit step test signal.

The nature of a system's response to a step input is determined by its poles, and the literature generally recognizes two broad classes of response: overdamped and underdamped. An *overdamped response* is one in which the final value is approached from below with no overshoot, like a simple exponential. An *underdamped system* is one whose response is oscillatory. Such a system responds with an output rising above the final value and then approaching the final value with a series of decreasing under- and overshoots. Figures 5.4 and 5.5 show these two types of responses normalized to unit final value and marked with various labels to be defined in the next two paragraphs.

A term often used in electronics is *rise time*. Ideally, the response to a step input is a step output, which means an immediate rise to the final value. Since a real system can't respond instantly and we need to quantify this shortcoming in order to compare alternatives, a definition of rise time should be agreed on. The most common definition is that rise time equals the time taken for the response to go from 10 to 90% of the final value. The 10 and 90% times are marked in Figs. 5.4 and 5.5 as t_{10} and t_{90}. So

$$t_{\text{rise}} = t_{90} - t_{10} \tag{5.27}$$

This makes some physical sense, because most systems have a quick initial rise in comparison to the following part of their step response. Another definition sometimes used is that rise time is the time taken to go from zero to the first crossing of the final value level. For an overdamped system this is forever, but for an underdamped system it may not be much longer than the (5.27) time. This time is marked as t_c in Fig. 5.5. For a second-order system t_c is easier to calculate than (5.27) in terms of the gain and pole values. Another way to describe the response quickness is by the *delay time*. This is defined as the time the response

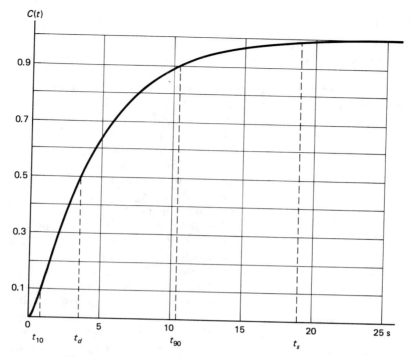

Figure 5.4 An overdamped step response.

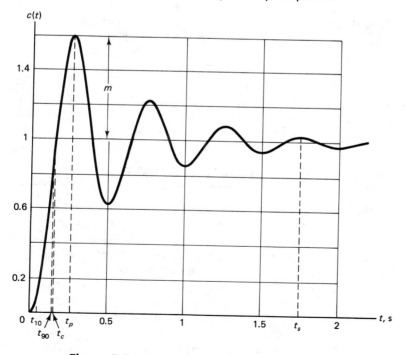

Figure 5.5 Underdamped step response.

takes to go from zero to 50% of the final value. The delay time is marked as t_d in Fig. 5.4 and not shown in Fig. 5.5 because it would make the time axis too crowded. In these figures the delay time is close to half the rise time, but in some systems there is a significant period during which nothing seems to be happening at all, so that the delay can be as long or longer than the rise time. The time before the rise is sometimes called *latency*, especially in biologically oriented literature.

Once the response is in the neighborhood of the final value, we are interested in how long it takes to settle down. Since in both the ideal (stictionless) and real cases the system never reaches the final value, settling time is defined as the time taken to go from zero to within some tolerance of the final value. The time taken to come within 2% of the final value is marked as t_s in each figure. For overdamped systems, knowledge of the initial response time (rise or delay or both) and the settling time tells everything. For underdamped systems though, overshoot is an additional important feature. A small amount of overshoot is often tolerated in order to produce a quicker rise. In Fig. 5.5, m is the per-unit overshoot, that is, the response peak minus the final value divided by the final value. It is often expressed as a percentage, which is $100m$. The time to the peak is also of interest and is labeled t_p.

A typical specification may include rise time and almost certainly will include a limit on overshoot and settling time. The next chapter will show that these three parameters cannot be arbitrarily chosen for a second-order system. Indeed, even with a higher-order system it may not be possible to meet a given set of limits on these three parameters. Some examples in later chapters will deal with this topic.

While Figs. 5.4 and 5.5 represent typical step response shapes, they don't cover all cases. For example, it sometimes happens that a system will have a mix of types in its step response. Figure 4.20 shows a response in which the ripple rides on the rising portion. With more zeros and poles in the system, more possibilities develop, but the range of possibilities, as the reachable sets of performance parameters, is limited by the gain phase requirements for stability. Again, examples in later chapters will illustrate this point.

5.3   COMPLETE PERFORMANCE MEASURES

In the first section we looked at the steady-state error, and in the second section we looked at the initial response to a step input. Both were examinations of a system's performance in detail. When one looks at alternative designs, one must have a basis on which to make a choice. Will you choose slightly less overshoot and a slower rise? How many integrators do you need? A continuing effort in engineering is to try to boil a system's performance down to a single number called a *figure of merit* or a *performance index*. Performance measures can be either specific: Move the specified load from point A to point B with a minimum expenditure of fuel, zero initial and final speeds, and taking no longer than T hours. Or they can be general: Minimize the mean squared error in response to a step input. I leave the mission-specific problem to a future course.

How should a measure be constructed? Without being mission-specific, what we have to work with is system type, test signal, error response, and time. Since we want an overall measure, we need a function that takes account of the response throughout its duration. This points to an integral over time. Error should be minimized, whether it is positive or negative, so the argument of the integral should be an increasing function of the error magnitude. Since the performance index is to help us make choices, it should be sensitive to parameter variations. The initial error is always the same, so a function which penalizes initial error lightly and later error more heavily might be more sensitive. This suggests including time in the integrand. Some of the simple choices for the integrand are error squared (ISE), absolute error (IAE), time times error squared (ITSE), and time times absolute error (ITAE). The easiest to deal with analytically is the error squared integrand, but the most sensitive to system parameters is the absolute error times time.

All these criteria have been used, studied, and reported in the literature. For a given closed-loop bandwidth, system type, and order, it is possible to determine the locations of the closed-loop poles that minimize the performance index. In the case of ISE, this can be done analytically. For the integrands involving absolute error, the index can be minimized by numerical methods. In 1953, Graham and Lathrop reported the results of an analog computer study in which they determined the pole locations to minimize the ITAE. Some of their results are given in Table 5.2. Notice that the entries in part b of the table are for the minimum ITAE for a ramp input, not a step input. Performance comparisons for a given system order between type-1 and type-2 systems and between different orders for a given type are quite interesting. I have generated plots of the step response for type-1 systems of orders 2 and 4 in Fig. 5.6, and for the step responses of fourth-order systems of type 1 and type 2 in Fig. 5.7. The most important effect of system order is that the time delay increases as the order goes up. The overshoot is quite small. These are examples of general properties you would observe if you ran step responses for the other cases. The step response of the type-2 system shown in Fig. 5.7 seems surprisingly quick. Upon reflection, it is reasonable for a system whose objective is to catch up to a rising input. Since the ramp is the integral of the step, the ramp response is the integral of the step response. The high ripple in the step response will be smoothed out by integration. Note that the ripple is entirely above the final value level. Figure 5.7 is an example of how strongly one's choice of system objective affects the performance in other tests. You might think of the 40% overshoot in the step response as a "side effect" of choosing optimum ramp response as the system objective.

Table 5.2 needs a little discussion. $T(s)$ is the closed-loop transfer function for a unity-feedback system. Therefore the open-loop transfer function can be calculated and is given in the table as $G(s)$. For the type-1 system G is an all-pole function. From Chap. 3 we know that a system with two integrators must have a zero whose frequency is lower than that of any of the other poles in G to be stable. This is why G has a zero at $-1/a_1$ in the type-2 system. Naturally, the zero turns up as a zero of $T(s)$ also. The coefficients in the second column are

Table 5.2   Transfer functions for minimum ITAE systems with $\omega_0 = 1$

(a) Type-1 systems with step input

$$T(s) = 1/D(s) = 1/(s^n + a_{n-1}s^{n-1} + \cdots + a_1s + 1) = G(s)/[1 + G(s)]$$

$$G(s) = 1/(D(s) - 1)$$

n	a_i	Poles of $T(s)$	Poles of $sG(s)$
1	1, 1	-1	—
2	1, 1.4, 1	$-0.7 \pm j0.7141$	-1.4
3	1, 1.75, 2.15, 1	$-0.7081, -0.521 \pm j1.068$	$-0.875 \pm j1.177$
4	1, 2.1, 3.4, 2.7, 1	$-0.626 \pm j0.4141, -0.424 \pm j1.263$	$-1.168, -0.466 \pm j1.447$
5	1, 2.8, 5, 5.5, 3.4, 1	$-0.8955, -0.5758 \pm j0.5339, -0.3764 \pm j1.292$	$-1.108 \pm j0.6598, -0.2916 \pm j1.3994$
6	1, 3.25, 6.6, 8.6, 7.45, 3.95, 1	$-0.5805 \pm j0.7828, -0.7346 \pm j0.2873, -0.3099 \pm j1.263$	$-1.2677, -0.8227 \pm j1.074, -0.1685 \pm j1.294$

(b) Type-2 systems with ramp input

$$T(s) = N(s)/D(t) = (a_1s + 1)/(s^n + a_{n-1}s^{n-1} + \cdots + a_1s + 1) = G(s)/[1 + G(s)]$$

$$G(s) = N(s)/(D(s) - N(s))$$

n	a_i	Poles of $T(s)$	Poles of $s^2G(s)$
2	1, 3.2, 1	$-2.849, -0.351$	—
3	1, 1.75, 3.25, 1	$-0.3643, -0.6929 \pm j1.505$	-1.75
4	1, 2.41, 4.93, 5.14, 1	$-0.2465, -1.2466, -0.4584 \pm j1.745$	$-1.205 \pm j1.865$
5	1, 2.19, 6.5, 6.3, 5.24, 1	$-0.2468, -0.4665 \pm j0.8797, -0.5051 \pm j1.957$	$-1.1866, -0.5017 \pm j2.249$
6	1, 6.12, 13.42, 17.16, 14.14, 6.76, 1	$-3.277, -0.2377, -0.9983 \pm j0.4329, -0.3044 \pm j0.9958$	$-3.1066, -2.112, -0.4506 \pm j1.3971$

due to Graham and Lathrop and can be found in many textbooks. I have recalculated the poles of $T(s)$ and $G(s)$ to avoid repeating possible past errors. If you reconstruct the polynomials from the poles, you will not get the exact coefficient values given, because I have rounded to the fourth decimal place. However, the effect on the ITAE value for a given system is negligible. All the values given are

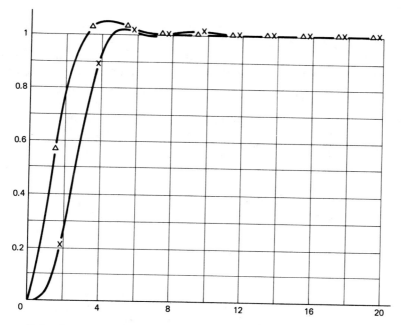

Figure 5.6 Step responses for type 1 minimum- ITAE systems. △, second-order system; ×, fourth-order system.

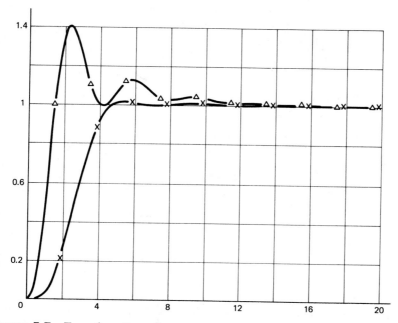

Figure 5.7 Two fourth-order system step responses. ×, Type 1 system designed for minimum ITAE for a step input; △, Type 2 system designed for minimum ITAE for a ramp input.

for $a_0 = a_n = 1$. This means that 1 rad/s is the geometric mean of the closed-loop pole values. To get the same performance, minimum ITAE, for another mean frequency, it is only necessary to multiply each pole value by that frequency. If you need to use the coefficients directly, the coefficient of s^k must be multiplied by ω_0^{n-k}, where ω_0 is the mean frequency. The table can be used in a number of different ways to help in SISO and MIMO design methods. The following example illustrates the design of a SISO system.

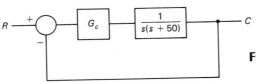

Figure 5.8 System for a constant-rate feed, Example 5.2.

■ Example 5.2

Suppose that a cutting tool is to be driven so that it follows the groove it is making with zero final velocity error and that Fig. 5.8 represents the amplifier, motor, and load as known at the present point of the design. G_c is the controller yet to be designed. A sensor measures the groove depth, and since position is the integral of velocity, we have a natural integrator in the motor. The armature corner frequency is an order of magnitude higher than the mechanical one, so it is omitted. In order to have zero final velocity error we need a second integrator, and in order to maintain stability we need a zero. So the controller transfer function must be, at least,

$$G_c = K\frac{1 + s/z}{s} \tag{5.28}$$

The overall forward transfer function is

$$G(s) = \frac{K(1 + s/z)}{s^2(s + 50)} \tag{5.29}$$

which is type 2 and third-order. From Table 5.2b, with the mean frequency included,

$$G(s) = \frac{a_1\omega_0^2 s + \omega_0^3}{s^2(s + a_2\omega_0)} = \frac{\omega_0^3(1 + a_1 s/\omega_0)}{s^2(s + a_2\omega_0)} \tag{5.30}$$

By comparison, $\omega_0 = 50/1.75 = 18.89$, $K = 356.8$, and $z = 18.89/3.25 = 5.8123$ rad/s. This controller can be made as a modified op-amp integrator, as you will see in Chap. 7. ■

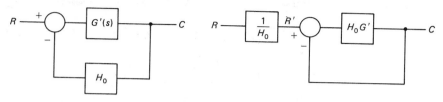

Figure 5.9 Transformation for a nonunity-feedback system.

Suppose, as is more likely to be the case, that you have to design for a system with nonunity transducer feedback. What then? Figure 5.9 shows a solution. By this transformation, the system from R' to C is a unity-feedback system which can be designed using Table 5.2. By letting $G(s) = H_0 G'(s)$ and using the procedure of matching forms, one can find the controller values. Since the system from R' to C has minimum ITAE performance, so also will the original system. Such is the scaling property of linear systems. Although the procedure looks simple and elegant, it is not always the best one to use for various complicating reasons which will be discussed in Chap. 8.

PROBLEMS

Section 5.1

5.1 Refer to Fig. 5.10. For the following systems determine if the final value theorem applies. Find the steady-state error for a unit-step input if it does, and determine (by calculating the time response) the maximum error if it doesn't.

(a)
$$G(s) = \frac{10}{s^2 + 25} \qquad H(s) = 7.5$$

(b)
$$G(s) = \frac{10}{s^2\,25} \qquad H(s) = 7.5\,(1\,+\,s)$$

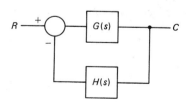

Figure 5.10 Generic system for probs. 5.1 and 5.2.

5.2 Find the steady-state error and output for a ramp input, $r(t) = 2t$, for the following systems. Again, Fig. 5.10 defines the system notation.

(a)
$$G(s) = \frac{30}{s(s + 3)} \qquad H(s) = 0.01$$

(b)
$$G(s) = \frac{2(1 + s)}{s(s + 20)} \qquad H(s) = \frac{500}{s + 5000}$$

Section 5.2

5.3 For the system in Fig. 5.10, let $G(s) = K/(1 + s/a)$, $H(s) = H_0$. Find the response $c(t)$ to a unit-step input. Define the final value as c_0, the final error as e_0, and the exponential time constant as T.
(a) Give expressions for c_0, e_0, and T as functions of K, a, and H_0.
(b) Give expressions for K, a, and H_0 in terms of c_0, e_0, and T.
(c) Find t_d, t_{10}, t_{90}, and t_{rise} in terms of c_0 and T.
(d) What performance parameters will completely define the system parameters K, a, and H_0? List alternatives, if there are any.

Section 5.3

5.4 The criteria discussed in Sec. 5.3 are not averages—the integrals don't have $1/T$ in front of them. One obvious reason is that it wouldn't make any distinction between either system or index choices to include this factor. Another reason is less obvious. To get at it, consider the system in Fig. 5.11.
(a) Calculate $e(t) = 1 - c(t)$.
(b) Calculate

$$I_A = \frac{1}{T} \int_0^T e^2(t) \, dt$$

(c) If $a > 0$, the system is type 0, and if $a = 0$ the system is type 1. Find, for each case,

$$\lim_{T \to \infty} I_A$$

(d) What general conclusions can you leap to about the relations between mean squared error, system type, and test signal?

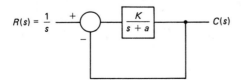

Figure 5.11 A simple system for testing performance measures.

5.5 Given the existence of an APL function (or a subroutine in some other language) for calculating the impulse response of a system, write a flowchart for a function (or a subroutine) for calculating the ITAE for that system. You may assume that a frequency domain description of the system is available. A vector of time values and another vector of impulse response values are the input and output of the function IMPULSE.

5.6 In Fig. 5.10, suppose $H(s) = 0.01$ and $G(s) = G_c(s)/[s(s + 1)]$. Find a $G_c(s)$ for minimum ITAE for a step input. Keep the system at type 1 and use as simple a controller as possible.

5.7 As with Prob. 5.6 except

$$G(s) = \frac{G_c(s)}{s(1 + s)(1 + s/5)}$$

Chapter 6

THE SECOND-ORDER SYSTEM: A MODEL FOR ALL SYSTEMS

The canonic second-order system consists of an integrator and a single-pole, low-pass element in the forward path, with unity feedback. Figure 6.1 shows the block diagram. This system is of special interest for three important reasons: Many practical systems can be modeled in this form, it performs well, and it is possible to obtain many exact and approximate results and design relations analytically. In this chapter, we will exercise your analytic tools to find these relations and display this behavior. We discussed modeling in Chap. 2, and in Chap. 8 we will discuss designing higher-order systems to behave as if they were of second-order.

6.1   STEP RESPONSE

We shall approach the step response through the inverse Laplace transform. Suppose the input is $r(t) = Au(t)$, where A is a real constant. The Laplace transform is $R(s) = A/s$. The closed-loop transfer function is

$$T(s) = \frac{Ka}{s^2 + as + Ka} \tag{6.1}$$

This form occurs so often in so many different contexts that it is customary to put it in a standard form as follows:

$$T(s) = \frac{\omega_0{}^2}{s^2 + 2\zeta\omega_0 s + \omega_0{}^2} \tag{6.2}$$

171

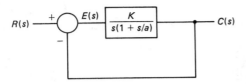

Figure 6.1   Canonic second-order system.

The standard form parameters must be related to the system parameters by

$$\omega_0 = \sqrt{Ka} \qquad \text{natural frequency} \tag{6.3}$$

$$\zeta = \frac{\sqrt{a/K}}{2} \qquad \text{damping factor} \tag{6.4}$$

You will see several more changes in notation to simplify the expression of relationships and display concepts as we go along.

As you know, the nature of a system's response is determined by the poles of its transfer function. In this case, we are interested in the underdamped condition. Designers generally favor some underdamping, as they feel it can give the quickest approach to the desired output value. The damping factor is less than 1, and the poles are

$$p = -\omega_0\zeta \pm j\omega_0\sqrt{1 - \zeta^2} \tag{6.5}$$

You can see that the real and imaginary parts form a right triangle whose hypotenuse is ω_0. The angle of the poles with respect to the real axis is an important parameter. For convenience, and because of their similarity to forms you have seen in the past, we denote the real and imaginary parts as

$$p = -\alpha \pm j\beta \tag{6.6}$$

Figure 6.2 shows the open- and closed-loop poles.

The output response, in the Laplace transform domain, is

$$C(s) = T(s)R(s) = \frac{A\omega_0^2}{s(s + \alpha + j\beta)(s + \alpha - j\beta)} \tag{6.7}$$

Using partial fraction expansion and the fact that $\alpha^2 + \beta^2 = \omega_0^2$, you will find

$$C(s) = A\left(\frac{1}{s} + \frac{\omega_0^2}{2j\beta(\alpha + j\beta)(s + \alpha + j\beta)} + \frac{\omega_0^2}{-2j\beta(\alpha - j\beta)(s + \alpha - j\beta)}\right) \tag{6.8}$$

Check that all the complex terms occur in conjugate pairs. This is a necessary, but not sufficient, condition for the work to be correct. Now you can go to the time domain:

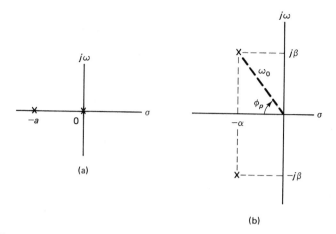

Figure 6.2 (a) Open-loop poles. (b) Closed-loop poles.

$$c(t) = Au(t)\left[1 + \frac{(\alpha - j\beta)e^{-(\alpha+j\beta)t}}{2j\beta} + \frac{(\alpha + j\beta)e^{-(\alpha-j\beta)t}}{-2j\beta}\right] \quad (6.9)$$

The exponential terms are begging to be turned into trigonometric functions. By factoring out the real quantities, using the exponential definitions for sine and cosine, the sine-of-a-sum identity, and the triangle in Fig. 6.2b, you can find

$$c(t) = Au(t)\left[1 - \frac{e^{-\alpha t}}{\sin\theta_p}\sin(\beta t + \phi_p)\right] \quad \phi_p = \tan^{-1}\left(\frac{\beta}{\alpha}\right) \quad (6.10)$$

Check the initial and final values. Are they what you would expect for a single-integrator system? When you examine (6.10), you will find that the response consists of a constant plus an exponentially damped sine wave. It is the very model of the transient response discussed in the last chapter, and we can proceed to find the performance parameters defined there.

Of the various definitions for rise time, the 0–100% rise time is meaningful for the underdamped response and easy to find. $c(t)$ crosses the final value line when the sine function is zero. The first crossing will occur when its argument is equal to π, as it doesn't start at zero. Therefore, the rise time is

$$t_r = \frac{\pi - \phi_p}{\beta} \quad (6.11)$$

To find the time to peak and the settling time, we need to find the values of t for maxima of c. You will find by straightforward differentiation that the maxima and minima occur at times

$$t_{pk} = \frac{k\pi}{\beta} \qquad k \text{ an integer} \qquad (6.12)$$

At the maxima,

$$c(t_{pk}) = A(1 + e^{-\alpha t}pk) \qquad k \text{ odd} \qquad (6.13)$$

$k = 1$ gives the highest peak,

$$c(t_{p1}) = A(1 + e^{-\pi/\tan \phi_p}) \qquad (6.14)$$

The per-unit overshoot is

$$m = \frac{c(t_{p1})}{A} - 1 = e^{-\pi/\tan \phi_p} \qquad (6.15)$$

You may have observed that (6.13) represents an upper envelope for c. If we take settling time as the time required for the envelope to approach within the specified tolerance δ of the final value, we have

$$\text{Upper envelope } (c) = A(1 + e^{-\alpha t}) = A(1 + \delta) \qquad (6.16)$$

from which

$$t_s = \frac{1}{-\alpha} \ln \delta \qquad (6.17)$$

■ Example 6.1

Suppose a system has a gain $K = 10$ and a corner frequency $a = 10$ rad/s. Then the natural frequency $\omega_0 = 10$ rad/s, and the damping factor $\zeta = 0.5$. The attenuation constant $\alpha = 5$ s^{-1}, and the ripple frequency $\beta = 5\sqrt{3} = 8.66$ rad/s. From Fig. 6.2b, $\phi_p = 60° = \pi/3$ rad. The per-unit overshoot $m = 0.163$, and the 100% rise time $t_r = 0.242$ s. The 2% settling time $t_s = -(\ln 0.02)/5 = 0.782$ s. The unit-step response is shown in Fig. 6.3. ■

6.2 OPEN-LOOP FREQUENCY RESPONSE

The open-loop transfer function has only one corner frequency, so there are two general possibilities for the unity-gain crossover. It is either above or below the corner frequency. This is illustrated in Fig. 6.4 with two straight-line Bode magnitude plots. You can find an exact expression for the crossover frequency and then for the phase margin (Prob. 6.5), but they are neither elegant nor illuminating.

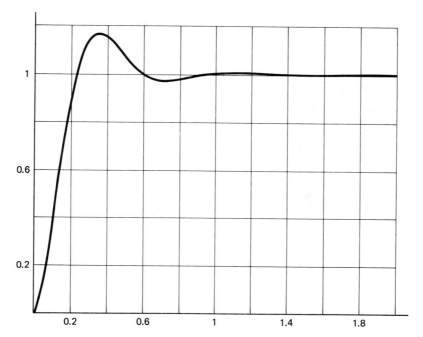

Figure 6.3 Step response for $K = a = 10$, Example 6.1.

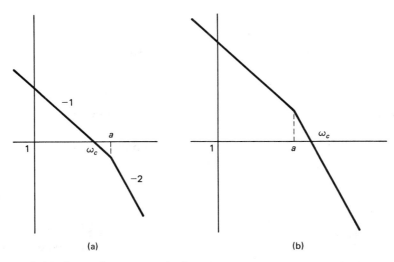

Figure 6.4 Open-loop straight-line Bode plots. (a) $\omega_c < a$. (b) $\omega_c > a$.

So we will use the straight-line diagrams to find approximate expressions. What is the gain margin?

From the approximate equations represented by Fig. 6.4, $K/\omega_c = 1$, $K < a$, and $K/(\omega_c^2/a) = 1$, $K > a$. These two cases will now be treated in separate paragraphs.

For $K < a$,

$$\omega_c = K \tag{6.18}$$

The crossover phase lag is

$$\phi_c = 90° + \tan^{-1}\left(\frac{\omega_c}{a}\right)$$

$$= 90° + \tan^{-1}\left(\frac{K}{a}\right)$$

$$= 180° - \tan^{-1}\left(\frac{a}{K}\right)$$

The phase margin is

$$\phi_m = 180° - \phi_c = \tan^{-1}\left(\frac{a}{K}\right) = \tan^{-1}(4\zeta^2) \tag{6.19}$$

For $K > a$,

$$\omega_c = \sqrt{Ka} \tag{6.20}$$

and, by analogy with the previous case,

$$\phi_m = \tan^{-1}(\sqrt{a/K}) = \tan^{-1}(2\zeta) \tag{6.21}$$

Is the approximate phase margin continuous at $K = a$? What are the phase margin and damping factor at $K = a$?

■ Example 6.2

A system has gain $K = 20$ and a corner frequency $a = 10$. Then, approximately, $\omega_c = 14.14$ rad/s and the phase margin is $\phi_m = 35.3°$. Exact expressions give $\omega_c = 12.5$ rad/s and $\phi_m = 38.7°$. We expect this direction of error because the exact curve lies under the straight-line diagram for an all-pole system. This makes the approximation more conservative in that it predicts a lower phase margin. The Bode plot is shown in Fig. 6.5.   ■

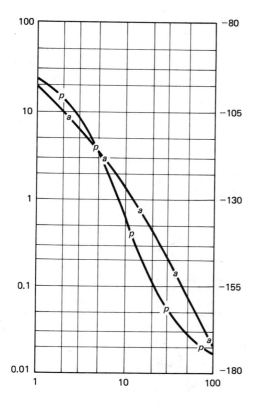

Figure 6.5 Bode plot for $K = 20$, $a = 10$, for Example 6.2.

6.3   CLOSED-LOOP FREQUENCY RESPONSE

You know from Sec. 3.8 that this system has unity gain at low frequency and rolls off at a -2 slope at high frequencies. Since the denominator of $T(s)$ is a quadratic, there is a possibility of resonance. Thus, the interesting questions are What frequency gives peak magnitude? What is the peak magnitude? What is the half-power bandwidth? You can answer these questions analytically by studying $|T(j\omega)|$.

$$T(j\omega) = \frac{1}{1 - (\omega/\omega_0)^2 + 2j\zeta(\omega/\omega_0)} \tag{6.22}$$

Since T is really a function of the frequency ratio, replace ω/ω_0 by ω_n.

$$|T|^2 = \frac{1}{(1 - \omega_n^2)^2 + 4\zeta^2\omega_n^2} = \frac{1}{D(\omega_n)} \tag{6.23}$$

Evidently, the peak occurs when D is a minimum. By differentiating D with respect to frequency, you will find the frequency for the peak is

$$\omega_{nM} = \sqrt{1 - 2\zeta^2} \qquad \zeta < 1/\sqrt{2}$$

$$= 0 \qquad\qquad \zeta \geq 1/\sqrt{2} \qquad\qquad (6.24)$$

After a little algebra, you will find the peak to be

$$M_p = \frac{1}{2\zeta\sqrt{1 - \zeta^2}} \qquad \zeta < 1/\sqrt{2}$$

$$= 1 \qquad\qquad \zeta \geq 1/\sqrt{2} \qquad\qquad (6.25)$$

The half-power frequency is that frequency for which $D = 2$. Direct solution gives

$$\omega_{nHp}^2 = 1 - 2\zeta^2 + \sqrt{2 - 4\zeta^2 + 4\zeta^4} \qquad 0 < \zeta < 1 \qquad (6.26)$$

$$= \omega_{nM}^2 + \sqrt{1 + \omega_{nM}^4} \qquad\qquad 0 < \zeta < 1/\sqrt{2} \qquad (6.27)$$

Remember that these are *normalized* frequencies. What are the ranges of values for the half-power and peak frequencies? Figure 6.6 shows a family of closed-loop frequency responses.

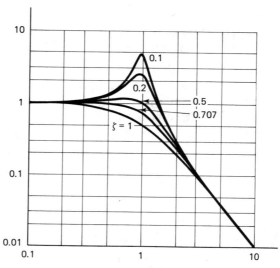

Figure 6.6 Closed-loop frequency responses for damping factors of 1, 0.707, 0.5, 0.2, and 0.1.

If you think back over the first three sections of this chapter, you will see that there are common tendencies and relations among the three responses. For example, frequency parameters are proportional to ω_0, and times are inversely proportional to ω_0. By analogy with filter theory, let us consider the frequency-normalized case, that is, let us set $\omega_0 = 1$. Now all the previous results become functions of just one parameter, which I will choose to be the angle of the closed-loop poles ϕ_p in Fig. 6.2b. I will denote normalized parameters with an additional subscript n, as in Sec. 6.3.

Step response:

Time to peak,

$$t_{np} = \frac{\pi}{\sin \phi_p} \tag{6.28}$$

Rise time,

$$t_{nr} = \frac{\pi - \phi_p}{\sin \phi_p} \tag{6.29}$$

Settling time,

$$t_{ns} = \frac{-\ln \delta}{\cos \phi_p} \tag{6.30}$$

Ripple frequency,

$$\beta_n = \sin \phi_p \tag{6.31}$$

Per-unit overshoot,

$$m = e^{-\pi/\tan \phi_p} \tag{6.32}$$

Open-loop frequency response:

Phase margin,

$$\phi_m = \tan^{-1}(2 \cos \phi_p) \qquad \phi_p < 60°$$
$$= \tan^{-1}(4 \cos^2 \phi_p) \qquad \phi_p > 60° \tag{6.33}$$

Unity-gain crossover frequency,

$$\omega_{nc} = \frac{1}{2 \cos \phi_p} \qquad \phi_p < 60°$$
$$= 1 \qquad \phi_p > 60° \tag{6.34}$$

Closed-loop response:

Frequency for peak,

$$\omega_{nM} = \cos(\pi - 2\phi_p) \qquad \phi_p > 45°$$
$$= 0 \qquad \phi_p < 45° \tag{6.35}$$

Magnitude peak,

$$M_p = \frac{1}{\sin(2\phi_p)} \qquad \phi_p > 45°$$
$$= 1 \qquad \phi_p < 45° \tag{6.36}$$

179

Half-power bandwidth,

$$\omega_{nHP}^2 = -\cos(2\phi_p) + \sqrt{1 + \cos^2(2\phi_p)} \qquad 0 < \phi_p < 90° \qquad (6.37)$$

Clearly, if one specifies the closed-loop pole positions, all the response descriptors are completely determined. Do you see any inverse relationships, that is, any two quantities whose product is independent of the pole angle? The only obvious pair on the list is time to peak and ripple frequency. I hope you are surprised that the time-bandwidth inverse relation so often mentioned in other courses does not exist in exact form. Observe that, in fact, settling time and crossover frequency are proportional to each other for large pole angles. A useful exercise is to construct a table of limiting values for the response parameters as the pole angle moves from 0° to 90°.

Of course, once the pole angle is fixed, then all times are inverse to ω_0, and all frequencies are proportional to ω_0. Fixing the pole angle determines the shape of the time and frequency responses, and then the scales are determined by the natural frequency. In the following example, we will calculate the responses for a special case.

■ Example 6.3

Consider the case of $\phi_p = 45°$. This is the boundary between a drooping closed-loop frequency response and one with a peak. The magnitude peak is 1, and the unit-step response overshoot m is 0.0432. These values do not depend on the natural frequency. They are shape functions. The times are

100% rise time, $$t_r = \frac{3.332}{\omega_0}$$

Time to peak, $$t_p = \frac{4.443}{\omega_0}$$

2% settling time, $$t_s = \frac{5.543}{\omega_0}$$

Ripple frequency, $$\beta = 0.7071\omega_0$$

Phase margin, $$\phi_m = 54.74°$$

Unity-gain crossover, $$\omega_c = 0.7071\omega_0 = \beta$$

Half-power bandwidth, $$\omega_{HP} = \omega_0$$

Note that, since the peak overshoot is about 4%, the settling time is not much longer than the time to peak. Also, from (6.3) and (6.4), the corner, straight-line crossover, and ripple frequencies are all equal.   ■

The designer of a second-order system faces a number of questions that are different in character from those discussed in the preceding paragraphs. You, as the designer, have two parameters to control, at most. These are the gain and corner frequency. This means that you can satisfy exactly, at most, two specified values of performance descriptors. However, in most cases, the specifications are given in terms of inequalities, or ranges of values, and more than two parameters are so specified. Sometimes, the customer may want performance numbers that are incompatible. A part of your job is to look at the requirements and determine if this is the case. We will go into this in more detail in Chaps. 7 and 8. In the meantime, it will be useful to recast the previous results in terms of K and a. You should verify the following results.

Step response:

Ripple frequency,
$$\beta = \left(aK - \frac{a^2}{4} \right)^{1/2} \tag{6.38}$$

Time to peak,
$$t_p = \frac{\pi}{\beta} \tag{6.39}$$

Peak overshoot,
$$m = e^{-(\pi a/2\beta)} \tag{6.40}$$

100% rise time,
$$t_r = \frac{\pi - \tan^{-1}(\beta/a)}{\beta} \tag{6.41}$$

Settling time,
$$t_s = \frac{-2 \ln \delta}{a} \tag{6.42}$$

Closed-loop frequency response:

Frequency for peak,
$$\omega_M = \left(aK - \frac{a^2}{2} \right)^{1/2} \qquad K > \frac{a}{2}$$
$$= 0 \qquad K < \frac{a}{2} \tag{6.43}$$

Magnitude peak,
$$M_p = \frac{K}{\beta} \qquad K > \frac{a}{2}$$
$$= 1 \qquad K < \frac{a}{2} \tag{6.44}$$

Half-power bandwidth,
$$\omega_{HP}^2 = aK - \frac{a^2}{2} + \left[1 + \left(aK - \frac{a^2}{2} \right)^2 \right]^{1/2} \tag{6.45}$$

■ Example 6.4

A second-order system is to have a 1% settling time of less than 10 s, a 100% rise time less than 1 s, and an overshoot not greater than 5%, in response to a unit step. Choose values of a and K to satisfy these requirements.

From (6.42), you see that the settling time depends on the corner frequency alone. Thus,

$$a > \frac{-2 \ln \delta}{t_s} = \frac{-2 \ln 0.01}{10} = 0.921 \text{ rad/s}$$

From (6.38) and (6.40),

$$K = \frac{a}{\pi}\left(1 + \frac{\pi}{\ln^2 m}\right) = 0.4835$$

for the minimum value of a. From (6.38) and (6.41), the rise time for these values will be $t_r = 3.12$ s. This is too long for the specification. If you multiply K and a by 3.2, you will preserve the overshoot value and decrease the times by this factor. In this way, the settling time will be less than one-third of the maximum allowed, and the other two requirements will just be met. ■

PROBLEMS

Section 6.1

6.1 Express the per-unit overshoot as a function of the damping factor.

6.2 If $a = 10$ and $m = 0.2$, find K.

6.3 If $a = 10$ and $m = 0.1$, find the ripple frequency both in radians/second and hertz.

6.4 If $t_s = 2$ s for 2% settling and $m = 0.05$, find K and a.

Section 6.2

6.5 Write the expression for the square of the magnitude of the open-loop gain. Set it equal to 1 with $\omega = \omega_c$. Find the phase at ω_c and then the phase margin. Calculate ω_c and ϕ_m for $a = 1$, $K = 0.5$, 1, and 5, and compare these values with those obtained from the approximate expressions.

Section 6.3

6.6 For the system in Example 6.2, what are M_p and the half-power bandwidth?

6.7 A second-order system is to have a half-power bandwidth of 10 kHz and a frequency response peak of 1.1. Find K and a.

6.8 If $m = 0.3$ and $t_s(2\%) = 5$ s, what are M_p and ω_{HP}?

6.9 Verify the half-power bandwidth expressions (6.26) and (6.27).

Section 6.4

6.10 Suppose $K > a/2$. Is it possible to find K and a to satisfy arbitrary choices of $m > 0$ and $M_p > 1$?

6.11 A system is to have a time to peak <1 s and an overshoot <0.2. Find K and a to satisfy these conditions with minimum bandwidth.

6.12 Write a program to find ϕ_p as a function of δ to satisfy $t_{np} = t_{ns}$. Generate a plot of ϕ_p versus δ.

Part II

Chapter 7

DESIGN OF SINGLE-INPUT
SINGLE-OUTPUT SYSTEMS

COMPENSATION NETWORKS

7.1   INTRODUCTION

In Chaps. 3 and 4 you saw some examples of how system performance can be
improved by adding poles or zeros to the transfer function to reshape the frequency response or the root locus. In Chap. 2 you read a number of examples
and worked some problems on circuits which have simple transfer functions with
names such as lead and lag networks. This chapter reformulates and adds to the
network results in preparation for their application in the next two chapters. In
Sec. 7.2 I will give some examples of electrical networks in which I will relate
their part values to the corner frequencies in the transfer functions and then discuss some properties of the transfer functions that are of interest for system
design. In Sec. 7.4 I will give mechanical network examples for the same kinds
of transfer functions so the discussion of transfer function properties applies to
these networks as well. The examples are only a few out of many ways to realize
a given function. If you can't come up with reasonable component values using
one circuit, work up another. This topic will be discussed further in Chap. 9.

7.2   ELECTRICAL NETWORKS

In this and the following section I will assume that you are going to use *RC*
networks in conjunction with operational amplifiers. Therefore, all networks will
be considered to be driven by voltage sources and terminated by either open
circuits for voltage output or short circuits for current output.

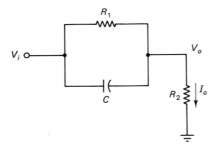

Figure 7.1   A general lead network.

7.2.1   Low- and High-Pass Networks for Real Factors

Figure 7.1 shows a semi-high-pass network which is known in control engineering as a lead network. This name refers to the property that the network adds phase lead to the system transfer function over the frequency range in which the capacitor's reactance is comparable in magnitude to the resistances. The transfer function for voltage is

$$\frac{V_o}{V_i} = \frac{a_1}{a_2} \frac{1 + s/a_1}{1 + s/a_2} \tag{7.1}$$

where $a_1 = 1/(R_1 C)$ and $a_2 = (R_1 + R_2)/(R_1 R_2 C)$. The transfer function for current output is

$$\frac{I_o}{V_i} = \frac{1}{R_1 + R_2} \frac{1 + s/a_1}{1 + s/a_2} \tag{7.2}$$

Figure 7.2 is a sketch of the Bode plot for (7.1). The magnitude has odd symmetry, so the phase must have even symmetry. The maximum magnitude is 1, since no transformer is present, so a low-frequency attenuation equal to the corner frequency ratio is introduced and probably will need to be corrected by additional system gain. The phase lead is due to the $+1$ slope section. Since, by symmetry, the maximum phase shift occurs at the middle of the slope, $\omega_M = \sqrt{a_1 a_2}$, it is

$$\phi_M = 90° - 2 \tan^{-1} \sqrt{a_1/a_2} \tag{7.3}$$

Table 7.1 shows some properties of the phase function. If the $+1$ slope were infinitely long, the phase would be 90°, so the function approaches this value as an asymptote for large corner frequency ratio. The $90° - \phi_M$ column shows this behavior clearly. The $\Delta \phi_M$ column illustrates the diminishing-return nature of the relationship.

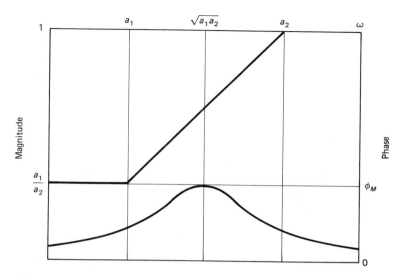

Figure 7.2   Bode plots for a lead network transfer function.

Table 7.1   Phase lead as a function of corner
frequency ratio

a_2/a_1	ϕ_M	$90 - \phi_M$	$\Delta\phi_M$
2	19.5	70.5	—
5	41.8	48.2	22.3
10	54.9	35.1	13.1
20	64.8	25.2	9.9
50	73.9	16.1	9.1
100	78.6	11.4	4.7
200	81.9	8.1	3.3

Figure 7.3 shows a general lag network form. This circuit gets its name for a reason similar to the lead network, it introduces phase lag. Its main importance, however, is that it provides a way to produce attenuation over an intermediate frequency range in order to reduce the magnitude at high frequencies without changing the low-frequency loop gain. This application will be demonstrated in Chap. 8. When it is used to produce a voltage output transfer function R_3 is usually ∞. For this case,

$$\frac{V_o}{V_i} = \frac{1 + s/a_2}{1 + s/a_1} \tag{7.4}$$

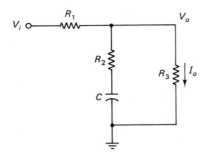

Figure 7.3 A lag network.

where $a_1 = 1/((R_1 + R_2)C)$ and $a_2 = 1/(R_2C)$. The high-frequency attenuation produced is

$$\frac{a_2}{a_1} = 1 + \frac{R_1}{R_2} \tag{7.5}$$

For current output, with $G_i = 1/R_i$,

$$\frac{I_o}{V_i} = \frac{1}{R_1 + R_3} \frac{1 + s/a_2}{1 + s/a_1} \tag{7.6}$$

where this time $a_1 = 1/(CR_2[1 + G_2/(G_1 + G_3)])$. Again, the high-frequency attenuation is

$$\frac{a_2}{a_1} = 1 + \frac{G_2}{G_1 + G_3} \tag{7.7}$$

Figure 7.4 shows a sketch of the Bode plot for the voltage output case. Most of the attenuation takes place between the two corner frequencies, so that the system transfer function has the original shape for frequencies above a_2, shifted down by the desired attenuation. If the system response has a good shape in the first place, say a section with an adequate length of -1 slope, the lag network can be used to reduce the gain so that unity-gain crossover occurs on the -1 slope instead of at some higher frequency where the slope is also higher. The phase behavior is the negative of that for the lead network.

7.2.2 The Bridged T for Complex Zeros

The simplest way to produce complex zeros is with an *LCR* circuit. The physical action is that the *L* and *C* elements provide paths through which the signal is phase-shifted in opposite directions so that a cancelation takes place where the paths come together. As pointed out in Chap. 2, the inductor is a very bulky

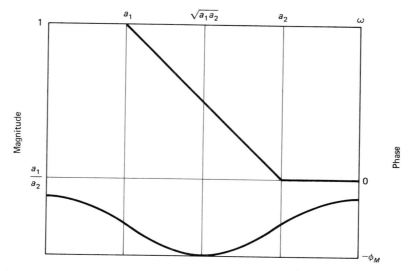

Figure 7.4 Bode plots for a lag network transfer function, $G_3 = 0$.

component at audio and subaudio frequencies, so that another way to accomplish this action is desirable. From the previous subsection, you can see that RC networks can also be used to produce opposing phase shifts. One such network is the bridged T shown in Fig. 7.5. For $R_L = \infty$, the voltage transfer function is

$$\frac{V_o}{V_i} = \frac{s^2 + 2\zeta\omega_0 s + \omega_0{}^2}{(s + a_1)(s + a_2)} \tag{7.8}$$

where $\omega_0{}^2 = 1/(R^2 C_1 C_2)$, $\zeta^2 = C_1/C_2$, and

$$a_1, a_2 = \zeta\omega_0 \left[1 + \frac{1}{2\zeta^2} \pm \left(1 + \frac{1}{(2\zeta^2)^2} \right)^{1/2} \right]$$

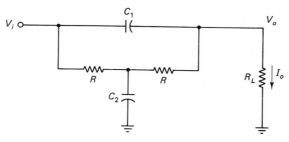

Figure 7.5 A bridged-T network.

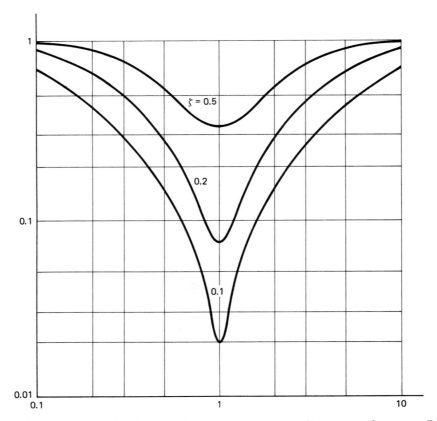

Figure 7.6 Magnitude vs. ω for the bridged-T voltage transfer case. Damping factors are 0.5, 0.2, and 0.1.

For sufficiently small ζ the poles go to $a_1 \approx \zeta \omega_0$ and $a_2 \approx \omega_0/\zeta$. Evidently, a low-frequency pole comes along with lightly damped zeros. In most applications, one would have to correct this with an additional lead network. Some typical frequency responses are shown in Fig. 7.6. By examining either the circuit or the transfer function, you can see that the high- and low-frequency gains go to 1 but the figure shows attenuation at both ends of its two-decade frequency range. This is due to the low-frequency pole. Without it, the plots would look the same as the quadratic factor plots in Fig. 3.27, since the high-frequency pole's effect doesn't show much in the frequency range in Fig. 7.6. A pole-zero map for this case is given in Fig. 7.7. Note that the low-frequency pole is equal to the real part of the zeros.

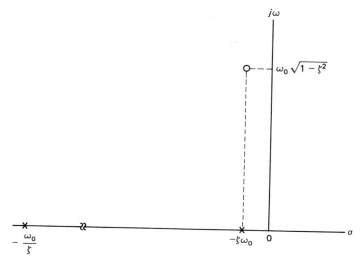

Figure 7.7   Pole-zero map for the bridged-T voltage transfer case. Note that the pole at $-\omega_0/\zeta$ is really off the scale compared to the other pole and the zeros.

For current output, with $R_L = 0$, the transfer function is

$$\frac{I_o}{V_i} = \frac{G}{2} \frac{1 + (2\zeta s/\omega_0) + (s/\omega_0)^2}{1 + s/(2\zeta\omega_0)} \tag{7.9}$$

There is no high-frequency pole, and the low-frequency pole is twice the real part of the zeros. Figures 7.8 and 7.9 show frequency responses for $G = 2$ and a pole-zero map.

7.3   OPERATIONAL AMPLIFIER USAGE

The operational amplifier was originally used in analog computers to study dynamic system behavior. Its advantages as a system component were recognized, and since integrated circuit versions became available it has become very widely used. Its advantages for control engineering are its very low output impedance, which provides network isolation, and its ability to realize transfer functions through passive-element feedback. The second property is a simple application of the general advantage of feedback in systems. These properties are applied in the next subsection. I assume you are already acquainted with the op amp's use as a gain block, sum and difference amplifier, and integrator, so I will start the treatment from a fairly general level.

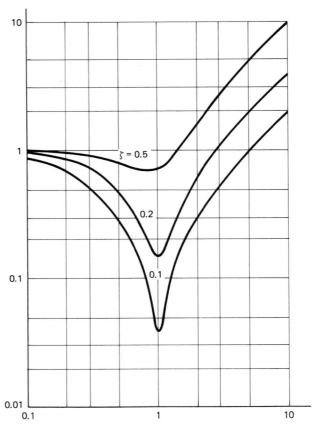

Figure 7.8 Magnitude vs. ω for the bridged-T current output case. Damping factors are 0.5, 0.2, and 0.1.

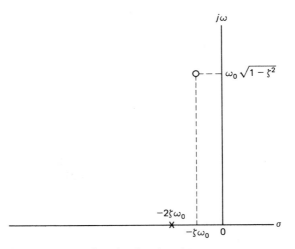

Figure 7.9 Pole-zero map for the bridged-T current output case, $\zeta \ll 1$.

The op amp is a real device, so it has limitations. Gain, input and output impedance effects have already been discussed in Example 5.1. I will discuss the effects of electrical disturbances in the following subsections. My treatment of all these topics is superficial, but I hope it will serve to alert you to consider these aspects when you do a practical design. There are many books available on operational amplifiers.

7.3.1 Transfer Function Realization

The generalization of a gain block is a transfer block. In designing a gain block as an inverter, one takes advantage of the property that no current flows into the $(-)$ terminal, so that input and feedback currents must balance. This leads to the idea of using transadmittance, voltage-to-current transfer functions, as input and feedback networks. Figure 7.10 illustrates this arrangement. Since the $(+)$ terminal is grounded, the $(-)$ terminal will be held at ground by the amplifier. Thus each network is operating into a short circuit. We have $I_1 = G_1(s)V_1$, $I_2 = G_2(s)V_2$, and $I_1 = -I_2$. Therefore,

$$V_2 = \frac{-G_1(s)V_1}{G_2(s)} \tag{7.10}$$

The consequence of this result is that a three-terminal network can be used in the input and its transfer function will appear unchanged, while another three-terminal network can be used in the feedback path to invert its transfer function.

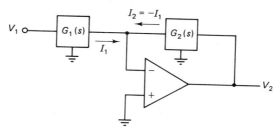

Figure 7.10 General inverting-mode operational amplifier topology.

■ Example 7.1

An ordinary two-terminal admittance is a special case of a transadmittance. Suppose we need a lead network with a transfer function of $-(1 + s)/(1 + s/20)$. An R in parallel with a C has an admittance $G + Cs$. Using one such circuit in the input will give us the zero, and another such circuit in the feedback path will give us its inverse, the pole. This leads to Fig. 7.11.

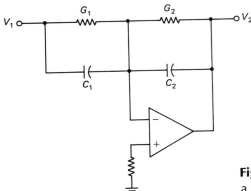

Figure 7.11 A circuit that can be used as either a lead or a lag network.

$$\frac{V_2}{V_1} = -\frac{G_1 + C_1 s}{G_2 + C_2 s} = -\frac{G_1(1 + sC_1 R_1)}{G_2(1 + sC_2 R_2)} \tag{7.11}$$

To get the needed values, $G_1 = G_2$, $R_1 C_1 = 1$, and $C_2 = C_1/20$. We are free to choose the impedance level. Since the Cs are inverse to the Rs, large values of R give smaller, more compact, cheaper Cs, but too high choices will produce loading by the input terminal, excessive noise, or both. ∎

■ Example 7.2

The circuit in the previous example has the disadvantage that it loads the previous stage as the frequency increases. The step response of the two stages will initially be limited by the rate at which they can charge the capacitors, and this may cause a nonlinear response if the step is big enough to cause one of the stages to saturate. One alternative is to put the lead network entirely in the input path and use an appropriate feedback resistor to set the low-frequency gain to 1, as shown in Fig. 7.12. To realize the transfer function of the previous example, $R_3 = R_1 + R_2$, and from (7.2), $R_1 C = 1$ and $R_1 = 19 R_2$. Again, the impedance level is up to you. ∎

The voltage follower is a common example of an op amp used in the non-inverting mode. In this mode, the input drives the $(+)$ terminal and the output, fed back, drives the $(-)$ terminal so as to have the same voltage. The noninverting gain block is another such circuit. In this case, a resistive voltage divider is used as the feedback network, and the inverse of its transfer function becomes the gain. Figure 7.13 shows a generalization of this idea. The $T_i(s)$ are voltage-to-voltage transfer functions implemented by three-terminal networks. $V_3 = T_1(s)V_1$, and $V_3 = T_2(s)V_2$. Therefore,

$$V_2 = \frac{T_1(s)V_1}{T_2(s)} \tag{7.12}$$

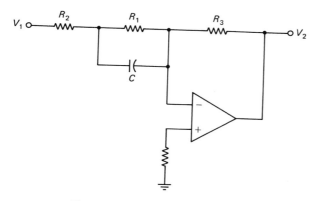

Figure 7.12 A lead network and gain block combination.

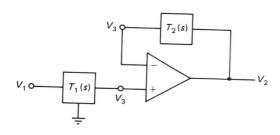

Figure 7.13   General noninverting operational amplifier topology.

■ Example 7.3

The lead network in the previous examples can be produced in noninverting form as shown in Fig. 7.14. The feedback voltage divider is designed so that the dc gain of the stage is 1. The part values could be the same as the corresponding parts in Example 7.2. ■

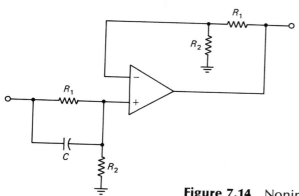

Figure 7.14 Noninverting lead network and gain block.

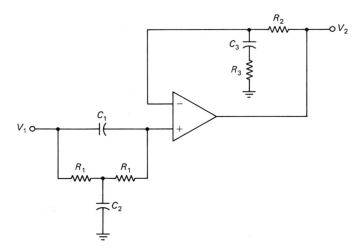

Figure 7.15   Circuit for complex zeros discussed in Example 7.4.

■ Example 7.4

Suppose we need to introduce a pair of complex zeros in the system, but we can't stand the low-frequency pole of the bridged T. We can use the bridged T as $T_1(s)$. To cancel the pole, we only need a pole of the same value as $T_2(s)$, as shown in Fig. 7.15. A lag network is used to limit the high-frequency gain and noise output of the stage.   ■

The two topologies discussed in this subsection are quite flexible and can be used to cover all the usual cases of interest in system design. As pointed out in Example 7.2 though, one should be cautious about side effects. In that example excessive capacitive loading was pointed out as a possibility. Another arrangement to avoid is one that blocks direct current in either the op amp feedback path or the system signal path. These blocks are illustrated in Fig. 7.16. The physical

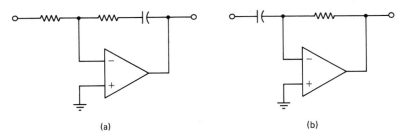

(a)   (b)

Figure 7.16   Two capacitor placements to avoid (a) dc open feedback (b) dc open input.

effect in both cases is to decouple the output from the input at low frequency. This allows a quasi-static disturbance to drive the output without any feedback correction, resulting in drift and decalibration or possibly saturation. This is discussed in a more analytical way in Sec. 7.3.3.

7.3.2 Offsets and Noise

Active devices produce noise above the simple thermal noise of their resistive components. The differential input stage of an op amp is subject to small voltage errors, and possibly bias currents and bias current differences, which have the effect of producing a dc output when there is no input. It is possible to calculate the effects of these error sources on the general topologies in the preceding subsection by representing the subnetworks in terms of their admittance parameters. My intent in this subsection is more illustrative and cautionary than synthetic; I believe the burden of such a general analysis would exceed these benefits. I will discuss the simpler situation in which the networks are represented by branch impedances as shown in Fig. 7.17. The offsets and noises can both be represented by the three generators shown.

To analyze this circuit, and any op amp circuit, start at the $(+)$ terminal: $V_+ = Z_3 I_{n2}$. The $(-)$ terminal must have the same voltage, so $V_- + E_n = Z_3 I_{n2}$. Writing a node equation for V_- gives $I_{n1} = (Y_1 + Y_2)V_- - Y_2 V_n$. Eliminating V_- yields the total output voltage:

$$V_n = (1 + Z_2 Y_1)Z_3 I_{n2} - Z_2 I_{n1} - (1 + Z_2 Y_1)E_n \qquad (7.13)$$

This is a transform relation which we can specialize for the two cases of interest.

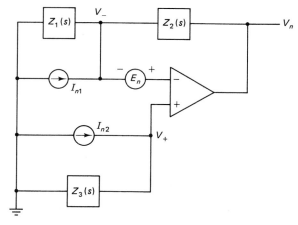

Figure 7.17 Op amp stage with equivalent noise and error sources.

For non-dc random noise sources, the total squared voltage out is the sum of the squares transferred from each noise source:

$$V_n^2 = |1 + Z_2 Y_1|^2 (|Z_3|^2 I_{n2}^2 + E_n^2) + |Z_2|^2 I_{n1}^2 \qquad s = j\omega, \text{ V}^2/\text{Hz} \quad (7.14)$$

The dc values are transferred by the dc values of the impedances in the usual straightforward algebraic way. They are given by (7.13) evaluated at $s = 0$.

One important conclusion that is usually drawn here is that, given equal dc bias currents, their effect can be nulled at the output by making the resistance from each amplifier input terminal to ground the same value. From (7.14) we can also conclude that, for fixed $Z_2 Y_1$, the noise from the current sources is reduced by using smaller Z_3 and Z_2 values. Circuits in which either Z_2 or Y_1 becomes large as frequency goes down (in the former case) or up (in the latter case) will have large amounts of voltage noise. Indeed, if Z_2 is a capacitor, the voltage offset will charge it to the supply limit, while if Y_1 is a capacitor, the amplifier is effectively running open loop above some frequency, and if it has a lot of gain it will produce a lot of noise.

7.3.3   System Effects of Offset and Noise

If an op amp stage is driven by a source whose impedance is low compared to the stage's input impedance, then we can model the stage as an ideal circuit with the error voltage of the last subsection in series with its output. This is so because a condition of the analysis is that the input signal lines are grounded, i.e., of zero impedance. Figure 7.18 illustrates this model in a general system. Let the transfer function of the op amp stage be $-G_c(s)$. Then the transfer from V_n to the output is

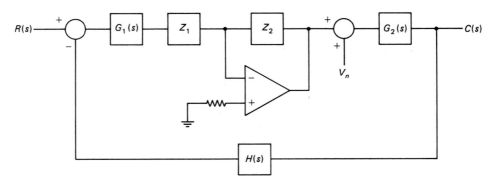

Figure 7.18   A system with an op amp noise source.

$$T_n(s) = \frac{G_2}{1 + G_2HG_1G_c} \tag{7.15}$$

The output signal-to-noise ratio is

$$\frac{S}{N} = \frac{|T|^2 R^2}{|T_n|^2 V_n^2} = |G_1G_c|^2 \frac{R^2}{V_n^2} \qquad s = j\omega \tag{7.16}$$

If V_n is proportional to the stage gain, one should back it up in the model by dividing it by the gain and putting it at the input rather than the output. In any case, (7.16) is a reaffirmation of the result from communication theory that noise is less important the further it is from the input. This has a counterpart in control literature which says that the system should have as much gain before a disturbance input as possible.

Next suppose that the dc error is dominated by the amplifier's voltage offset and our subject stage is an integrator. See Fig. 7.19. The output for this situation is

$$C(s) = \frac{E_{no}}{s} \frac{(1 + 1/(RCs))G_2}{1 + G_1G_2H/(RCs)}$$

$$= \frac{E_{no}}{s} \frac{(s + 1/(RC))G_2}{s + G_1G_2H/(RC)} \tag{7.17}$$

If G_1 has no integrators, the steady-state output is

$$C_{ss} = \frac{E_{no}}{G_1(0)H(0)} \tag{7.18}$$

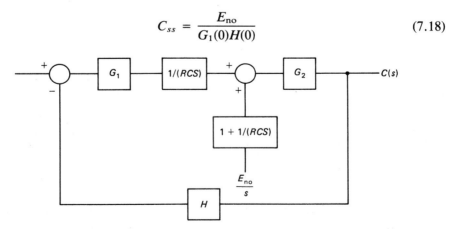

Figure 7.19 Voltage offset in an integrator stage.

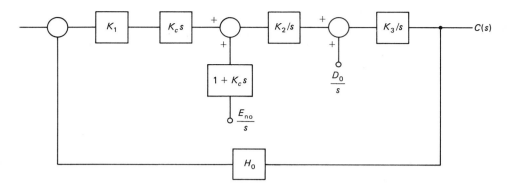

Figure 7.20   Canceling an origin pole with a differentiator.

It is possible for C_{ss} to be greater than E_{no}, for example, if the first stage is a unity-gain subtractor and the feedback is a transducer with a conversion gain less than 1. Normally, one thinks of an integrator as a way to make the steady-state step response errorless and to cancel a constant disturbance at the output, but this shows that the integrator can introduce its own constant error at the output. The system feedback has the effect of helping the integrator. We saw in the last subsection that the offset will drive an isolated integrator to the supply limit.

Finally, suppose our stage has a capacitor input which we want to cancel one of two natural integrators in the system. Consider Fig. 7.20. I have represented the blocks and signals by their low-frequency values, and an external disturbance signal between the two integrators. The response to the offset is

$$C(s) = \frac{E_{no}}{s} \frac{(1 + K_c s)K_2 K_3}{s^2 + K_2 K_3 H_0 K_1 K_c s}$$

$$C_{ss} \to \infty \tag{7.19}$$

The effect of the offset is the same as that of a dc disturbance appearing in front of the two integrators. The response to D_0 will be a constant. Without the differentiator, the responses would be a constant and zero, respectively. A real-axis zero is preferable to a zero at the origin.

I need to mention a few details now. I absorbed the minus sign from the inverting stage in the other system blocks, which is why you didn't see it in the equations. In all this talk about noise, I haven't mentioned the thermal noise of the network resistors. These are, of course, sources independent of the op amp active devices and make their own contribution to the output noise power. Analysis of their effects depends on the particular circuit, and I have put one case in the problem set.

7.4   MECHANICAL NETWORKS

The properties of the transfer functions for the various compensators used in control systems were presented in Sec. 7.2, so in this section I will give some examples of mechanical networks which realize those transfer functions.

Figure 7.21 shows the three types of transfer functions realized with linear motion spring-and-damper networks. I am assuming that the admittance of the following network is negligible. If you want to use one of these networks in practice, you will have to analyze the load to see what conditions are needed to make this assumption a good approximation.

Lead network:

$$\frac{X_2}{X_1} = \frac{a_1}{a_2} \frac{1 + s/a_1}{1 + s/a_2} \tag{7.20}$$

where $a_1 = K_1/B$ and $a_2 = (K_1 + K_2)/B$.

Lag network:

$$\frac{X_2}{X_1} = \frac{1 + s/a_2}{1 + s/a_1} \tag{7.21}$$

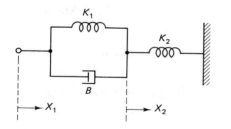

(a) Lead network

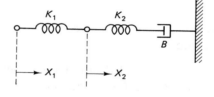

(b) Lag network

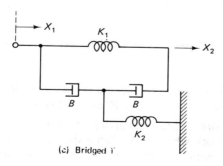

(c) Bridged T

Figure 7.21 Some mechanical compensators.

where $a_1 = K_1K_2/[B(K_1 + K_2)]$ and $a_2 = K_2/B$.

Bridged T:

$$\frac{X_2}{X_1} = \frac{s^2 + 2\zeta\omega_0 s + \omega_0^2}{(s + a_1)(s + a_2)} \tag{7.22}$$

where $\omega_0^2 = K_1K_2/B^2$, $\zeta^2 = K_1/K_2$, and a_1, a_2 are given by (7.8).

These networks are analogous to the voltage-to-voltage transfer functions in Sec. 7.2. In the case of a mechanical equivalent to an operational amplifier, one may wish to use networks with a displacement-to-force transfer function. Possible networks for this use are given in the problems.

FURTHER READING

There are many books on operational amplifiers and linear integrated circuits. They range from cookbooks, many of which have inadequate schematic labeling which prevents one from easily filling in their inadequate analysis, to texts which describe design procedures for the IC itself as well as for its applications. I'm sure your library has several. The one I used for some of the data in the examples is intermediate in level.

Morley, Michael S., *Linear IC Handbook*, Blue Ridge Summit, PA, TAB Books, 1986.

The following two items give design help on bypass, grounding and shielding of electric circuits.

Wilson T. C., Jr., "Life After the Schematic: The Impact of Circuit Operation on the Physical Realization of Electronic Power Supplies", Proc. IeEE vol. 76 no. 4, April, 1988.

Morrison, Ralph, *Grounding and Shielding Techniques*, New York, John Wiley and Sons, New York, 1986.

PROBLEMS

Section 7.2.1

7.1 Design an *RC* circuit to give the voltage output transfer function $(s + 2)/(s + 30)$. Choose values so that $5 \leq R_i \leq 100$ kΩ and $C \leq 100$ μF.

7.2 A lead network is desired which will give a 45° phase shift at 15.9 Hz. Design a current output network which will require the minimum makeup gain, with a maximum resistor value of 1 MΩ and a minimum capacitor value.

7.3 A lag network is required which will produce an attenuation of 25:1, ending at 0.1 rad/s. Use the constraints of Prob. 7.2.

Section 7.2.2

7.4 In (7.8), what does $a_1 a_2$ equal? What are a_1 and a_2 in terms of the circuit part values?

7.5 In the discussion following (7.8), what do you think "sufficiently small ζ" means? Find and tabulate ζ values for error in the pole values from 20% to 1%.

7.6 Design a current output bridged T to give a resonant frequency of 31.8 Hz and a damping factor of 0.14. The maximum allowed R is 470 kΩ. Use C values as small as possible.

7.7 Figure P7.7 shows another form of bridged T which has equal-valued capacitors. Find the transfer functions for $R_L = \infty$ and $R_L = 0$. For a given transfer function and a maximum R value, which circuit will require larger capacitor values?

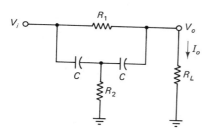

Figure P7.7   Another bridged T.

Section 7.3.1

7.8 Realize the transfer function in Example 7.1 by using a lag network in the feedback path of an op amp. The maximum C allowed is 1 μF. Minimize the R values.

7.9 Figure P7.9 shows a circuit whose purpose is to get the specified gain while using more uniform input and feedback resistor values. Suppose a gain of

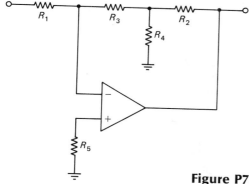

Figure P7.9   A circuit to reduce input loading.

-100 is needed and $R_1 = R_2 = 47$ kΩ. Find minimum values for R_3 and R_4. Find R_5 equal to the net resistance on the $(-)$ terminal with the input grounded.

7.10 A generalization of the difference stage is possible using transadmittances on the $(-)$ terminal and a voltage output network connected to the $(+)$ terminal. However, since the voltage on these terminals is not zero, the transadmittance networks would have to be represented by their Y parameters. A somewhat simpler situation is shown in Fig. P7.10, where Y_1 and Y_2 are two-terminal admittances. Find the transfer functions for this circuit.

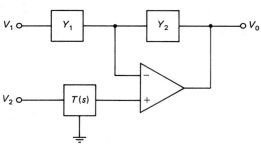

Figure P7.10 A two-input network.

7.11 Design a circuit on the models in Figs. P7.9 and P7.10 to give $V_0 = (V_2 - V_1) (1 + 10s)/(1 + 0.3s)$. Use resistor values as close to 100 kΩ as possible.

Section 7.3.2

7.12 Design the lead stage in Fig. P7.12 to have the transfer function $(1 + s)/ (1 + s/10)$. Suppose the op amp has an equivalent input noise source of 10 μV/$\sqrt{\text{Hz}}$. Choose component values so that the output noise due to the resistors is equal to that due to the op amp at $\omega = 3.16$ rad/s.

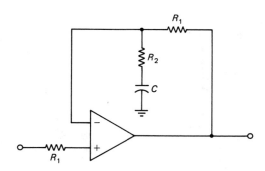

Figure P7.12 A lead stage.

Section 7.3.3

7.13 A shaft position control system is being designed. The plan is to drive a dc motor from a voltage-to-current amplifier. The load is pure inertia. A colleague proposes to use a differentiator in front of the amplifier to stabilize the system. Analyze the effect of a constant disturbance torque on the load with and without the differentiator.

Section 7.4

7.14 Design a spring-damper lead network to realize the displacement transfer function $(1 + 10s)/(1 + 2s)$. If the output is connected to a load mass of 50 g, what must the component values be to make the loading negligible in the range between the corner frequencies? Minimize the input force required.

7.15 A proposed compensation network is given in Fig. P7.15. Find and name its transfer function.

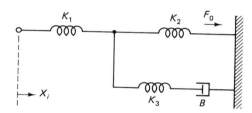

Figure P7.15 A displacement-to-force transfer network.

7.16 Find the transfer function for the bridged T given in Fig. P7.16 for the cases $Y_L = 0$ and ∞.

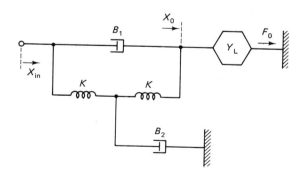

Figure P7.16 Another mechanical bridged T.

Chapter 8

COMPENSATION OF THE SYSTEM

8.1   OBJECTIVES

Once the major parts of a system have been chosen, it generally turns out that the dynamic performance is unsatisfactory. This is determined by modeling these components and plotting the system's frequency response or root locus. So what will you do? You are going to add transfer blocks to the system to make its frequency response or root locus into something that will satisfy the specifications, whether they are stated in the time domain or the frequency domain or both. This means you have to be able to relate, at least roughly, the specifications to the open-loop description which might meet them. Chapters 5 and 6 pointed to several ways to do this using either the ITAE optimum system or the second-order system as a model for translating between closed- and open-loop requirements. After meeting the specifications, your most important objective is simplicity in design. This goal of simplicity is not to make your life easier—it's to keep the cost down and enhance reliability. In fact, you may need to do quite a bit of work to achieve the simplest design. As a general guide, that design is best which seeks to alter the system least. If your system already has an integrator and a real pole much closer to the imaginary axis than any other pole or zero, it's easy to make it look like the second-order system that will meet the specs. If the system has several poles in a cluster, it may be best to redistribute them in an ITAE pattern. In between these situations you will find many that don't have an obvious best approach, so you will have to try several. In any case, once you have chosen a good model the methods of the following sections will help you convert the starting system to desired final form.

As far as the open-loop transfer function is concerned, it doesn't make any difference whether you put the compensating factors in the forward path or the feedback path. The closed-loop poles will be the same, too. However, the compensator location makes quite a difference to the zeros of the closed-loop transfer function and its time response. Figure 8.1 shows two fairly general arrangements for this comparison.

The N_x and D_x are polynomials in s, the subscript c is for compensator and the subscript p for plant. β is the transducer constant. For the cascade compensation case, the closed-loop transfer function is

$$T_a(s) = \frac{N_c N_p/(D_c D_p)}{1 + \beta N_c N_p/(D_c D_p)}$$

and for feedback compensation,

$$T_b(s) = \frac{N_p/D_p}{1 + \beta N_c N_p/(D_c D_p)}$$

When the loop gain is ≥ 1, these are, approximately,

$$T_a(j\omega) \approx \frac{1}{\beta}$$

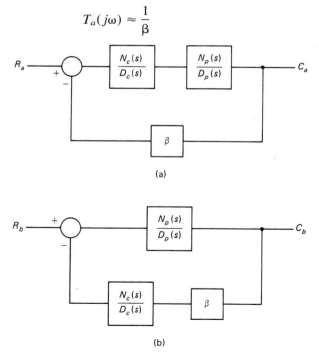

(a)

(b)

Figure 8.1   (a) Cascade compensation. (b) Feedback compensation.

$$T_b(j\omega) \approx \frac{D_c}{N_c\beta} \qquad \omega \leq \omega_c \qquad (8.1)$$

This means that the time response of the feedback-compensated system is apt to be similar to that of a system whose transfer function is the inverse of the compensator. It is often the case that many, if not all, of the poles and zeros are at lower frequencies than the unity-gain crossing. Those that are will show in T_b strongly.

■ Example 8.1

Suppose the plant is $G_p = 100/s[(1 + s/2)]$ and $\beta = 0.1$. The dc gain of this system is $1/\beta = 10$. At the corner between the -1 and -2 slopes $|G_p\beta| = 5/\sqrt{2}$, which means that unity gain is on the -2 slope. If the performance requirements say "little or no overshoot," we need to compensate this system. One way to do this is by adding a zero at -2 to cancel the pole. If the zero is cascaded in the forward path, the system will be first-order with a pole at -10 rad/s. This is also the unity-gain crossover frequency. From the analysis above in (8.1), putting the compensator in the feedback path will give a behavior similar to that of a system whose closed-loop transfer function is a simple pole at -2 rad/s. The step responses in both cases should be exponential, with time constants of 0.1 and 0.5 s, respectively. They are shown in Fig. 8.2.   ■

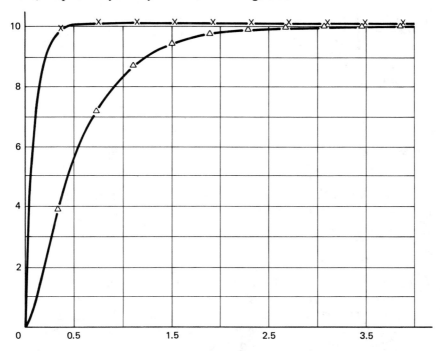

Figure 8.2 Step responses for Example 8.1. ×, Cascade compensator; △, feedback compensator.

■ Example 8.2

For reasons discussed in Chap. 7, we may not wish to use a simple zero compensator. Let's try a lead network, $G_c = (1 + s/2)/(1 + s/20)$. This time we have a net second-order system with a good phase margin (how much?). Cascade (forward-path) compensation gives a step response with a little overshoot. Again, putting the compensator in the feedback path will give a response dominated by the zero converted to a closed-loop pole. Figure 8.3 shows the two step responses. ■

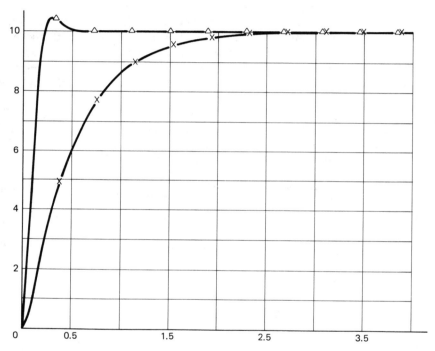

Figure 8.3 Step responses for Example 8.2. △, Cascade compensator; ×, feedback compensator.

■ Example 8.3

If we don't need such a fast response, another possibility is to use a lag network to bring the −1 slope down. Try the compensator $G_c = (1 + s/0.08)/(1 + s/0.0032)$. This will give $\omega_c = 0.4$ rad/s. If the compensator is put in the feedback path, the response will be dominated by the feedback-path pole at −0.0032 turned closed-loop zero. The low-frequency zero gives differentiator action and responds to the step with a pulse. Figure 8.4 shows this effect. The

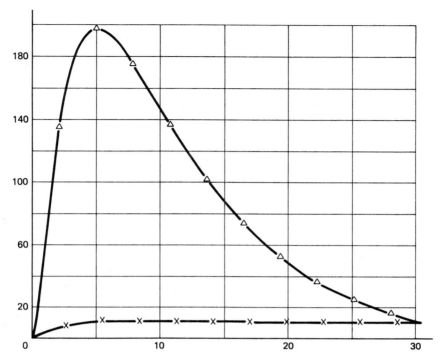

Figure 8.4 Step responses for Example 8.3. ×, Cascade compensator; △, feed-back compensator.

cascade-compensated system response looks small, but it's doing the right thing by easing up to the required final value of 10. ■

You may conclude from these examples that it's always the best policy to use the cascade arrangement. Certainly, putting a lag network in the feedback path isn't a good idea, but the other two examples were really not so bad. If the system can afford to be a little slower, it may be that getting the zero or the lead is more economical or desirable because of lower high-frequency noise, for example. As pointed out in Chap. 4, the commercially available motor-tachometer pair provides a rate signal which gives us a zero without the side effects of an op amp design. To use this zero as the compensator in Example 8.1, we need to move it off the origin. This is easily done by putting the tachometer signal in a parallel path, as I show in Fig. 8.5. The total feedback transfer is $0.1 + K_t s$. What should K_t be to produce the required zero?

Using the tachometer to get a real zero is an example of another kind of topology, minor-loop feedback. A general form using one minor loop is displayed in Fig. 8.6. If the connection points are accessible, the minor-loop approach can

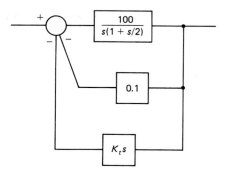

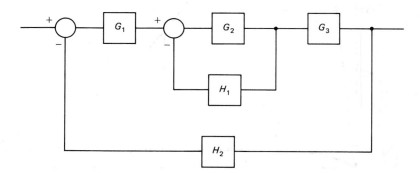

Figure 8.5 Tachometer feedback for the system in Example 8.1.

Figure 8.6 Compensation using a minor loop.

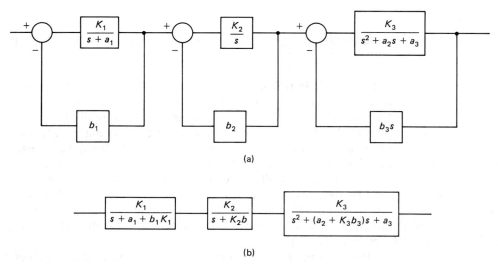

(a)

(b)

Figure 8.7 Using simple minor loops to adjust block transfer functions. (a) Original blocks with added loops. (b) Equivalent new blocks.

be used to tame a system in a step-by-step way. For example, if a system has two natural integrators and the specifications only point to one, a constant feedback around one integrator will convert its transfer to a real pole. Pole frequencies and quadratic damping factors can also be changed by simple minor-loop feedbacks, again I stress, if the connection points physically exist. Figure 8.7 shows some specific examples of this sort. State feedback is another special case of minor-loop feedback. It isn't as frequently useful in SISO systems as cascade compensation, so I defer discussion of state feedback to the chapters on MIMO systems.

One can think of the minor-loop operations in Fig. 8.7 as another way of achieving the desired open-loop transfer function with cascaded blocks in the forward path. Evidently the circuits needed are simpler than would be the case for a single network doing the whole job. Since minor loops are not always possible, and they are really a circuit alternative, the rest of the chapter is devoted to finding the transfer function for a single-cascade compensator.

8.3   CASCADE COMPENSATION

Each of the following subsections is intended to feature the application of a particular type of compensator transfer function. The instruction is mainly by example, and I make use of both Bode plot and root locus methods as alternative ways to get started on the problem. In every case, I assume that the low-frequency character of the system is to be preserved, that is, the number of integrators and the low-frequency gain constant are not to be changed.

8.3.1   Raising Bandwidth

This subsection features the lead network transfer function. Since the low-frequency gain is to be unchanged, using a lead network will raise the high-frequency gain and increase the bandwidth of the system.

■ Example 8.4

Suppose we have a second-order system. The design at this point consists of a plant with transfer function $G_p(s) = 100/[s(1 + s/2)]$ and a feedback of $H = 0.1$. The dynamic specification is simply to make the per-unit overshoot on a step input 0.02. From (6.32), the required pole angle is $\phi_p = 38.77°$, and $\zeta = \cos \phi_p = 0.7797$ and $\tan \phi_p = 0.803$.

Bode Plot Method

Figure 8.8 shows a straight-line Bode magnitude plot for the system as it stands. Since unity-gain crossover is above the corner frequency, on the -2 slope, the system will have more than trivial overshoot. From Chap. 6, the corner frequency-to-loop-gain-constant ratio is $a/K = 4\zeta^2 = 2.432$. Since the loop gain

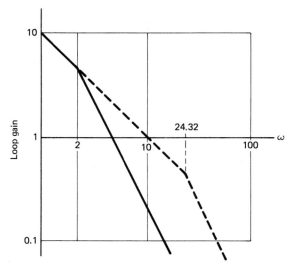

Figure 8.8   Bode plots for Example 8.4. Solid line, original system, dashed line, required system.

constant is 10, the corner frequency should be 24.32. The sketch is extended to show this correction. In order to get this new transfer function, I need to cancel the pole at -2 with a zero and add a pole at -24.32. Therefore, I will need a compensator whose transfer function is $G_c(s) = (1 + s/2)/(1 + s/24.32)$.

Root Locus Method

The root locus for the system at this point is depicted in Fig. 8.9. I need closed-loop poles with a tangent of 0.803. Inquiring of the LSP what the gain constant is for $s = -1 + j0.803$, I am told 8.224. Since the gain constant has to be 100, I must expand (rescale) the diagram by $100/8.224 = 12.16$. This means moving the real pole over to $-2 \times 12.16 = -24.32$. Thus, I will need a network to cancel the pole with a zero and add a new pole. The compensator will be $G_c(s) = (1 + s/2)/(1 + s/24.32)$.   ∎

∎ Example 8.5

Now let's look at a two-integrator plant. $G_p(s) = 100/[s^2(1 + s/2)]$, $H = 0.1$, and we need to meet the same overshoot requirement as in the previous example. Notice that these specifications must derive from steady-state ramp error and transient step response requirements. My approach will be to make the system look like a second-order one at high frequency. As it stands, we know from examples in Chaps. 3 and 4 that this system is unstable at any gain, so there's no use in making plots as it stands. Consider this though, since I want to have a system that looks like a single pole at the origin, I will need a relatively low-

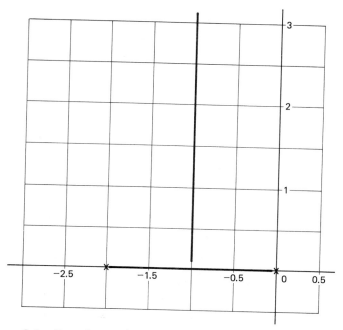

Figure 8.9 Root locus for the original system in Example 8.4.

frequency zero to mitigate the effect of one of the origin poles. This creates an equivalent gain constant for the equivalent second-order system. This is so because at frequencies above the zero, $G_p \approx (K/s^2)(s/z)(\ldots) = (K/z)(\ldots)/s$,

$$K_{eq} = \frac{K}{z} \tag{8.2}$$

where z is the corner of the zero. As in the previous example, the required corner frequency for the equivalent system is $a = 4\zeta^2 K_{eq}$, so

$$az = 4\zeta^2 K \tag{8.3}$$

For this problem, using the numbers from Example 8.4, I need $az = 24.32$, and $az = 25$ will do. The system has a third pole at -2 which I need to remove, since I already know from Example 8.4 that it is at too low a frequency. I am going to need two lead networks for this problem, one with a zero at $-z$ and one with a zero at -2. I'll need one pole at $-a$, and then there is an excess pole. Since I am trying to approximate a two-pole system, I will suppose that the excess pole is five times farther out than the desired corner, at $-5a$. Making some sketches will help you a lot in visualizing this discussion, but I didn't want to repeat the words for the devotees of each of the following methods.

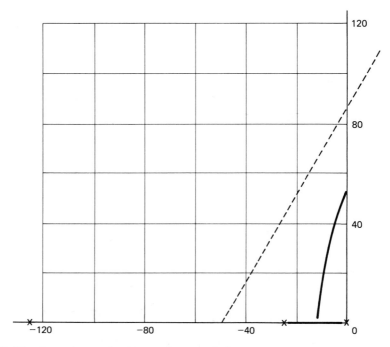

Figure 8.10   Root locus for Example 8.5. G_c has zeros at -1 and -2 and poles at -25 and -125.

Root Locus Method

The question now is, What should the ratio a/z be? If you think 25 should give a good approximation, then $z = 1$ and $a = 25$. Figures 8.10 and 8.11 show the root locus for this choice. Note the little circle near the origin in Fig. 8.11. Apparently, the origin poles circle around and come to the real axis near -2. Then they must split, with one going to the zero at -1 and the other meeting the pole coming from -25 at about -11. The poles meeting at -11 go off-axis, with a nearly vertical initial heading. If this were a true second-order system with one origin pole and a pole at -25, the poles would meet at -12.5 and go vertical. The matter of which poles dominate the system is a bit clouded, since there is a real pole very near the origin, but supposing the transient response to be dominated by the complex poles, I checked the forward-gain constant at $s = -11 + j9$ and the LSP gave me 112. So we seem to be on the right track. The step response for this plant and compensator is shown in Fig. 8.12. The overshoot is well above specifications.

Well, I suppose a/z wasn't large enough for a good approximation, so I tried 100. Choosing $z = 0.5$ and $a = 50$ gave me the root locus in Figs. 8.13 and 8.14.

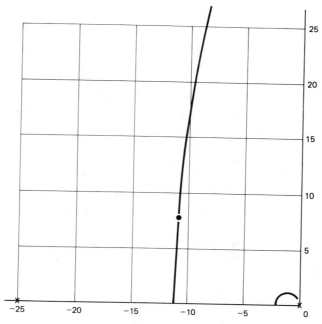

Figure 8.11 Expansion of previous root locus showing the approximation to a second-order system.

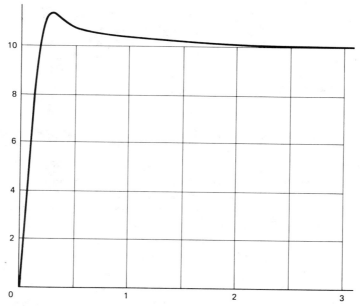

Figure 8.12 Step response for Example 8.5. G_c has zeros at -1 and -2 and poles at -25 and -125.

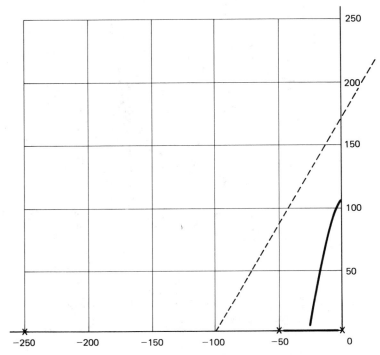

Figure 8.13 Root locus for Example 8.5. G_c has zeros at -0.5 and -2, and poles at -50 and -260.

The shape seems about the same, except that the little circle is too small to show up. Again, checking the gain at the poles with about the right tangent, $s = -23 + j18$ gave about 100. The step response is better, as you can see in Fig. 8.15, but still not good enough. Figure 8.15 shows an even greater spread, which again is an improvement, but doesn't meet the specification. Perhaps the problem is partly due to the excess pole. In consideration of it, maybe I should have chosen a larger product for az. Choosing $az = 50$ and going back to $z = 0.5$ gave me $a = 100$ and the step response in Fig. 8.16. This time I made it. I also checked $z = 1$ for the same poles, and you can see that this value doesn't make it either. Another possibility for meeting the specification with an az closer to 25 might be to raise the ratio for the excess pole. A figure of merit for comparing alternatives is the total amount of high-frequency gain required in the two lead networks for a compensator. In the present case, this is $100 \times 500/(2 \times 0.5) = 50,000$. Putting the 2 with the 500 and the 0.5 with the 100 gives one stage with a gain of 250 and the other with a gain of 200—reasonably even distribution. I leave the investigation of the excess pole ratio to you.

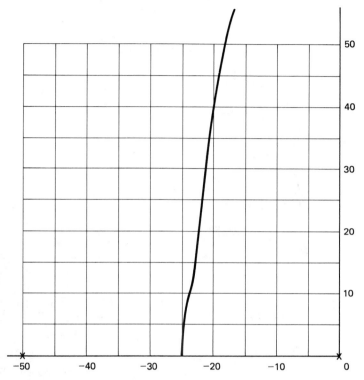

Figure 8.14   Expansion of the previous root locus, showing the local approximation to a second-order locus.

Bode Plot Method

How long should the -1 slope be to approximate the second-order system? I first tried 25:1, with $a = 25$ and $z = 1$. The Bode plots for the equivalent system and the compensated system are given in Fig. 8.17. The magnitude plots are quite close, especially for a decade around crossover, but the phase curves have a fair spread. The step response is shown in Fig. 8.12 and is not too good. Doubling a and halving z gives me a ratio of 100:1. The Bode plots for this are shown in Fig. 8.18. Again, the magnitudes are very close, but the actual system has more phase lag at all frequencies near crossover. Maybe I should increase the az product to compensate for the excess pole at $-5a$. Going to $a = 100$ and keeping $z = 0.5$ gave me the plots in Fig. 8.19. The equivalent system is the same as for Fig. 8.18. Now, the compensated system has phase lead over the equivalent system near crossover. The step response for the compensated system is in Fig. 8.16, and that for the equivalent second-order system is in Fig. 8.20. Pretty close!   ∎

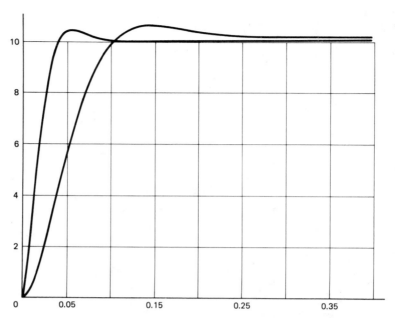

Figure 8.15 Step responses for Example 8.5. Slower curve for G_c with zeros at -0.5 and -2, poles at -50 and -250. Faster curve for G_c with zeros at -0.2 and -2, and poles at -125 and -625.

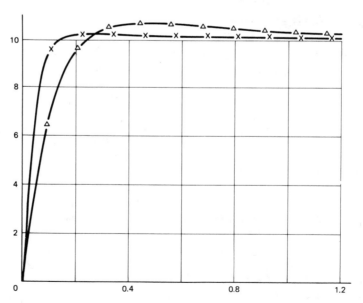

Figure 8.16 Step responses for Example 8.5. $\times$, G_c with zeros at -0.5, -2; $\triangle$, G_c with zeros at -1, -2. In both cases G_c has poles at -100, -500.

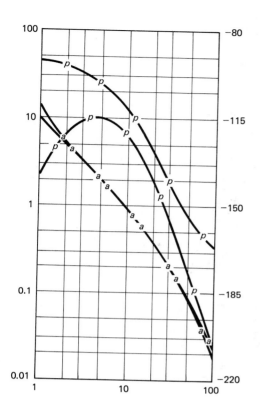

Figure 8.17 Bode plots for Example 8.5. On the left side, the upper (p)hase and (a)mplitude curves are for the equivalent second-order system, the lower curves are for G_c with zeros at -1, -2, and poles at -25, -125.

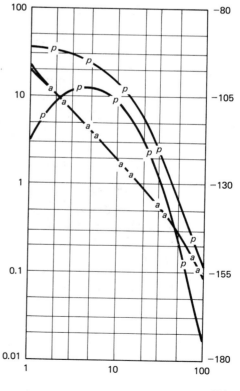

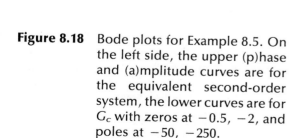

Figure 8.18 Bode plots for Example 8.5. On the left side, the upper (p)hase and (a)mplitude curves are for the equivalent second-order system, the lower curves are for G_c with zeros at -0.5, -2, and poles at -50, -250.

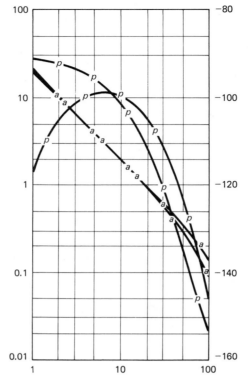

Figure 8.19   Bode plots for Example 8.5. On the right side, the upper (p)hase and (a)mplitude curves are for G_c with zeros at -0.5, -2, and poles at -100, -500. The lower curves are for the equivalent second-order system.

■ Example 8.6

Instead of the transient spec in Example 8.5, suppose the requirement is for an ITAE ramp response. Our system presently has one real pole, so it's already third-order without adding a lead network. Since we will need a zero for the ITAE and will likely have to shift the pole, the compensator will again have to be two lead networks. This gives us a fourth-order system as a minimum. However, as in the previous example, I'm going to make it approximate the third-order system by putting one of the poles far out on the negative real axis. The exact transfer function I want is

$$HG_pG_c = \frac{10(1 + s/z)}{s^2(1 + s/p)} = \frac{(10p/z)(s + z)}{s^2(s + p)}$$

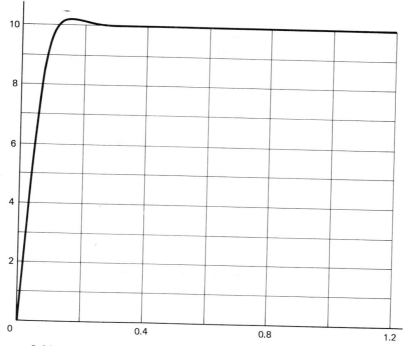

Figure 8.20   Step response for $G(s) = 200/[s(1 + s/50)]$ and $H = 0.1$.

From Table 5.2 the generic form for the third-order system is

$$G(s) = \frac{a_1\omega_0^2 s + \omega_0^3}{s^2(s + a_2\omega_0)}$$

Matching terms, $10p = \omega_0^3$ and $p = a_2\omega_0$ gives $\omega_0 = \sqrt{10a_2} = 4.1833$ and $p = 7.32$. Always, $z = \omega_0/0a_1$, which is 1.287 in this case. Since the required pole is at -7.32, I will put the extra pole at -100. This gives

$$G_c = \frac{(1 + s/2)(1 + s/1.287)}{(1 + s/7.32)(1 + s/100)}$$

I generated plots of the ramp response and error response for the exact ITAE system and the approximate design in Figs. 8.21–8.24. I felt the ramp response looked so smooth that it is hard to tell what's really going on. The error responses for the two systems are very similar, with a bit higher peak on the approximate design. Remember that these are actual errors not normalized to the desired value. From what time does the per-unit error stay below 0.02? ∎

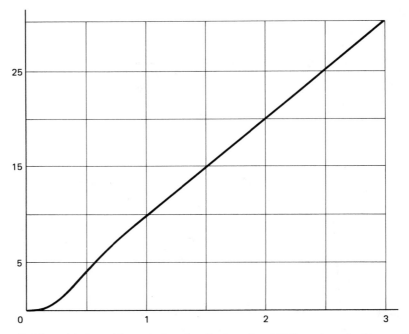

Figure 8.21 $c(t)$ for $r(t) = t$, for the 3rd-order ITAE system in Example 8.6.

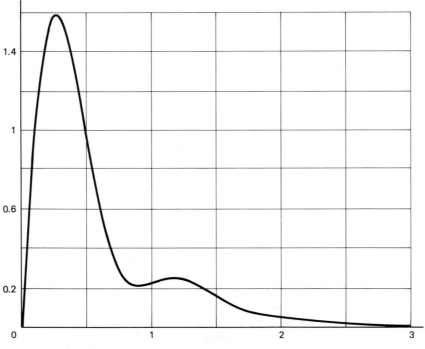

Figure 8.22 Error$(t) = 10t - c(t)$ for Fig. 8.21.

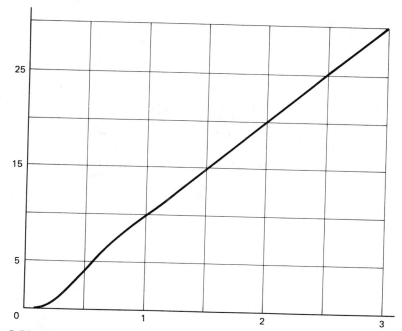

Figure 8.23   Ramp response for the approximate ITAE design, Example 8.6.

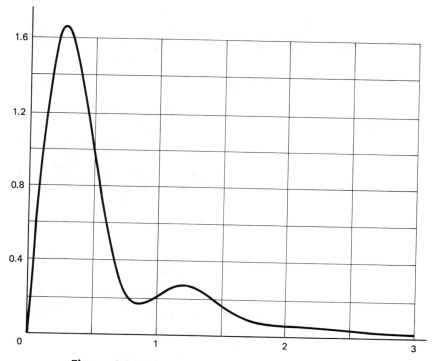

Figure 8.24   Error$(t) = 10t - c(t)$ for Fig. 8.23.

8.3.2 Lowering Bandwidth

In this subsection I will work through two examples which use the lag network transfer function to attenuate the high-frequency portion of the plant frequency response. Example 8.7 will use Bode plots, and Example 8.8 will use root locus plots. In each example, issues of criteria and design choice will be raised and illustrated. We cannot do a system design from scratch for a time domain specification because we don't have an analytical mechanism to find the time response directly in terms of system parameters. The advantage of frequency domain methods is that we do have a way to relate system parameters to frequency domain performance. Therefore, all design problems have to start with a translation from a desired time domain performance specification into a corresponding, approximate, frequency domain objective. The frequency-time relationships for the second-order system are a known guide for this translation, and if the design approximates second-order frequency response, it will also approximate second-order time response, as we saw in the last subsection. However, a single frequency domain objective can be satisfied by many different design solutions (infinitely many, in principle) which will have different costs and time domain performances. This is the point exemplified in the following solutions.

■ Example 8.7

$H = 1$ and $G_p(s) = 100/[s(1 + s)]$. The objective is to produce a 60° phase margin using a lag network. The plant's frequency response has a -1 slope cornering at 1 rad/s to a -2 slope. Since the gain constant is so high, unity-gain crossover occurs well down the -2 slope. The lag network will provide attenuation to bring the -1 slope down so that unity-gain crossover occurs below the corner. There are two procedures readily at hand. I can make the compensated system look like a second-order system by having the lag network's pole and zero at values well below crossover and therefore use the phase margin equations for a second-order system to find the needed equivalent gain. The other choice is to use the lag network to produce a long -2 slope, leaving a relatively short -1 slope and then the original -2 slope. This gives a system with a -2, -1, -2 slope sequence which approximates a two-integrator system, and if crossover is placed at the middle of the -1 slope section, I can use the phase results for a lead network to find the corner and crossover frequencies and then the required attenuation. I will work through both approaches.

Approach 1: Equivalent Second-Order Design

We are using the original corner frequency of the plant, so I need to find the gain constant that will give us the 60° phase margin. From the approximate relations in Chap. 6, $K \approx 0.7$. Using exact calculation, $K = 0.7$ yields $|G_{eq}| = 1\angle -121°$ at $\omega = 0.6$ rad/s. Trimming a little, $K_{eq} = 0.65$ yields $|G_{eq}| = 0.99\angle -119.7°$ at $\omega = 0.57$ rad/s. The required attenuation is $K/K_{eq} = 100/0.65$

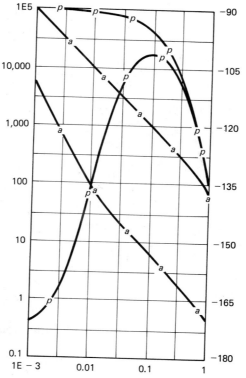

Figure 8.25   Bode plots for Example 8.7. On the left side, the upper (a)mplitude and (p)hase curves are for the original plant. The others are for the very low-frequency lag-compensated system.

= 154. We want a long −1 slope, so I choose the zero to be −0.01. The pole must be −0.01/154 = −6.5 × 10⁻⁵ = −65 μrad/s. Since this is a very low corner frequency, I refer to this network as the very low-frequency lag-compensated system. Figure 8.25 shows Bode plots for the original and compensated systems. Figure 8.26 shows the step response for the compensated system. We can see that the phase plots are close in the 0.1–1 decade and that unity gain occurs at a phase of −120°, as required. The compensator is

$$G_{c1} = \frac{1 + s/0.01}{1 + s/0.000065}$$

Approach 2: Approximate Two-Integrator Design

If I use the lag network to produce a long −2 slope, the effect is to produce a system which looks like a two-integrator system compensated with a lead net-

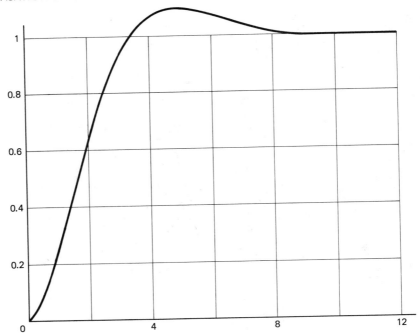

Figure 8.26   Step response for the very low-frequency lag-compensated system in Example 8.7.

work. The phase margin in such a case is the lead produced by the network. From Chap. 7, $\phi_m = 90° - 2 \tan^{-1} (\sqrt{a/b}) = 60°$, from which $a/b = (\tan 15°)^2 = 0.071797$. Since $b = 1$ rad/s, $a = a/b$. I want crossover to occur at the maximum phase lead frequency, so $\omega_c = \sqrt{a/b} = 0.26795$ rad/s. At this frequency, $|G_p| = 360.49$, so this is the required attenuation. The corner frequency of the lag network zero is a, so its pole corner is at $a/360.49 = 0.071797/360.49 = 0.0001992$ rad/s. This is about three times the pole frequency found for the first approach. The Bode plots for the original and compensated systems are shown in Fig. 8.27. The phase curve peaks at $-120°$ as the amplitude crosses unity at just below 0.3 rad/s—quite different from the second-order system approach. The step response is shown in Fig. 8.28.

The overshoot for the first approach is about 10%, and that for the second is about 18%. In both cases, the response is smooth. Note the slow rise times compared to the lead-compensated examples. This is the time domain change equivalent to the bandwidth change for the two types of compensators. Neither of these is a simple system; their responses have something of the character of both single- and double-integrator systems. The second approach would probably give a better transient response to a ramp input than the first, or follow a triangle or short-term ramp better, because it has the high-frequency character of a two-

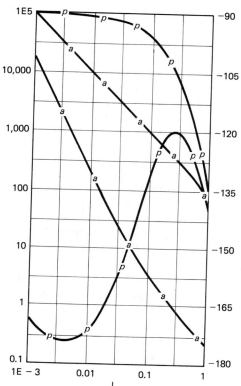

Figure 8.27 Bode plots for the low-frequency lag-compensated system in Example 8.7. On the left, the upper (p)hase and (a)mplitude curves are the original system.

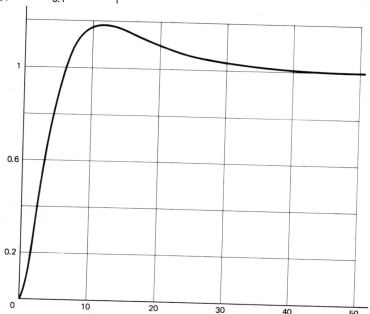

Figure 8.28 Step response for the low-frequency lag-compensated system in Example 8.7.

integrator system. The quasi-second-order design of the first approach could be tuned to give lower overshoot if that is desired. In either case, one should choose the approach that will lead to a closer fit to the original time domain specification. Once the frequency domain design has gotten the parameter values and performance in the neighborhood, the parameters can be adjusted (tuned) to meet the time domain requirements more closely. ■

■ Example 8.8

The root locus method is best suited to the placement of closed-loop poles, so a specification which is translated into a closed-loop pole location is the best frequency domain objective for this method. Suppose then that $H = 1$ and $G_p = 100/[s(1 + s)]$ and that a lag network is to be used to place the complex closed-loop poles at an angle of 30° off the negative real axis. Without compensation, the conventional locus for gain variation is the real axis between -1 and 0 and vertical tracks leaving from -0.5. You may verify that the pole angle is much steeper than 30° for the required gain. Adding a lag network doesn't change the net pole count, so there will still be two vertical asymptotes. But now there is a real pole close to the origin so the track will leave the axis and swing over to the asymptote, while the pole at -1 goes to the zero. I have two parameters to adjust, and using the locus for gain would be a clumsy process of guessing pole and zero values, plotting, finding K at the required pole angle, and guessing new values to start over. The better way to go since the gain constant is already fixed, is to work with the locus for one of the compensator parameters. The open-loop transfer for the compensated system will be

$$G_c G_p = \frac{100(1 + s/z)}{s(1 + s)(1 + s/p)}$$

The locus equation is

$$s(1 + s)\left(1 + \frac{s}{p}\right) + 100 + \frac{100s}{z} = 0$$

I must choose either p or z to make a locus plot. Since $1/z$ is part of the simpler polynomial, I will choose it. Multiplying through by p and rearranging,

$$s^3 + (1 + p)s^2 + ps + 100p + \left(\frac{100p}{z}\right)s = 0$$

or

$$\frac{s^3}{100p} + \frac{1 + p}{100p}s^2 + \frac{s}{100} + 1 + \frac{1}{z}s = 0$$

I need to choose a value of p, factor the cubic, and then find a locus for $1/z$ to see what the shape of things might be. For the lag network, $p < z < 1$, so I will start by choosing $p = 0.01$. This gives poles at -1.4675 and $+0.2288 \pm j0.732$. The locus for this case is plotted in Fig. 8.29. The $p = 0.01$ choice can't give $\phi_p = 30°$ for any value of $1/z$ because the imaginary part of the complex pole is too high with respect to the asymptote position. I need to find a p which sets the complex poles closer to the real axis compared to the distance between the real pole and the imaginary axis. This leads naturally to the desire to look at the root locus of the cubic as a function of p. Forming this into a locus equation I have

$$s^3 + s^2 + p(s^2 + s + 100) = 0$$

The zeros are at $-0.5 \pm j9.9875$. Figures 8.30 and 8.31 show the locus. Between Fig. 8.29 and 8.31, I can see that the asymptote is near -0.5 and, for small p, the complex pole is very near the imaginary axis. Trying a pole at $+0.02 + j0.2$ gives a value $p = 0.00042$. When I ran the locus for $1/z$, it had a dip toward the real axis and went low enough so that two values would give the needed pole

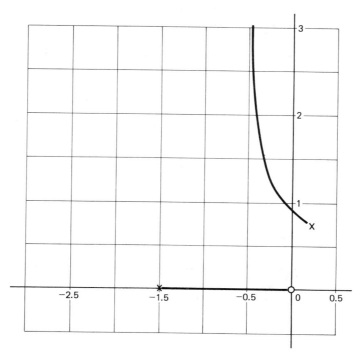

Figure 8.29   Root locus for $1/z$ with $p = 0.01$, Example 8.8.

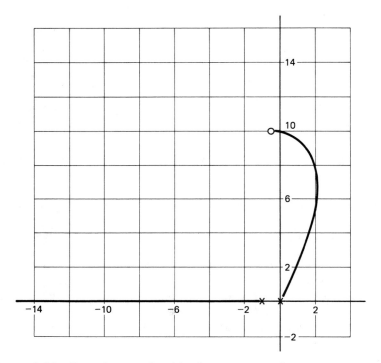

Figure 8.30 Root locus of cubic denominator for *p*, Example 8.8.

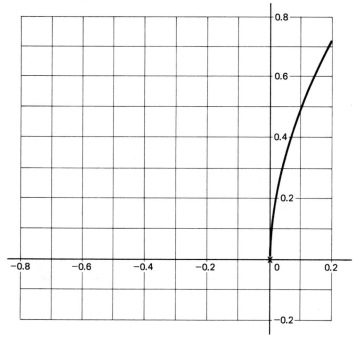

Figure 8.31 Expansion of Fig. 8.30 to show locus for small *p*.

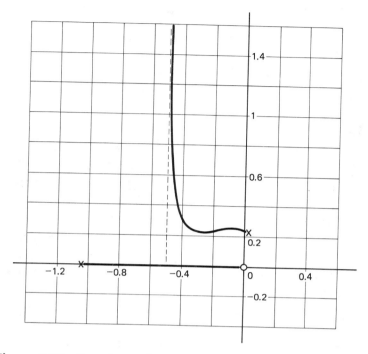

Figure 8.32   Root locus for 1/z with $p = 0.00055$, Example 8.8.

angle. In the absence of any other requirement, I decided to look for the largest value of p which would satisfy the pole angle requirement. This corresponds to a minimum-cost solution, which is a point I will discuss in the next chapter. After a few cycles in which I set the poles of the system model in the LSP equal to the output of the factoring function for a guess at p, I found the locus in Fig. 8.32 for $p = 0.00055$. This looks about right for tan 30°, which 1 square up giving 1.7 squares left near the low point of the curve. Checking $s = -0.34 + j0.2$ gives a gain $1/z = 6.98$ with a phase error of 0.98°. I'll settle for this. Then $z = 0.143$ and $p = 0.00055$ are the compensator values at the end of the frequency domain design. Figure 8.33 shows the step response. Because the network zero is relatively close to the plant pole, the system has the transient response of a two-integrator system. If the overshoot is not acceptable, the value of p can be reduced and the smaller of the two values of z (two solutions for 30° pole angle) taken for the network. At this point then,

$$G_c = \frac{1 + s/0.143}{1 + s/0.00055}$$

∎

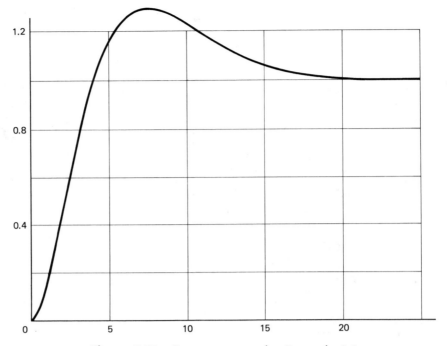

Figure 8.33 Step response for Example 8.8.

8.3.3 Improving Performance With Fixed Bandwidth

In the previous subsections, the examples used had rather loose performance specifications—just an overshoot or a phase margin or a pole angle. In this subsection you will see a case in which a tighter specification in combination with the given plant forces the use of both lead and lag networks.

■ Example 8.9

Again, $H = 1$. The plant is $G_p = 8/s(1 + s/2)(1 + s/10)$. The requirements are a 0–100% rise time less than 0.5 s and an overshoot less than 5%, in response to a step input. As it stands, the system is barely stable. I will make it look like a second-order system which meets the specifications. The overshoot requirements sets the pole angle or damping factor. From Chap. 6, $\tan \phi_p = -\pi/\ln m$, so $\phi_p = 0.80916$ rad $= 46.36°$. The natural frequency is set by the rise time, $\omega_0 = (\pi - \phi_p)/(t_r \sin \phi_p) = 6.4458$ rad/s. The gain and corner frequency for the equivalent second-order system are given by $aK = \omega_0^2 = 41.548$ and $0.5\sqrt{a/K} = \zeta = \cos \phi_p$, which yields $a/K = 1.905$. From these two values I have $a = \sqrt{(41.548)(1.905)} = 8.8965$ rad/s and $K = 4.67$.

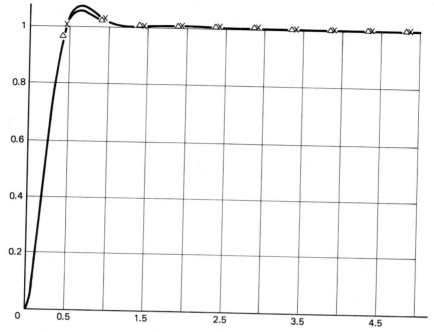

Figure 8.34 Step responses for Example 8.9. ×, two lead solution; △, one lead solution.

A rigorous reworking of the system would require two lead networks to cover the plant's poles at -2 and -10, put in a new pole at $-a$ and another high-frequency pole at, say, -100. Then a lag network would be needed to turn the high-frequency gain down from 8 to 4.67. From the previous subsection, one might choose the lag's zero at -0.08, about 100 times lower than the new plant pole, and the lag's pole at $-0.08 \times 4.67/8 = -0.0467$ rad/s. These choices produce a compensator whose transfer function is

$$G_c = \frac{(1 + s/0.08)(1 + s/2)(1 + s/10)}{(1 + s/0.0467)(1 + s/8.8965)(1 + s/100)}$$

This gives the step response shown in Fig. 8.34. It meets the rise time but has too much overshoot. As before, we can blame this on the extra pole needed for the lead network. Since we have to make a change, and the corrected system doesn't exactly duplicate the equivalent second-order system, we might as well take out one lead network and let the corner frequency be the original one at 10. The remaining lead network covers the pole at -2 and has its own pole at -100. The step response with this solution is also shown in Fig. 8.34. It is better, but still not good enough.

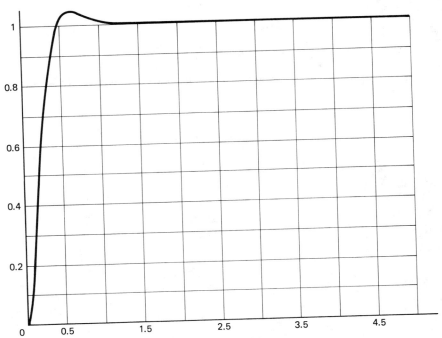

Figure 8.35 Step response with the farthest pole moved to -200, Example 8.9.

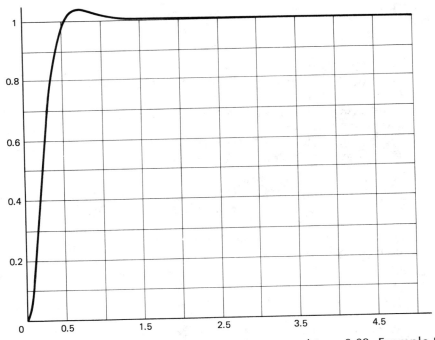

Figure 8.36 Step response with the lag zero moved to -0.09, Example 8.9.

Among the several possibilities that appear at this point are tinkering with the lead's extra pole or moving the lag zero slightly to produce more high-frequency attenuation. Doubling the lead's pole frequency to -200 rad/s meets the specifications as you can see in Fig. 8.35. Moving the lag's zero up to -0.09 (with the lead's pole still at -100) makes the step response a little too slow, as in Fig. 8.36. The best choice seems to be

$$G_c = \frac{(1 + s/0.08)(1 + s/2)}{(1 + s/0.0467)(1 + s/200)}$$

This solution is good because it meets the performance requirements, and it is fairly simple because it uses an original plant pole which happens to be close to the desired equivalent second-order system pole.   ■

8.4   SOME ADDITIONAL OBSERVATIONS

The general process for compensation design is as follows:

1. Pick a frequency domain model to which you will shape the system.
2. Translate the time domain transient performance into frequency domain parameters which are appropriate both for the model and for the method you wish to use in the frequency domain.
3. Design the compensator transfer function.
4. Tune the compensator by running the transient response and adjusting the network parameters as in the examples in this chapter.

This list assumes that the low-frequency parameters have already been set by steady-state error requirements and choices made on the actuator and transducer(s) to be used. If the steady-state has not been completely specified, then one is free to choose the number of integrators and/or the gain constant to simplify the compensator design. Maybe you *can* turn the gain down.

In all these examples, I worked the problem until the requirements were just barely met. As I have said several times, an important advantage of feedback is that it can make an accurate control system out of an inaccurate actuator or plant. This often means that the plant parameters aren't known to any better precision than 5 or 10%, or worse, and sometimes means that the parameters vary with time or temperature or changing loads. What will happen to the step response if the power amplifier gain changes by 20%, or if the unit-to-unit variation on the motor torque constant is 10%? Precision passive-feedback components and the number of integrators only guarantee the steady-state error, not consistency in the transient response. In most cases, the transient performance will be given in terms of inequalities and you will want to design to meet them with considerable margin. Some practice in this course and experience later will help you decide how much margin to aim for on the first pass. In such cases, it is not worthwhile to nail down the design part values to more than two significant figures. If you have parameter ranges in which a design must fit, e.g., $0 \le m \le 0.01$, you will wind up specifying part values to three figures and probably either making them up from assemblies of precision parts or incorporating adjustable parts into the system. This runs the cost up rapidly, both for components and labor. The system engineer who specified the range had better have a good reason for doing it.

PROBLEMS

Section 8.2

8.1 Derive the step response for the tachometer feedback system shown in Fig. P8.1. Find K_t for critical damping.

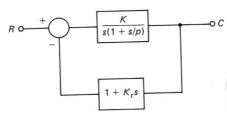

Figure P8.1 Tachometer-damped position control system.

8.2 Derive the closed-loop transfer function for the system in Fig. P8.2. Relate it to the standard form for a second-order system given in Chap. 6. Given $H_0 = 0.0375$ and $K_2 = 28$, find b and K_1 to give a step response with a rise time of less than 0.1 s and an overshoot of less than 10%.

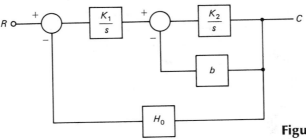

Figure P8.2 Minor-loop design.

Section 8.3.1

8.3 $G_p = 25/[s(1 + s)]$ and $H = 0.8$. Find a lead network to keep the step response overshoot under 3%.

8.4 $G_p = 25/[s(1 + s)(1 + s/40)]$ and $H = 0.8$. Find the lead network(s) with the minimum total high-frequency gain to give a step response overshoot less than 5%.

8.5 An alternate approach to the design in Example 8.6 is to use an exact fifth-order ITAE solution. Since the fifth-order system has two complex roots, a bridged T in the feedback path of an inverting stage can be used. This network will have a real zero, likely in the wrong place. Thus, a network will be needed to cover the bridged-T zero and the plant pole, and a lead network will be needed to give the required zero and pole for the ITAE transfer function. Find all the transfer functions needed. Sketch minimum part count op amp circuits to realize the compensator. There should be an even number of inversions between input and output. Compare the step and ramp responses with the quasi-third-order solution given in the example.

Section 8.3.2

8.6 A speed control system has $H_0 = 0.01$ and $G_p = 10,000/[(1 + s)(1 + s/10)]$. Use a lag network to get a step response with a 10–90% rise time less than 1 s and an overshoot less than 10%. What combinations of pole and zero values will satisfy these requirements?

8.7 A premium version of the system in Prob. 8.6 is to be designed. Add an integrator and redesign the lag network. If the specifications conflict, meet the $\leq 10\%$ overshoot requirement with the quickest possible rise time.

Section 8.3.3

8.8 Redesign the system in Prob. 8.6 for a maximum rise time of 0.05 s, $\leq 10\%$ overshoot, and no integrator.

8.9 Although the subsection titles mention bandwidth, there has been no real examination of the question in the examples. Run plots and make a table listing the following properties for all the system designs, before and after compensation, where applicable: open-loop unity-gain frequency, closed-loop frequency response peak frequency, closed-loop 3-dB down frequency, step response rise time.

Chapter 9

SYSTEM DESIGN

9.1   UNDERSTANDING THE PROBLEM STATEMENT

The nature and quality of a problem statement depend strongly on its source. If you are an engineer working on a subsystem which is part of a larger system, then presumably a system engineer has defined your problem to the level of stating the load to be controlled, the format and level of the control signal, and the steady-state and dynamic performance requirements. An example of this sort is the position control for one axis of a robot system. The robot's mission has been defined by an end user or a marketing group acting as the end user; the mission has been translated into a set of loads, movements, degrees of freedom, sizes of parts, and so forth, by a team of system engineers. Each degree of freedom needs its own control system with performance specifications compatible with the other parts and the mission of the entire system, so when it gets to you, your problem should be quite well defined.

On the other hand, you may receive only a mission statement such as, "We need to keep this circuit at a constant temperature. Please design a small oven package to do this." This means you will have to ask a series of questions and may have to help the end user get to the answers to questions, such as, "At what temperature, and to what tolerance? What ambient temperature range do you expect? How fast does the temperature have to stabilize (how long a warmup can be tolerated before the circuit has to perform to its spec)? etc. In addition, it is quite likely that you will have to solve, at least roughly, the heat transfer problem, which means you will have to use either your company's library or the technical library of the nearest engineering school. As a practicing control engineer with an electrical background, I often had to consult texts on heat transfer and fluid flow to help model, and sometimes design, such components. Ask around. In a bigger company there are many resources, the most important of which is its people. It is often more challenging than looking in the telephone directory to find the senior engineer or specialist who has the datum you need or who happens to know that a heat transfer software package exists and who knows how to use it.

Drawing pictures is an important aid in clarifying your understanding of the problem and an important means of communicating with the customer. If no pictorial representation exists, sketch one as you talk, so you can be sure you both agree on the problem. Sketch a preliminary block diagram of the system parts as they are known and add in blocks for things that you reasonably can expect will be needed, such as amplifiers and sensors.

9.2   FILLING IN THE PHYSICAL BLOCK DIAGRAM

Figure 9.1 shows a hypothetical block diagram. To begin with, you will have information only on the left- and rightmost blocks—the input signal description and the load description. Maybe you won't have even that much. At any rate, the load description should be the first block you fill in. Using the various performance requirements, you should work away from the load, as I will show in the next section. The point now is to simply list the kind of item and what you need to know about it, in each block. You should fill in the blocks in Fig. 9.1 with something a little more specific. Given a load, what kind of actuator will you use? If it's a mechanical load, you might use either a hydraulic piston or an electric motor. If you choose a hydraulic piston, the power amplifier block could represent a solenoid-driven spool valve. The summing amplifier might be an op amp with a current-boosting stage to drive the solenoid. The sensor block depends on what you are trying to control in the load's motion. If you are controlling position, the sensor might be a rotary potentiometer with a dc voltage applied, or it might be a capacitance transducer. There are many different position sensors, rotary and linear, available on the market, and you will have to spend some time deciding which one gives the required performance, with negligible loading and design impact, for the least cost.

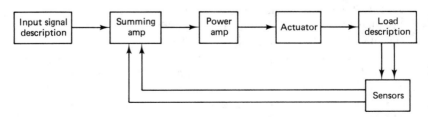

Figure 9.1   A hypothetical physical block diagram.

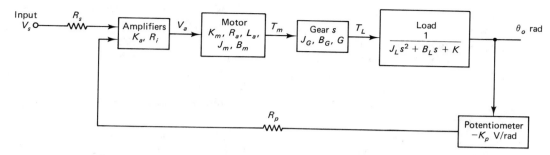

Figure 9.2 A physical block diagram for a position controller.

Figure 9.2 shows a physical block diagram for an angular position system with some choices filled in. Each block also contains specific items you will either be given initially or have to find out in order to model and complete the system design. The load is given as a parallel inertia, friction, and spring. For reasons which I will discuss in the last section, I have chosen to use a dc motor and a gearbox to drive this load. The gearbox needs to be described in terms of its ratio G and its equivalent inertia and friction. The motor has five electrical and mechanical parameters to be established for its model. Then I assume there will be amplifiers, which are represented by their overall gain and input resistance(s). I am showing the input sources and position feedback potentiometer are shown as equivalent voltage sources with series resistances. This is not quite accurate and serves mainly as a reminder to me that I have to take the source resistances into account when I prepare the detailed electrical design. The input source may be using a different voltage scale to represent shaft angle than the potentiometer, so the gain from each will have to be adjusted if an equivalent unity-gain system is required.

9.3   ESTIMATING THE REQUIRED ACTUATOR SIZE

How much power or torque or pressure is the system expected to deliver and still be linear? Once an element in the system saturates (or cuts off), the specifications can't be met because the output is effectively disconnected from the input. The output proceeds at a fixed rate or is held at a fixed value determined by the maximum capability of the saturated element, regardless of the input command. This condition will exist until or unless the error signal diminishes and the element comes out of saturation. This may happen either because the input command has changed or the output has caught up to the input. In a case where the output is in constant acceleration during the saturation period, it may come out of saturation with so much initial speed that it will enter saturation in the other direction. This large-signal nonlinear behavior can imitate lightly damped oscillation and even be unstable. In cases where the system is expected to be linear only for small errors but large ones are possible, graceful recovery from the non-linear condition is an important design objective. I will assume that we are dealing with a design which is expected to be linear always. In the absence of a statement to the contrary, one should assume that linear operation is expected through the full specified ranges of the input and output variables. There are many situations where this is not the case. For example, a linear tool position control system may have a travel range of 10 cm, but the largest change command might be less than 1 cm.

The maximum effort the actuator will have to make comes when the command input makes the maximum step change allowed. We have already looked at step responses for second- and higher-order systems, and they are not simple to calculate, nor is it possible to tell before the compensator is designed exactly what the step response is going to be. Furthermore, actuators come in discrete size increments. If a quarter-horsepower motor isn't big enough, the next size to be considered may be half-horsepower. In some cases the actuator will be custom-designed and made for the system, so more sizing accuracy is justified than in the purchased actuator case.

The size of the actuator often has an important effect on the dynamic description of the system. Thus, we are faced with a circular problem; we can't calculate the maximum effort until we know the dynamic description, and we can't know the dynamic description until we know the size of the actuator and design the compensator. To break out we need to make an estimate.

To begin with, we know that the output has to reach close to the desired value in less than the specified rise time. The simplest way to get an estimate is to assume maximum effort during the rise time. The following two examples illustrate this method.

■ Example 9.1

Suppose a temperature control system is being designed. The equation relating the heater output to the controlled temperature might be

$$Q = \frac{1}{C}\frac{dT_c}{dt} + k(T_c - T_a)$$

where Q is the heat, T_c is the controlled temperature, T_a is the ambient temperature, t is time, C is the bulk thermal capacity, and k is the bulk thermal conductance of the system being heated. Suppose that the maximum desired value is T_0 and the largest command is at turn-on with the minimum ambient temperature T_m. If we neglect the heat loss term, then

$$T_0 - T_m = CQ_Mt_r$$

where Q_M is the maximum heat rate and t_r is the rise time. If I actually used this to find the heater size, it would be an underestimate because I neglected the loss during warmup and because, in a linear system, maximum effort may not occur right away and the effort diminishes as the error does. For these reasons, I usually double Q_M from this result.   ■

■ Example 9.2

An angular position control system load can be simplified to just one term. Suppose that inertia is the dominant load component. The response to a step input has to accelerate the load to a maximum speed and then decelerate it so that zero speed is reached when the desired position is reached. A torque program that will do this is a square wave, as shown in Fig. 9.3. Here, T_M is the maximum torque, t_r is the rise time again, J is twice the load inertia so as to include an estimate of

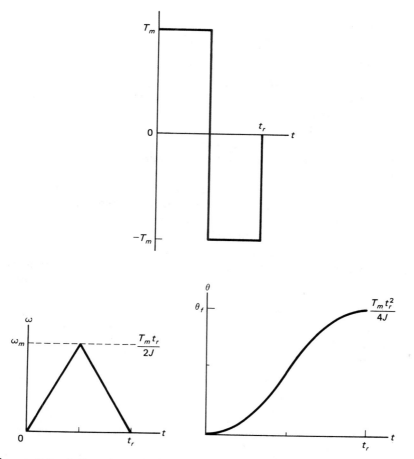

Figure 9.3   Estimating maximum torque by assuming a square wave.

the motor and possible gear contribution, ω is the angular speed, and θ is the output. The speed is $1/J$ times the integral of the torque. This gives the triangle wave shown in Fig. 9.3. The output position is the integral of the speed, which gives two parabolic sections. The final value relations are shown in the figure. The maximum torque is $T_M = 4J\theta_f/t_r^2$. Again, this is probably optimistic, so initially a factor of 2 should be used.   ■

9.4   THE DESIGN CYCLE

The discussion to this point has really been an introduction to the preliminary steps in the design process. Figure 9.4 shows a block diagram of one possible form of the sequence of steps which one can follow designing a control system. It includes specifically the possibility of loops for correcting problems discovered in later stages by going back to earlier stages with new information. Some kinds of corrections are not shown. For example, it may not be discovered until one is well along in the design that a conflict exists between some of the initial requirements. Such a conflict can occur either for physical reasons or because of exorbitant cost. This sort of problem may be more apparent than real, though. All reasonable alternatives at the physical diagram level should be tried before saying the specs need to be reworked. The inclusion of loops is what makes the process a cycle. It may also be part of a larger system design cycle. If you are creating a subsystem design, it may be that the results of your effort cause a change in the overall system design for reasons of size or cost. Also, a change in some other part of the larger system may cause a change in your subsystem's performance requirements.

The first three blocks in Fig. 9.4 were discussed in Secs. 9.1–9.3. The fourth block is obvious. Chapter 5 is used to block five. The material in Chaps. 5, 6, and 8 address the translation task in block 6. The task in block 7 is grounded in Chap. 2, and block 8 is based on Chap. 3 and/or 4 and uses a software package. Block 9 returns to Chap. 8 for a transfer function design. The tasks in blocks 10–13 are treated later in this chapter, and the circuit design in block 11 is based on Chap. 7. Now you may wonder about the order of presentation of the material in this book. Synthesis is often analysis reversed, and sometimes it is analysis iterated. In either case, analysis has to be learned first. Thus Chaps. 2–6 mainly consider analysis with some hints at design which provide justification for the design topics of Chaps. 7 and 8. This chapter tries to put everything together and in perspective.

9.5   VALIDATING THE ESTIMATE

Once the dynamic analysis and compensator design are finished, the next step is to determine what actual maximum effort the actuator needs to supply. This is done by finding the unit step response for the variable representing effort and

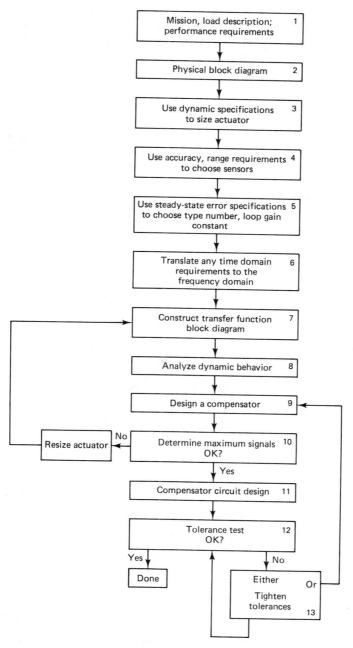

Figure 9.4 A design sequence.

then scaling (multiplying) by the largest expected input step change. If the system is specified for some other type of input, say a ramp or a trapezoid, then the response for that type of input must be found and scaled. So how do you do that? It depends on the software available. In some systems you can identify the individual blocks and indicate whether you want their outputs. In others, you have to give a state variable description of the system and you can specify the effort as one of the states and request it as part of the output. The LSP offers two choices. You can describe the system in terms of state variables, which is added work because you used the pole-zero description to prepare the design, or you can manipulate the block diagram to obtain the effort as the output and redescribe the forward and feedback paths to the LSP, which is also more work. I will present both methods in the next two examples.

■ Example 9.3

As promised, I will now revisit Example 4.5. That example concerned a linear position controller driven by a dc servomotor through a screw actuator. It was a study example and did not have dynamic specifications or a maximum excursion requirement. Its block diagram is shown in Fig. 4.17 and redrawn as

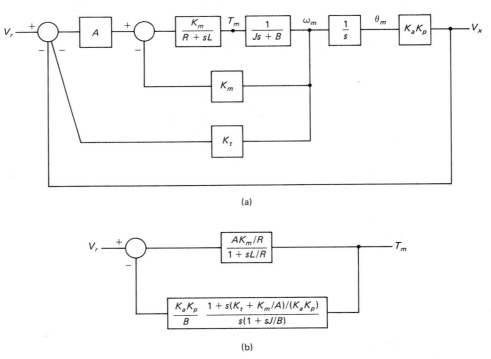

(a)

(b)

Figure 9.5 Block diagrams for the positioning system of Examples 4.5, 9.3, 9.4, and 9.6.

part of Fig. 9.5. Before dealing with the question of effort, I first want to redesign the system to have a standard output response. I shall choose to find the gain and rate feedback constants that will give the closed-loop poles the ITAE locations. The open-loop transfer function is not in the standard form for Table 5.2, so I will form the closed-loop transfer function using the notation in Fig. 4.17b.

$$\frac{V_x}{V_r} = \frac{K}{s(1 + as + bs^2) + K(1 + K_t's)}$$

$$= \frac{K/b}{s^3 + \dfrac{a}{b}s^2 + \dfrac{1 + KK_t'}{b}s + \dfrac{K}{b}}$$

We only know a and b in this equation, so I'll start there. From Table 5.2, for a type-1 system with $n = 3$, $a/b = a_1\omega_0$. So $\omega_0 = a/(ba_1) = 7.9/(0.019922 \times 1.75) = 226.6$ rad/s. Next, $K/b = \omega_0^3$, so $K = 231.8$. Finally, $1 + KK_t' = ba_2\omega_0^2 = 2.1993$ and $K_t' = 0.005174$. Figure 9.6 shows the output step response for these values. Note that the zero in the feedback path reduces the small overshoot one would normally see in an ITAE optimum system.

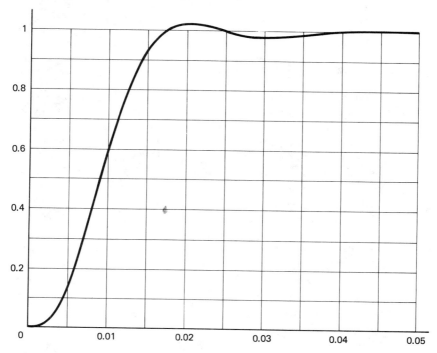

Figure 9.6 Step response of the position sensor voltage for the system of Example 9.3.

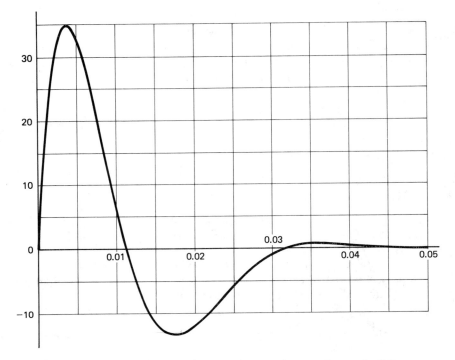

Figure 9.7 Torque for a unit step input, Example 9.3.

Figure 9.5a is a redrawing of Fig. 4.17a with the intermediate variables identified. I need to rearrange this diagram to place the motor torque T_m at the output and to form a single-loop system. First, I put all the feedback paths in parallel by moving the motor's internal voltage back to the input summing junction. Next, I add these blocks together and multiply by the mechanical load block. Last, I put the poles and zeros in the Bode plot standard form, which is used by the LSP. The result is Fig. 9.5b. The following values are from Example 4.5 and Table 2.1: $A = K/0.177 = 1309.6$, $K_t = cK_t' = 102.96$ μV-s, $B = B_L + B_m = 1923.35$ μN-m-s, $J = 55$ μN-m-s^2, $B/J = 34.97$ rad/s, $R = 3.04$ Ω, $L = 8.41$ mH, $R/L = 361.47$ rad/s, $K_m = 0.1318$ V-s, $K_aK_p = 0.0199$, $K_aK_p/(K_t + K_m/A) = 97.74$ rad/s, $K_aK_p/B = 10.35$, and $AK_m/R = 56.78$. The LSP gave me the torque response in Fig. 9.7. The peak is about 35 N-m/V. The maximum pulse torque for the motor is 2.5 N-m. Since the torque response looks like a pulse about 10 ms wide, I will assume it qualifies, so the maximum step in voltage is about $2.5/35 = 0.07$ V. At 5 V/in. this corresponds to 0.014 in. = 14 mils. ∎

■ Example 9.4

Figure 9.5a can be used to write state equations for the system. Although torque is not an energy storage variable in this system, it is proportional to current which is, so I will use torque, shaft speed, and angle as state variables. Rewriting the transfer blocks into a preliminary form for state representation,

$$(sL + R)T_m = -K_m(K_m + AK_t)\omega_m - AK_aK_pK_m\theta_m + AK_mV_r$$

$$(Js + B)\omega_m = T_m$$

$$s\theta_m = \omega_m$$

In standard state equation form,

$$sT_m = -\frac{R}{L}T_m - \frac{K_m}{L}(K_m + AK_t)\omega_m - AK_aK_p\frac{K_m}{L}\theta_m + \frac{AK_m}{L}V_r$$

$$s\omega_m = \frac{T_m}{J} - \frac{B}{J}\omega_m$$

$$s\theta_m = \omega_m$$

Putting in the numbers from Example 9.3, the state matrix is

$$\mathbf{A} = \begin{bmatrix} -361.5 & -4.179 & -408.42 \\ 1,8182 & -34.97 & 0 \\ 0 & 1 & 0 \end{bmatrix}$$

The input matrix is

$$\mathbf{B} = [20,524 \quad 0 \quad 0]^\tau$$

The output matrix isn't important because I want to see the states, so I chose $C = [1 \quad 0 \quad 0]$. One can select each state for an individual plot by setting the element in $\mathbf{C}$ which selects that state to 1 and the others to zero and asking for an output plot. When I ran the state step response, the speed was so large it scaled the other states down too much for detail. I rescaled the speed by dividing the second row of $\mathbf{A}$ by 100 and multiplying the second column by 100. This is equivalent to replacing the speed by $100X$, where X is the new state variable. Figure 9.8 shows the results. The angle output has the same shape as Fig. 9.5, scaled by 1/0.0199. The speed peaks earlier, as it should, about halfway through the position rise. The torque has both positive and negative peaks and looks the way it did in Fig. 9.7. If the position is an approximation to a step, then the speed is an approximation to a pulse and the torque is an approximation to a double pulse. The relation would be more perfect and symmetrical if it weren't for the large friction component in the load. Another limit to think about (later) is that of the amplifier voltage required to supply the torque at the speed peak. If 0.07 V is the maximum step size, the 3600 rad/s peak speed represents about 33 V in generated back emf. ∎

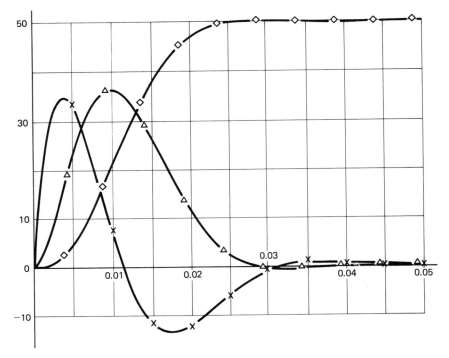

Figure 9.8   State response for Example 9.4. x, torque; △, speed/100; ◇, angle.

9.6   ELECTRONIC DESIGN: RATINGS AND REASONABLE VALUES

Once the maximum actuator effort has been found, one must find the input conditions that have to be supplied by the power amplifier. This involves more than just the maximum current. The maximum voltage and the maximum power dissipated by the amplifier itself need to be found. Consider again a dc motor application. If the load and command function are such that maximum torque is generated when the load is near maximum speed, the current delivered will be at a low voltage drop across the power stage, so that the power dissipated in this stage will be small compared to the power delivered to the load. On the other hand, if maximum torque is required at very low speed, the power stage will be delivering maximum current while blocking most of the supply voltage, therefore dissipating much more power than the motor is delivering to the load. This is a terribly inefficient mode of operation which can be avoided by using a gear train that allows the motor to run faster. This solution is okay unless the load is occasionally required to move much faster. Since the motor is presumably running near the maximum speed allowed by the power supply, an increase by a factor of 2 isn't possible. Another solution is to pulse the motor using any of several pulse modulation schemes embodied in commercial dc motor speed controllers. These controllers also have problems—response nonlinearity and hysteresis—but they do operate efficiently over a wide speed range and should be considered for such applications. This discussion points to the need to study the actual mission profile for the system.

The simplest thing to do about the power amplifier is to say that it must deliver the maximum voltage corresponding to maximum speed, the maximum current corresponding to maximum torque, and be able to dissipate their product. This approach can lead to a greatly oversized and needlessly expensive unit. For example, in Fig. 9.8 you can see that the peak torque occurs at about one-half peak speed. The power amplifier is dropping only about one-half the supply voltage at maximum current. This observation suggests, but doesn't guarantee, that a stage rated at less than one-half of the product of the maxima may be enough. Time is another important factor. The transient is over in about 30 ms. It would not be hard to generate the voltage drop and current time functions from the state response and integrate their product over the transient period to find the total heat generated. One can then divide by the transient duration to get an average dissipation value. This value can be used as the steady-state rating for the amplifier, or as the peak rating provided the peak is defined to include duration covering the transient period and the duty cycle (time on–time off) is sufficiently low. The pulse power ratings and duty cycle are given by some manufacturers or, if this is a high-investment product an amplifier can be designed to meet the load's service requirements. If the transient is long compared to an amplifier's ability to get rid of the heat, the peak of the load current–amplifier voltage drop function will have to be chosen as the power rating. ■

■ Example 9.5

Suppose a current-controlled heater is to be designed as part of a temperature-controlled box for a small circuit. It must operate from a 15-V supply, and the peak power required in heat is 10 W. If the stage in question consists of a power transistor in series with a resistance heating element, so that they both heat the circuit, what are the ratings the transistor will have to meet?

The maximum heater current is 10 W/15 V = $\frac{2}{3}$ A. The heater resistance should be nominally 15 V/($\frac{2}{3}$ A) = 22.5 Ω. A proposed circuit which uses an op amp to control the heater current is shown in Fig. 9.9. For the purpose of this example, I will assume the feedback current is negligible compared to the load current, so $I_R = I_c$. The total power dissipated in the resistor and the transistor is $P_T = V_{cc}I_c$. The power dissipated in the transistor itself is $P_D = V_cI_c$, and the collector voltage is $V_c = V_{cc} - I_cR_H$. Then the dissipated power as a function of I_c is $P_D = I_c(V_{cc} - R_HI_c)$. I want the maximum transistor dissipation, so I differentiate the power with respect to the current and set the slope to zero. $P_D' = V_{cc} - 2R_HI_c = 0$ and the current for maximum dissipation is $I_{cM} = V_{cc}/(2R_H)$. Then the maximum dissipation is $P_{DM} = V_{cc}(V_{cc}/2)/(2R_H) =$

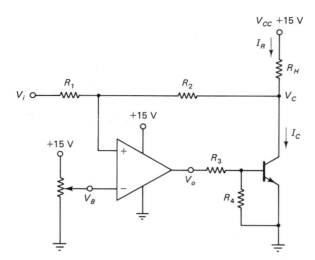

Figure 9.9 Part of a temperature controller.

$V_{cc}^2/(4R_H) = \frac{1}{4}P_{T,\max}$. Therefore the power rating on the transistor must be at least 2.5 W. The room-temperature rating will have to be higher, because the case temperature will be higher in the box than the average room temperature; otherwise there would be no room for control when the ambient temperature goes up. This consideration says that the case temperature, which is the circuit temperature, should be at least as high as the maximum ambient. If the case temperature is 70°C, the designer will have to find a power transistor with a derated dissipation rating of 2.5 W at 70°C, not 25°C. ∎

■ Example 9.6

Since I already have the state description, I thought it would be interesting to generate the power amplifier dissipation time function for Example 9.4. The data from a screen plot are saved in global variables, *YH* for the vertical values and *XH* for the horizontal values. I reshaped *YH* into a three-row matrix **X**. Each row of **X** is the value of a state variable as a function of discrete time. Following Example 9.4, I initially took the value of the input step to be 0.07 V and the power supply to be 33 V. Since I want the power amplifier dissipation, I need to calculate $P_d = |I_a(V_{cc} - V_a - V_b)|$, where I_a is the armature current, $V_{cc} = 33$ V, V_a is the armature impedance voltage drop, and V_b is the back emf. Since $K_m = 0.1318$, $I_a(kT) = 0.07 \times X(1, kT)/0.1318$. *T* is the time step, 0.25 ms in this case.

$$V_b(kT) = K_m\omega_m(kT) = 0.1318 \times 0.07 \times 100 \times X(2, kT)$$

$$V_a(kT) = RI_a(kT) + \frac{L}{T}[I_a((k + 1)T) - I_a(kT)]$$

I used the forward difference to approximate the derivative to find the inductance voltage. When I plotted the total motor voltage I was shocked (almost electrically) to find the armature drop so large that a 142-V supply would be needed. I decided to scale the step down to 24 V, which allows the supply to be 50 V. The voltages and power are plotted in Fig. 9.10. The maximum armature current is 6.35 A. This value times the supply voltage is 318 W. The actual maximum dissipation is 173 W. The second peak in the curve is due to the motor braking the load to decelerate to zero speed at the end of the step. When the motor brakes, the torque and the current go negative and the motor acts as a generator driving the supply through the amplifier. Without the magnitude sign on the expression for P_d, one would see negative dissipation because the equation only assumes power flow to the motor.

As I mentioned in the discussion earlier in this section, it is possible to use a lower-power amplifier if the transient heat can be removed before the transistor's junction temperature rises too high. To see what might be gained by such an approach, I need the heat energy dissipated in the transient:

$$J_d = \int_0^t P_d(\tau)\, d\tau \approx T \sum_{m=0}^{m=k} P_d(mT)$$

The result of this calculation is shown in Fig. 9.11. The final value of J_d is 2.69 J. Dividing this value by 25 ms gives an average over the transient of about 108 W. If the thermal resistance of the power transistors is low enough and a heat sink is chosen large enough, one can use a power amplifier rated at less than $\frac{1}{3}$ of the crude 318-W estimate.

The peak motor voltage is another value of interest. It is 48.44 V. This is interesting because it shows that the amplifier doesn't have to go to the rail (supply voltage). Since the motor must be able to reverse, the amplifier must have either a two-transistor output stage operating from ± 50 V or a bridge output stage operating from a single-voltage supply. ■

The gain and drive requirements for earlier stages are determined by considerations similar to those for the power amplifier. The stage that drives the power amplifier must be able to deliver enough drive current and the voltage range required at the power amplifier's input. Eventually, the forward-path gain requirement must be met. Usually, the power amplifier is a low-gain stage designed with voltage gain as a minor consideration. The number of gain stages needed depends on the bandwidth of the rest of the system and the gain required. For example, if the highest pole frequency is -50 Hz, the gain stages should corner at 5 kHz in order to have a negligible effect on the compensation design. Suppose the gain required is 900 and we want to use op amps that have a unity-gain frequency of 1 MHz. Approximately, the stage's corner frequency is the unity-gain frequency divided by the stage gain, so if we do it all in one stage, the corner will be 1 MHz/900 = 1.1 kHz, which isn't good enough. Two stages, each with a gain of 30, will each corner at 1 MHz/30 = 33 kHz. This is good enough, even considering that we are adding two poles at -33 kHz instead of one at -5 kHz. Of course, the designer can test this by adding the new poles to the dynamic

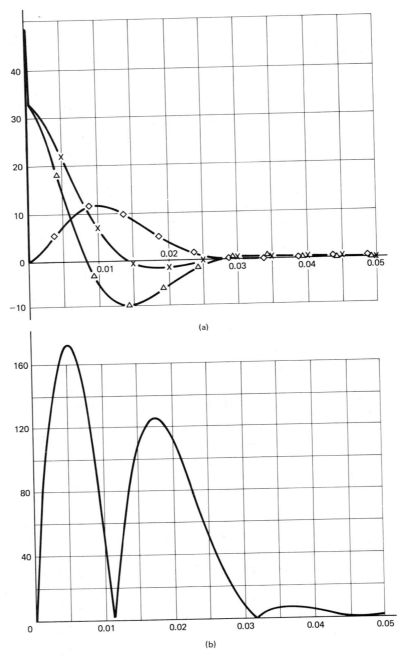

Figure 9.10a Voltages during the response to a 0.05 V input step. ×, Motor terminal voltage; △, armature impedance voltage drop; ◇, back emf. **b** Power amplifier dissipation, in watts, during the 0.05 V input step.

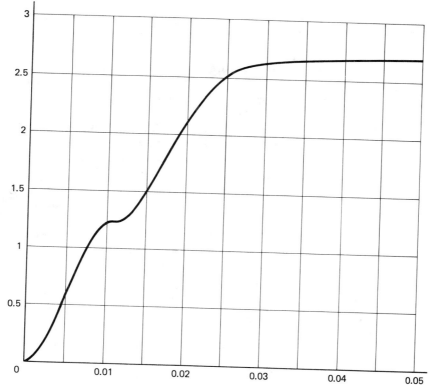

Figure 9.11   Heat energy generated in the power amplifier, in joules, during the 0.05 V input step, Example 9.6.

model. Using two stages allows a lower-feedback resistor, produces less noise, and is more immune to amplifier parameter variation. Why not use more stages? Economy.

Choosing resistor and capacitor values can be a problem. Generally, one should have resistor values between 10 kΩ and 1 MΩ to avoid loading the op amp on the one hand and to avoid either input loading or excessive noise on the other hand. However, this policy can force one to try to use excessively large capacitors. Suppose a network design calls for an RC product of 1000. If $R = 1$ MΩ, then $C = 1000$ μF. This is commercially available, but is larger and more expensive than a capacitor an order of magnitude smaller. There are at least four possible courses of action:

1. Look for a network which realizes the transfer function with smaller RC products.

2. Use a larger R value and do a noise analysis or test to see if it matters. A more expensive op amp may also be necessary to mitigate loading.

3. Failing items 1 and 2, use the large C or

4. Use a mechanical network. It is quite possible for very low frequencies that a mechanical network will be smaller and can be designed into either the actuator or the sensor.

Capacitance values above 1 μF are generally only available as electrolytics. They have value tolerances of ±10% or worse and are mainly polarized. If your design can stand the tolerance and you can't find a suitable nonpolarized version, the electrolytic will have to have a dc bias applied to it to maintain a total voltage with the polarity marked on its package. In a lag network, this can be done by connecting the appropriate end of the capacitor to either the plus or minus supply voltage line. These capacitors also have substantial leakage currents, however, which will cause an offset voltage in the stage that may not be acceptable. For example, a leakage current of 1 μA will cause a 10 mV drop across a 10 kΩ resistor. This can be reduced by a balanced arrangement as depicted in Fig. 9.12. If the capacitors are ±10% to begin with and they come from the same production run, their leakages may be very close, reducing the offset voltage by an order of magnitude or more. This arrangement also has the advantage that the leakages will track as the temperature changes. The factors of 2 shown in Fig. 9.12 are with reference to network design formula values. Other ways of balancing the leakage current are with a series potentiometer or a transistor collector with adjustable base current. If the system needs an offset adjustment anyway, these may be better choices.

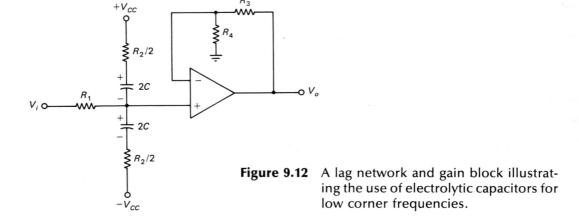

Figure 9.12 A lag network and gain block illustrating the use of electrolytic capacitors for low corner frequencies.

9.7   PERFORMANCE VARIATIONS AND CHOOSING TOLERANCES

I want to begin this section by continuing the design of the heater in Fig. 9.9. The following example doesn't deal with tolerances, but it shows some effects of parameter variation and illustrates a subsystem design.

■ Example 9.7

In this example, I proceed alternately with parts of the static and small-signal designs because the choices made in one part affect the next. From Example 9.5, the heater resistor should be 22.5 Ω. I will choose standard values for the parts in this circuit, so let $R_H = 22\ \Omega$. Nominally, the transistor has to be able to dissipate 2.5 W at 70°C and handle 700 mA. Transistors are normally derated to 0 W at 200°C and specified at 25°C. The derating factor is 2.5 W/130° = 0.02, so the 25°C rating must be 0.02 × 175° = 3.5 W. In perusing the selection chart in a (probably obsolete) data book the nearest suitable transistor I found is the 2N5148. It has ratings of 4 W at 100°C case temperature and a maximum current of 2 A. If it is available, it should be a $1 part. It has a maximum base current rating of 1 A, and an h_{FE} ranging from 30 to 90, typically 60. If the op amp is going to drive the transistor to 700 mA and the minimum current gain is 30, the op amp must be able to supply 700/30 = 23 mA. Again looking through another selection chart in a handbook on linear ICs, I found the LM13080 at $1.29 to be the cheapest unit for the job. Its maximum current is actually a lot more than I need at 250 mA, but its supply rating just qualifies at ±7.5 V. I have already set it up as a raised reference circuit in Fig. 9.9 to run from the single 15-V supply. The handbook also gives the minimum gain-bandwidth product as 1 MHz. I assume a fixed corner frequency of 10 Hz and a gain range of 10^5–10^6. The amplifier and the transistor are the big sources of variation in this circuit, with a potential forward-gain variation of at least 30 to 1. Now for some small-signal analysis.

A small-signal model is drawn in Fig. 9.13. I have neglected the transistor's base resistance and feedback capacitance as elements which are not necessary for the degree of accuracy of this illustration. As it turns out, the RC products for the transistor are on the order of 10^{-8} s and the base resistance is likely very

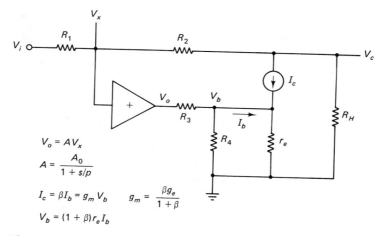

Figure 9.13   Small-signal model for the heater in Fig. 9.9.

small compared to R_3, so their inclusion would represent only minor corrections. To keep the algebra simple, I will assume that R_2 doesn't load R_H, so that $V_c = -R_H I_c = -g_m R_H V_b$. From the amplifier output, $G_3 V_o = [G_3 + G_4 + g_e/(1 + \beta)]V_b$, which gives the collector voltage as

$$V_c = \frac{-R_H \beta g_e G_3 V_o}{(1 + \beta)(G_3 + G_4) + g_e} = \frac{-\beta R_H V_o}{(1 + \beta)(1 + R_3 G_4)r_e + R_3}$$

If r_e were a reasonably constant value, the gain variation could be reduced by making the $\beta G_4 r_e$ product large compared to 1. However, there's no telling where the operating point is going to be for this package, and r_e is inversely proportional to the collector bias current. In fact, in this system it's a little hard to tell the signal from the bias. The current gain is a function of log (I_C) and changes only about 30% for a decade current change. So, I choose $G_4 = 0$ and to make R_3 large enough to just allow the op amp to saturate the transistor. If 0.8 V is the saturation base voltage, then $R_3 = (15 - 0.8)/0.023$ A $= 617 \ \Omega$. Choose the next lower standard value, 560 Ω.

To study the variation effects, I choose $I_C = 300$ mA as the operating point. At 70°C, this makes $r_e = kT/(qI_C) = 0.1 \ \Omega$. For $\beta = 60$, this makes the r_e term in the denominator 6. If the operating point drops to 30 mA, this term becomes about 40.

Returning to small-signal analysis, let the gain from V_o to V_c be A_H. Writing the current balance for V_x, $G_1 V_i + G_2 V_c = (G_1 + G_2)V_x = -(G_1 + G_2)V_c/(AA_H)$, which can be put in the form

$$\frac{V_c}{V_i} = -\frac{R_2}{R_1}\left(1 + \frac{1 + R_2 G_1}{AA_H}\right)$$

Substituting in the amplifier's transfer function and doing some more algebra with a view to separating the desired gain, gain error, and pole term into separate factors, I found

$$\frac{V_c}{V_i} = -\frac{R_2}{R_1}\frac{F}{F + 1}\frac{1}{1 + s/(p(F + 1))}$$

with $F = A_0 A_H/(1 + R_2 G_1)$.

F is the ratio of the open-loop gain to the closed-loop gain in the noninverting mode. This is, in fact, a rather general result for operational amplifiers. The actual terminals used are reversed from usual because of the inversion through the transistor stage. Assuming perfect resistor values, the gain error is $1 - F/(F + 1) = 1/(F + 1)$. Large F, which corresponds to low closed-loop gain, raises the pole frequency and reduces the gain error, which are both desirable effects. However, the required system forward gain has to be made up somewhere. For the range of β $1.16 < A_H < 3.5$, so $1.16 \times 10^5 < A_0 A_H < 3.5 \times 10^6$. If I choose $R_2/R_1 =$

22, the overall transfer is $I_c/V_i = 1$ A/V and the heat out is $Q/V_i = 15$ W/V. Then $5 \text{ k} < F < 15.2 \text{ k}$. The minimum corner frequency is 50 kHz, and the maximum gain error is 2×10^{-4}. In the usual course of things, the design can probably tolerate an order of magnitude reduction in the corner frequency and an increase in the gain error, which says that a voltage gain of 220 could be used if necessary for the system performance. ∎

The preceding example showed how feedback on a highly variable plant controls the closed-loop gain constant, but the pole frequency has the plant's basic variability. Also, keep in mind that the main variation discussed in the example is unit-to-unit variation. A given circuit will only have gain variations due to temperature effects, and sometimes due to aging, but if the unit is operated at nearly constant temperature, these are small. If the corner frequency can't be nailed down, then it should be at such a high value compared to the network corner frequencies that it has no effect. The next example looks at the question of tolerances on the components that determine the dominant corner frequencies in the system. I will again examine the position control system in Examples 9.4 and 9.6.

∎ Example 9.8

This example again examines plant unit-to-unit variations, this time for the dc servomotor in the position control system. Electro-Craft gives the tolerance on the motor constant as $\pm 10\%$ and that on the armature resistance as $\pm 15\%$. I will assume that the inductance has the same tolerance as the resistance, and that they track. Examining the numbers in the calculations for earlier examples, I find that K_m^2 dominates Q_m and, for the purpose of tolerance estimation, $a \approx JR/K_m^2$ and $b \approx JL/K_m^2$. Assuming a $\pm 5\%$ tolerance on J, I see that a and b also track and have a total range of $\pm 40\%$. At the high end, the open-loop poles are -135 and -265, and at the low end they are $-200 \pm j208$. The open-loop gain constant is $K \approx AK_aK_p/K_m$, and the feedback zero is $-K_aK_p/K_t$. So far as the motor constant error is concerned, the higher-frequency poles correspond to the low end of K_m, which also causes the gain to be higher. The screw and position sensor errors have the same effect on the gain and the zero's magnitude. The amplifier gain affects only the loop gain, so its error is not mitigated by any helpful tracking by another parameter.

The effects of these variations on the step response are shown in Figs. 9.14–9.16. In Fig. 9.14 the gain and rate constants are at their design values, while the poles are set at the design values and the two extreme sets given above. The real poles are the ones "slow" in the figure caption, and the complex ones are "fast." The design values for K and K_t' are 231.8 and 1/193.3, respectively. The design values for the poles are $-198.3 \pm j104.3$. The zero is quite close to the real part of the poles in both the nominal and fast cases, whereas it is between the real poles in the slow case. In the slow system the relatively high zero corner frequency accounts for the nearly 20% overshoot. The response by the fast-pole system is

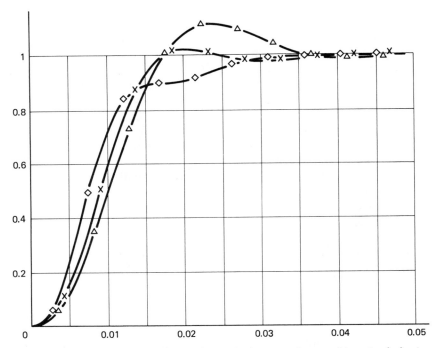

Figure 9.14 Step responses for pole variations only. ×, Nominal design; △, Slow poles; ◇, Fast poles.

initially quick but seems to almost dwell at 0.9 for 5 ms before rippling up to the final value. This response is due, as we have seen before, to the zero being in the feedback path causing a simple pole-like response, with the underdamped poles causing the ripple to ride on it. If one has an amplifier gain constant variation range of ±10%, the extreme values can either be canceled or added to by the motor constant error. Thus, for the fast poles the gain range is from nominal to +20%, and for the slow poles it is from nominal to −20%. Figure 9.15 shows the responses for the high and low extremes of gain. The general shapes are the same, and the main change is a wider range in rise rates. Notice that I didn't say "rise times." The peculiar shape of the fast-pole response makes it unclear as to what definition of rise time fairly describes this performance. Figure 9.16 displays the effect on the slow poles of a ±10% range in the zero's value. It has a small effect on the rise time and a major effect on the overshoot.

What should be done with this system? A little experimenting with step responses will soon show you that there is no combination of amplifier gain and zero position that will give a shape approximately like the nominal system's for the extreme pole value cases. One simple solution is to provide an adjustment on the rate feedback signal to move the zero to a good location during production

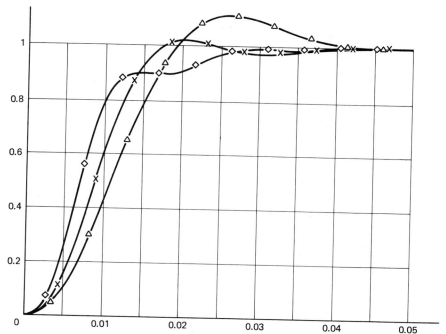

Figure 9.15 Step responses for pole variation and adding gain errors. ×, Nominal design; △, slow system; ◇, fast system.

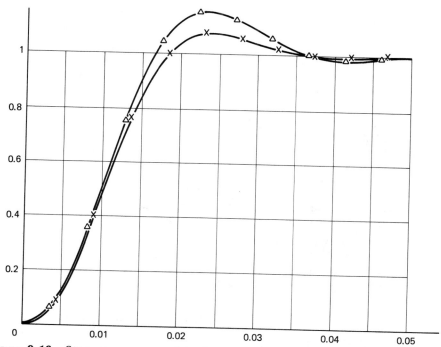

Figure 9.16 Step responses with slow poles and nominal gain. ×, zero at −173; △, zero at −213.

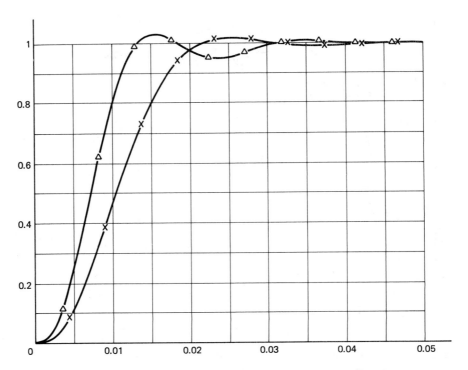

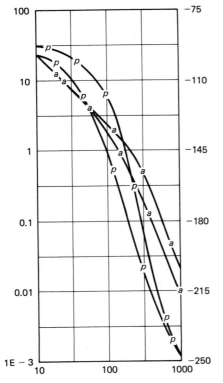

Figure 9.17 Step responses for adjusted rate feedback, $K = 232$. $\times$, Slow poles, zero at -150. $\triangle$, fast poles, zero at -290.

Figure 9.18 Bode plots for the forward path. Upper p and a curves are for the fast poles.

and test. A reasonable criterion is to set the overshoot to lie between 0 and 4%, and Fig. 9.17 shows the result of this strategy. An easily managed 2:1 adjustment range is sufficient to meet this criterion and to give reasonable shapes to the responses. This test and adjustment can be done either manually or by an automatic test setup.

∎

■ Example 9.9

Nowhere is it written that the rise time shall be under 25 ms, or should be anything in particular, for the system in Example 9.8. Figure 9.18 has Bode plots for the forward path which show that the pole variation has little effect on the amplitude transfer befow 50 rad/s. Indeed, if I use a lag network to get a 10:1 attenuation, unity-gain crossover will be about 20 rad/s with a phase lag between 90 and 100°. With our previous experience in mind, I tried a lag network with the zero at −2 rad/s (Fig. 9.19) and at −1 rad/s (Fig. 9.20). In both cases the step response shape is good. This is an easier design to test, and no adjustment is needed. The step response test specs could read:

1. 0–90% rise time less than 100 ms

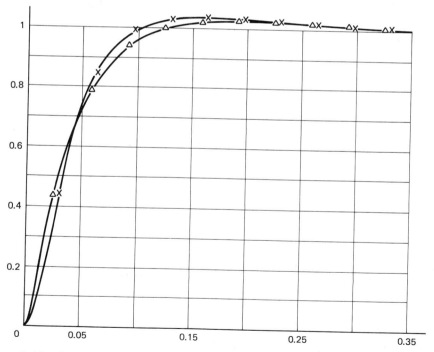

Figure 9.19 Step responses with a lag network $(1 + s/2)/(1 + s/0.2)$, $K = 232$ $K_t = 0$. ×, Slow poles; △, fast poles.

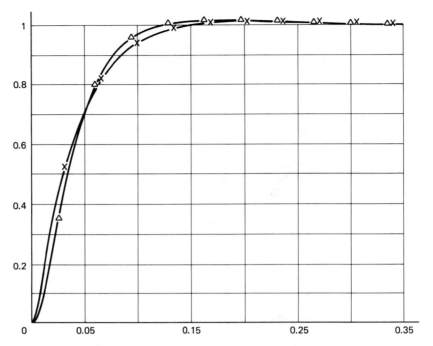

Figure 9.20 Step responses with a lag network $(1 + s)/(1 + s/0.1)$, $K = 232$, $K_t = 0$. $\triangle$, Slow poles; $\times$, fast poles.

2. Maximum output less than 1.04

3. Level shall be $1 \pm$ (static accuracy) for time beyond 0.4 s.

Since the location of the lag zero is not critical, a low-precision capacitor can be used. The crossover frequency and therefore the rise time are determined by the lag network resistor ratio and the variations in amplifier, motor, and screw constants. If these are each allowed to be in a wide range, say $\pm 10\%$, a greater attenuation and a lower crossover frequency will have to be chosen to avoid the variation in shape caused by the motor poles. I leave the study of this point to you in the problems. ∎

9.8 AN EXAMPLE: A STEERING SYSTEM

The last few examples had the purpose of illustrating ways of solving certain problems in the design process. They lacked a feeling of completion, because they were usually worked against incomplete specifications. In this section, I will start with a more complete problem description and guide you through the design cycle odyssey.

An aircraft nose wheel steering system is to be designed. It will be driven by a dc motor and have attached spring and shock absorber assemblies to return the wheel to straight-ahead in the event of a power failure. The wheel and steering plate have an inertia of 1 ± 0.1 N-m-s^2/rad.

Some items have already been chosen. The pilot's steering wheel potentiometer and the nose wheel position-sensing potentiometer are both connected to a 28V supply and range through a half-turn. The pilot's wheel has stops at 60° away from center. The steering motor and other electronics must also operate from the 28-V supply.

The performance requirements are

1. Linear operation from stop to stop.
2. If the power fails, the nose wheel will settle to 2% of the initial displacement in less than 1.2 s.
3. For a step input, time to peak shall be less than 1 s, and the overshoot between 0 and 50%.

The Physical Diagram

This is a low-speed motor application, so I expect a gear stage will be needed. Steady-state accuracy isn't specified, but the system is driven by a human operator who likely won't require better than 1°. Initially then, I assume a voltage amplifier, a power amplifier, a motor, a gear train, the load, and input and feedback potentiometers. I am simply reversing the power supply connections to the input potentiometer to get the difference signal. Figure 9.21 is a sketch of these ideas.

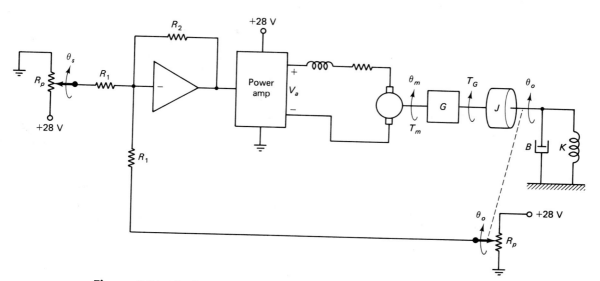

Figure 9.21   Preliminary diagram for the nose wheel steering system.

Suspension Design

The spring and damper values need to be chosen before the dynamic description of the whole system can be developed. We know the load inertia, but the spring has to move the connected motor and gears as well. A reasonable starting point is to assume the added inertia is equal to the load. This is a recommended objective for maximum efficiency in general applications. With the motor and gear properties reflected to the output shaft, the output mechanical admittance can be written as

$$\frac{\theta_o}{T_G} = \frac{1}{Js^2 + Bs + K} = \frac{1/J}{s^2 + (B/J)s + K/J} \qquad (9.1)$$

where $J = J_L + J_G + G^2 J_m$ and $B = B_L + B_G + G^2 B_m$ and J_L is the load inertia (1 N-m-s^2/rad), J_G is the gear train inertia, J_m is the motor inertia, B_L is the load damping to be chosen, B_G is the gear train friction coefficient, B_m is the motor friction coefficient, and K is the spring constant to be chosen. I will choose $J = 2$ N-m-s^2 and hope it turns out that way. By comparison with the standard quadratic form in Chap. 6, $B/J = 2\zeta\omega_0$ and $K/J = \omega_0^2$, so that $2\zeta = (B/J)\sqrt{J/K} = B/\sqrt{JK}$. Again from Chap. 6, the settling time is

$$t_s = -\frac{\ln \delta}{\zeta\omega_0} = -\frac{2J}{B} \ln \delta \qquad (9.2)$$

I choose the settling time to be 1 s to give a little margin to the requirement. This gives me $B = -4 \ln 0.02 = 15.65$ N-m-s. K can be chosen to set either the natural frequency or the damping factor. I think the damping factor is more pertinent, and I choose K to be the smallest value that still gives quadratic behavior—the value which makes $\zeta = 1$. This yields $K = 30.625$ N-m/rad.

Torque and Speed Estimates

The system is supposed to be linear from stop to stop, which is $\pm\pi/3$ rad. At an angle other than zero, the motor is opposed by the spring, and if it is going from $-\pi/3$ to $+\pi/3$, it will get help on the first half from the spring. For the sake of simplicity and because the motor has to fight the spring all the way, I will consider a step from 0 to $+\pi/3$ as the case for sizing the motor. Again for simplicity, I assume (estimate) that the motor has to produce the square wave pattern to accelerate and decelerate the load plus the torque to compress the spring. Even though this may seem conservative, I will take the required torque to be the sum of the absolute values of the maxima of each part. The square wave needed to drive the inertia is given by $\theta_f = T_M t_f^2/(4J)$. To allow some margin, I set $t_f = 0.8$ s. This makes $T_M = 13.09$ N-m. The spring force is $K\pi/3 = 32.076$, which gives a total maximum motor torque, as seen at the load end of the gears, of 45.17 N-m $= 399.75$ lb-in. The peak load speed is $\omega_M = T_M t_f/(2J) = 2.618$ rad/s $=$

25 rpm. I am looking for a motor with a speed-torque product of $25 \times 400 = 10K$ lb-in.-rpm = 160K oz-in.-rpm. The last two numbers are in the units used in the catalogs. As a matter of interest, this corresponds to 118 W = 0.159 hp.

Choosing the Motor

There are two kinds of speed limits to consider when looking for a dc motor. One is the machine's maximum rated speed, and the other is the maximum speed possible with the power supply voltage. In looking though the Electro-Craft catalog, I see that most motors come in a choice of windings, each with a different back-emf constant K_E. For the motors I am interested in, it turns out that $28/K_E$ is a much smaller number than the rated maximum speed in most cases. Another consideration I have in mind is keeping the gear ratio low. Gears are more efficient at lower ratios, and the match between reflected motor inertia and load inertia turns out to be better for a larger-torque motor than for a smaller-torque, higher-speed motor. I settled on the E703-B. At 500 rpm the safe continuous torque limit is 400 oz-in., which gives a product of 200,000 oz-in.-rpm. The K_E for this winding is 0.0405 V/rpm, so the supply-limited value is $28/0.0405 = 691$ rpm. Since I need 25 rpm at the load, I choose a gear ratio of $G = 20$. This gives me some room on both the speed and torque limits.

The motor parameters are

- ☐ Inertia: $J_m = 0.2$ oz-in.-s^2; $G^2 J_m = 0.565$ N-m-s^2
- ☐ Damping: $B_m = 0.01$ oz-in./rpm; $G^2 B_m = 0.27$ N-m-s
- ☐ Motor constant: $K_m = K_E 30/\pi = 0.38675$ V-s/rad
- ☐ Armature resistance: $R_a = 1 \ \Omega$
- ☐ Armature inductance: $L_a = 6$ mH
- ☐ Motor weight = 11 lb. = 5 kg.

Dynamics

Assume that the gear inertia is enough to bring J up to 2 N-m-s^2/rad. The gear loss should be small enough so that a damper will still need to be added. The block diagram in Fig. 2.24 gives the model for a motor-gear-load arrangement. I incorporated it into a system block diagram given as Fig. 9.22a. I reduced this to a unity-feedback system by bringing the two K_p blocks ahead of the summing junction and closing up the inner loop. The result is in Fig. 9.22b.

The next step is to evaluate the coefficients and find the poles of the cubic. $KR = K = 30.625$, and $K_m G/(KR) = 0.2526$ is the motor-load gain constant. Thinking of the denominator as $b_0 s^3 + b_1 s^2 + b_2 s + 1$,

$$b_0 = \frac{JL}{KR} = 3.918 \times 10^{-4}$$

$$b_1 = \frac{RJ + LB}{KR} = 0.06837$$

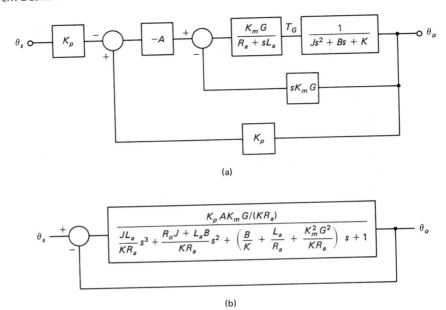

Figure 9.22 Preliminary block diagrams for the nose wheel system. (a) Direct. (b) Reduced.

$$b_2 = \frac{B}{K} + \frac{L}{R} + \frac{K_m^2 G^2}{KR} = 0.511 + 0.006 + 1.954 = 2.471$$

The poles are -0.4094, -50.41, and -123.7.

There is nothing in the performance requirements that says we need an integrator, but when I tested the step response with various gains, I found a steady-state error of 10–20% for values that gave the desired transient response. For the canonic second-order system the time to peak is $t_p = 2\pi/a$, where a is the open-loop corner frequency. For $t_p = 0.8$ s, $a = 7.86$ rad/s. To compensate the system I choose to use an integrator to drive the steady-state error to zero (or to the accuracy of the potentiometers) and a lead network to shift the first pole to -8. Since I want to avoid excessive overshoot, and the next two poles are fairly close, I initially chose a gain constant of 6. This gave a little quicker rise than needed, so I tried 4 next. The time to peak is right on target. To test the lower end, I tried 3 next. This gave $t_p = 1$ s. These results are plotted in Fig. 9.23. Choosing 4 as the overall gain constant means that $K_pA = 4/0.2526 = 15.84$.

Validation of Maxima Estimates

Now I need to find the speed and torque responses to see what the peaks really are. The speed is simply the derivative of the position, so it is easy to alter the system description to get speed as the output. All I need to do is add an origin

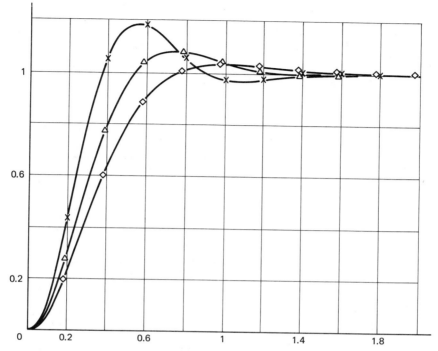

Figure 9.23   Step responses for $G_c = (1 + s/0.41)/[s(1 + s/8)]$. Forward gain is $\diamond$, 3; $\triangle$, 4; $\times$, 6.

zero to the forward path to produce the derivative, and an origin pole to the feedback path to compensate. This gives the correct transient but gives a nonzero steady-state value for the speed. I suspect that the program is having trouble with the 0/0 limit caused by the s and $1/s$ terms, so I simulate the zero and pole at the origin with very low-frequency values, a zero and a pole at -0.00001. I simulate s with $10^{-5}(1 + s/10^{-5})$ and $1/s$ with $10^{5}/(1 + s/10^{-5})$. This fixes the problem and gives the response shown in Fig. 9.24. The peak translates to a peak of 540 rpm for the motor with a 60° step input. This is acceptable because it is under the limit set by the supply voltage.

Rearranging the block diagram to bring torque to the output takes a fair amount of algebra. First, I return to the diagram in Fig. 9.22a and redraw it with K_p and G_c in the forward path in Fig. 9.25a. I then move the inner summing junction back to the input in Fig. 9.25b. Next, I substitute $G_c = (1 + s/z)/s(1 + s/p)$ and work the factors into the form for which the lowest-order term is 1, as in Fig. 9.25c. You may be interested to know that the zeros in the feedback path are -0.4479 and $-3.776 \pm j4.724$ rad/s. The armature L and R produce a forward-path pole at -166.7 rad/s. The gain constant in the feedback path G/K

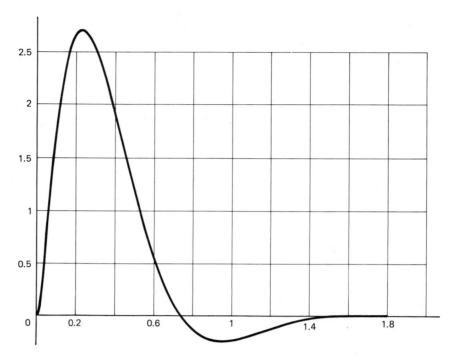

Figure 9.24 Speed response for a unit-step input. The peak of 2.7 rad/s corresponds to 25.78 rpm. This, in turn, means a peak of 27 rpm for a 60° step.

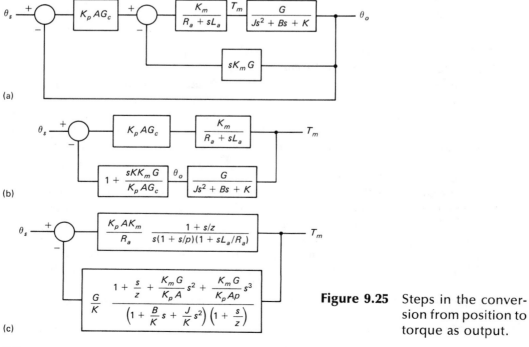

(a)

(b)

(c)

Figure 9.25 Steps in the conversion from position to torque as output.

sets the steady-state level. It checks with what we would expect, since the spring torque will be $K\theta$ and the motor torque will be this value divided by the gear ratio $K\theta/G$.

The peak of the torque response shown in Fig. 9.26 is too high by almost 25%. This means I will have to go back and consider choosing another motor or another gear ratio.

Recycling

Since I had room to move to a higher motor speed, I decided to try increasing the gear ratio to $G = 25$. Now I have to start back at Fig. 9.22b and recalculate everything. J, B, and K stay the same because they are design choices, and b_0 and b_1 stay the same because they don't depend on G. The value of b_2 becomes 3.57, an increase. The poles change to -0.2816, $-87.11 \pm j38.4$ rad/s. The step response curves for gains of 4 and 3, in Fig. 9.27, are shifted slightly to the left of 0.8 and to the right of 1 s for their peak times. Sticking with 4 as the forward-gain constant, $K_pA = 12.67$.

The speed response peak shown in Fig. 9.28 is almost the same as in the earlier case. Since G is now 25, this translates to a motor speed peak of 675 rpm

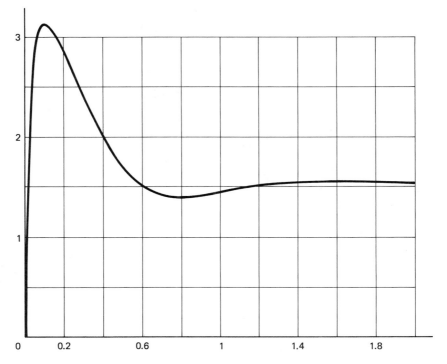

Figure 9.26   Motor torque response to a unit step input. The peak of 3.2 N-m corresponds to 453 oz-in., which will be 475 oz-in. for a 60° step.

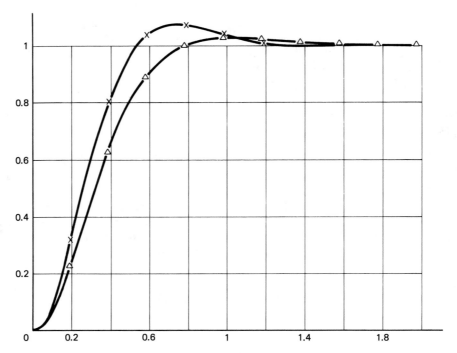

Figure 9.27 Step responses with $G = 25$, $G_c = (1 + s/0.28)/[s(1 + s/8)]$. Forward gain ×, 4; △, 3.

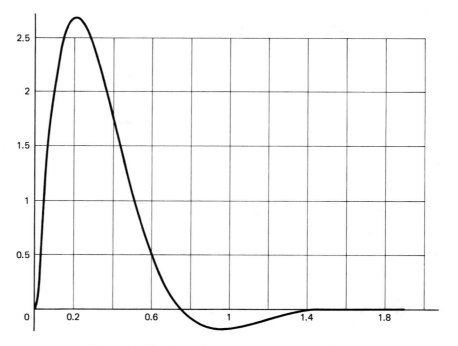

Figure 9.28 Speed response to a step input.

for a 60° step. The motor torque peak of 2.6 N-m from Fig. 9.29 translates to 386 oz-in. for a 60° step. Both of these values are within the motor safe operating area and under the power supply speed limit, so I will settle for these choices. $K_p = 28$ V/π rad = 8.913, so $A = 12.67/K_p = 1.42$.

Variations and Tolerances

The moment of inertia of the load has a ±10% tolerance, and the motor and gears may have the same. The tolerance on the motor constant K_m also has a ±10% range. The damper is mostly under design control, so I look at the effects of J and K_m on the transfer function and step response. K_m affects both the gain constant and b_2, and J affects b_0 and b_1. These are linear relations except for b_2. The term $K_m{}^2$ appears in b_2 and that term is about 85% of b_2, so that a 20% change in $K_m{}^2$ will make about a 17% change in b_2. Table 9.1 presents the poles of the motor-load transfer cubic for all the cases of high and low values for J and K_m.

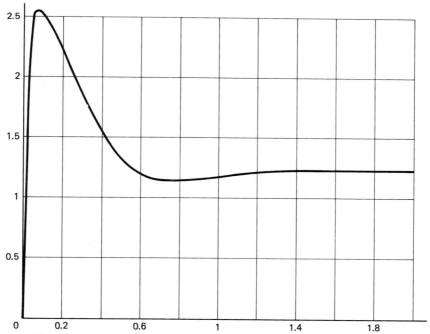

Figure 9.29 Motor torque response to a unit step input.

Table 9.1 Motor-load poles

J	K_m	Poles
High	High	-0.2405, $-87.13 \pm j45.36$
Low	High	-0.2403, $-87.13 \pm j64.9$
Low	Low	-0.3399, $-87.08 \pm j27.58$
High	Low	-0.3404, -114.8, -59.38

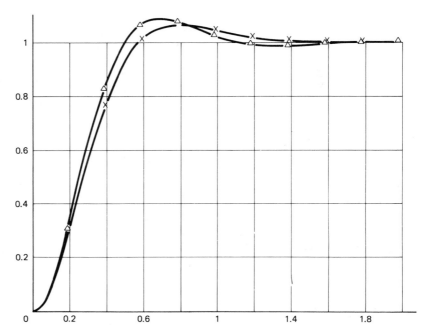

Figure 9.30 Step responses for motor-load parameter variations, high J. $\times$, high K_m; $\triangle$, low K_m.

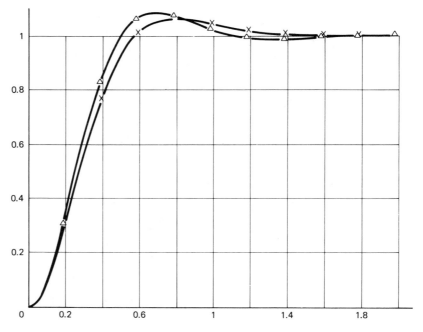

Figure 9.31 Step responses for motor-load parameter variations, low J. $\times$, high K_m; $\triangle$, low K_m.

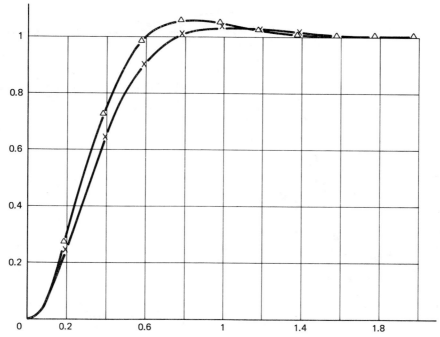

Figure 9.32   Nominal gain reduced to 3.5. Step responses for high J and K_m, other gains 10%. ×, low; △, high.

The step responses for these pole sets are shown in Figs. 9.30 and 9.31. The dominant effect is the position of the low-frequency pole, which is mainly a function of K_m. The two sets of curves lie right on each other, which is why I show them in two figures instead of one.

Since $t_p \leq 0.8$ s for all cases, I decide to reduce the gain constant to 3.5 and consider the effect of a ±10% variation in the parameters other than k_m. High values of J and K_m give the slowest system, and Fig. 9.32 shows that the variation in peak time due to just gain variation is tolerable.

It turns out that the positions of the pole near −0.3 and the compensator zero, with respect to each other, have a strong effect on the crossover frequency and the step response time. Since a precision capacitor is more expensive than a precision resistor, I consider the effect of tolerance in C with perfect resistors. Assuming a single-capacitor network is used to realize the lead function, the ratio of the pole to the zero will be constant, so changing C will move them both in proportion. A ±10% variation in C gives too big a variation in t_p. I next try ±5%, and this is just barely acceptable, as you can see in Fig. 9.33. If gain variation is allowed on top of this, the performance will likely overstep the torque limit on the high-speed end and the 1-s peak time limit on the low-speed end. This is as far as I can go without getting into evaluating the cost of various alternatives. If

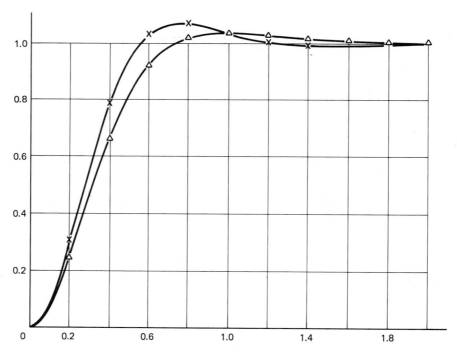

Figure 9.33 Nominal gain at 3.5, 5% variation in C. ×, high C, low J and K_m; △, low C, high J and K_m.

a larger, therefore more expensive, motor is used, the tolerances on the R and C components can be looser. Will the increment in motor size be more expensive than using a 2% C and 1% R? Suppose we use a 5 or 10% C, 1% R, and the present motor, with an adjustment which is only used if the unit fails inspection? To evaluate this last choice, one would have to make a probability analysis and know the cost of the adjustment. In any case, these matters are beyond the scope of this text, so I just point to them as possibilities.

Electronic Design

The peak current taken by the motor is 7.04 A. I already had in mind a bridge-type power amplifier to be able to run the motor in both directions, and this current peak suggests transistors with a 10-A rating. This is too much for any reasonably priced operational amplifier, so I drew a schematic for a hybrid of op amps and power transistors in Fig. 9.34. The bridge permits current flow in either direction through the motor, and either Q_1–Q_4 or Q_2–Q_3 will conduct. Suppose that the output of A1 is higher than that of A2 and the motor is delivering power to the load. Then Q_1–Q_4 will conduct. If the input changes such that A1 must reduce, the motor's back voltage will hold up the Q_1–Q_3 emitters while A1 drops their base voltage. The op amp will step down through the base voltage dead zone,

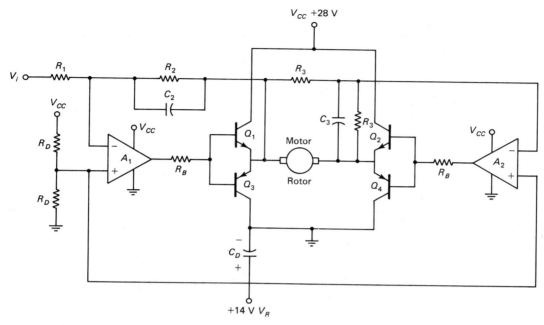

Figure 9.34 A bridge power amplifier to supply the full $\pm V_{cc}$ range to the motor.

turning off Q1 and turning on Q_4. A_2 performs the inverse action of A_1 because it is connected as a unity-gain inverter driven by the Q_1–Q_3 emitters. The op amps are biased at $+14$ V on their noninverting inputs, so that they act as if they were operating from a dual 14-V supply. The capacitors C_2 and C_3 serve the purpose of rolling off the gain at high frequencies to reduce noise. The base resistors are there to protect the bases from overdrive, if they need it. For some choices of power transistor and op amp, it will not be possible for the op amp to overdrive the bases.

The signal-processing circuit is shown in Fig. 9.35. As a sensor, a potentiometer is a relatively high impedance source. One must either use a resistor in series with the wiper which is large in value compared to the total pot. resistance, or use an op amp buffer stage, which I have shown here. This method is more linear than the series wiper resistor approach and simplifies the design of the following lead network. The compensator must have both an integrator and a lead network. C_5 is the integrating capacitor, and R_6 serves the purpose of bleeding its charge off should that be needed. R_6 converts the integrator into a low-frequency pole, but this conversion is acceptable as long as the low-frequency gain is much higher than the reciprocal of the accuracy of the sensors. The two R_4s, R_5, and C_4 form a combination summing and lead network. Physically, you can see that it's a lead network because the impedance at low frequency is higher

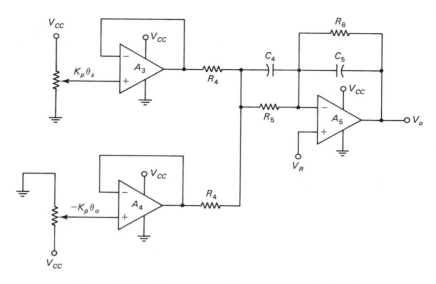

Figure 9.35 Summer and compensator circuit.

than that at high frequency when C_4 shorts R_5. At little circuit analysis will show you that $z = 1/(R_5C_4)$, $p = (2R_5 + R_4)/(R_4R_5C_4)$, $p/z = 1 + 2R_5/R_4$, and the gain constant including the integrator is $1/[C_5(2R_5 + R_4)]$. In analyzing this circuit, remember that you are looking for the transfer from one input with the other one held constant.

With the system's forward gain reduced to 3.5, $K_pA = 11$. And $A = 11\pi/28 = 1.234$. The value of $p/z = 8/0.28 = 28.57$, so $R_5/R_4 = 13.79$. I start by choosing $R_5 = 150$ kΩ, which makes $R_4 = 10.877$ kΩ, close to 11 kΩ. $C_4 = 1/0.28R_5 = 23.81$ μF. To find C_5, $2R_5 + R_4 = 310{,}877$ Ω, and for a gain of 1, $C_5 = 3.217$ μF. If I assign all of A to this stage, $C_5 = 3.217/1.234 = 2.607$ μF.

In the section on variations, I pointed out that one possibility requires C_4 to have a tolerance of 5%, and a tight gain tolerance requires C_5 to be 1% or both of them to be 1%. To obtain these accuracies, one must go to smaller values of C, by at least two orders of magnitude. Thus, dividing the C by 100 requires multiplying R_4 through R_6 by 100. Going to resistors this high forces A_5 to be a field effect transistor (FET) input amplifier and a little more expensive. Also, though, with R_4 over 1 MΩ, the buffers can probably be eliminated. If the larger motor option is chosen, looser tolerances can be used, larger Cs and smaller Rs in this stage.

As it is, the transistors have to dissipate quite a bit of power. I use the LSP to generate the torque and speed data for the current nominal design and then follow the procedure given earlier to find the power dissipation shown in Fig. 9.36. There is a sharp peak of almost 85 W near the start, and the steady state is 75 W holding the wheel against the spring with no back voltage generated. Since

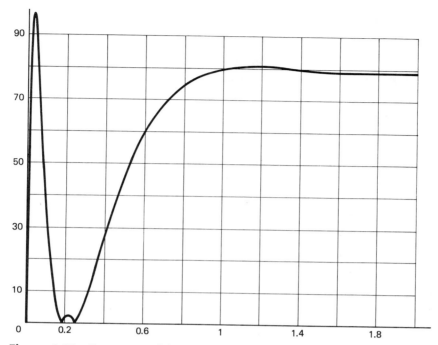

Figure 9.36 Power amplifier dissipation response, in watts versus seconds.

two transistors are on at a time to split the voltage drop, each transistor should be chosen and heat-sunk to handle 50 W. This gives some margin for unbalance in the bridge and a mismatch in the transistors.

This concludes the example. As you can see, one can traverse the design cycle several times before a project is really finished. I have made two major passes and indicated a few branches that need evaluation on the basis of cost. I have not made the detail calculations for several parts, as they are ordinary. The network for V_R should have small enough resistors to swamp variations in the op amp bias currents and should be bypassed to ground with a suitably large capacitor. Likewise, the power supply lines to each stage should have a decoupling network at the stage, or at least one for the small-signal stages and another for the power stage. Supply lines are not the zero-impedance sources we always model them to be, so a large capacitor must be at the stage connection point to give this simulation.

My main purpose in this example has not been to give you a demonstration of electromechanical system design. It is more important that you understand that the design process for a control system is a learning process. In your years in school and even later, you won't be able to learn the specifics of every possible kind of plant. Therefore each new system will be an experience in which you learn the nature of the plant and use the specialists who designed it to help develop a dynamic model. In many situations you will have to go to the specialists to ask for a redesign because the performance requirements can't be met on the first try. You will need a positive attitude toward studying and searching the literature and learning from and with other people.

FURTHER READING

The usefulness of catalog literature for actuators and sensors is quite variable. Too many manufacturers just give static ratings and physical dimensions. A couple of better sources are given here.

DC Motors, Speed Controls, Servo Systems: An Engineering Handbook, Electro-Craft Corp., 1600 Second Street South, Hopkins, MN 55343.
Measurement Handbook Series, Omega Engineering Inc., P.O. Box 4047, Stanford CT 06907-0047.

PROBLEMS

Section 9.3

9.1   Suppose that a mass M is to be moved through a distance x_M, starting and finishing at zero speed. The transit time is to be not more than t_f seconds. If the speed function is a triangle, what are the maximum speed and force required?

9.2   For the requirements of Prob. 9.1, suppose the speed function is a symmetrical trapezoid with the constant-speed section lasting $t_f/2$ seconds. What are the maximum speed and force needed?

9.3   For the friction load and screw actuator in Examples 4.17 and 9.3 estimate the maximum motor torque if the load must move
(*a*) 10 in. in 1 s
(*b*) 1 in. in 1 s
(*c*) 0.1 in. in 0.1 s
starting and finishing at rest.

9.4   A plant has the transfer function $(1 + 0.1s)/(1 + 0.1s + s^2)$. Estimate the maximum input amplitude required to move the output through a change of 10 in 2 s, with initial and final rest conditions.

Section 9.5

9.5   For the plant in Prob. 9.4, find a controller transfer function to give an ITAE type-1 system with a 100% rise time of 2 s in a unity-feedback arrangement. Find the maximum controller output for a step input of 10.

Section 9.6

9.6 Suppose the controller in Prob. 9.5 is electronic. Design it for minimum part count. Assume that ± 12 V supplies are to be used, that the plant has a 120 Ω input resistance, and that the command and sensor voltage sources have 10 kΩ output impedances.

9.7 Suppose that the circuit in Figs. 9.9 and 9.13, discussed in Examples 9.5 and 9.7, must have gain such that a 5 V change in V_i corresponds to the full 10 W range in output heat. Further, a lead network with transfer $(1 + 10s)/(1 + s)$ must be included in the stage. Do the electronic design with the op amp's input resistance of 1 MΩ minimum. If you need a capacitor larger than 1 μF, it must be a polarized electrolytic. Use two to balance the leakage currents, if possible, and bias it or them correctly.

Section 9.7

9.8 Figure P.9.8 shows the block diagram of a temperature control system which uses the circuit in Fig. 9.9 and Prob. 9.7. The nominal values of the circuits are $K_I = 5$, $K_H = 2$, $z = 0.1$, $p = 1$. Suppose that all four parameters have an independent variation with an equal percentage range. What should the tolerance be to keep the overshoot and settling time within $\pm 20\%$ of the nominal values?

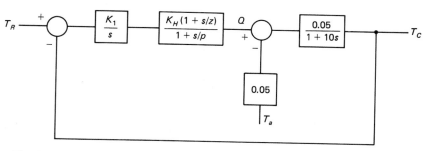

Figure P.9.8   Temperature control system for Prob. 9.8 and 9.10.

9.9 Continuing from Example 9.9, suppose the screw gain has a tolerance of 2% and the amplifier gain and motor constants each have 10% tolerance. Redesign the lag network with 10% parts and restate the test specifications so that the step response shape will be more-or-less uniform over the total range of variations.

Section 9.8

9.10 A temperature control system is to be designed to hold 70 $\pm$ 2°C in the face of an ambient which varies from 0 to 60°C. The circuit must settle to within

2° in 2 s after turn-on. The block diagram in Fig. 9.37 and the circuit in Fig. 9.9 have already been chosen.

(a) Since the thermal properties of the transistor-resistor heater are not significant compared to those of the box, an estimate of the effort is not required. Determine the gain and controller parameters needed to nominally meet the specs.

(b) Obtain the step response of Q for the largest reference temperature change. Assigning the peak heat entirely to the resistor, determine the transistor's heat as a function of time. Using these step responses for the transistor heat Q_T and T_c, plot Q_T against T_c. Fit a derating function of the usual shape (flat to 25°C, straight line down to 0 W at 200°C) over this plot to determine the transistor power rating.

(c) Choose a suitable power transistor and operational amplifier. Design the circuit for the gain and compensator stage. Assume that the required overall gain is divided equally between K_I and K_H.

(d) Tolerance your design. Using current catalog prices, distribute the tolerances to minimize the cost, meet the specifications, and not overheat the transistor.

9.11 Figure P.9.11 shows another possibility for the compensator in Sec. 9.8. Derive the design equations. Discuss the effects of the parts' tolerances on the network's performance parameters.

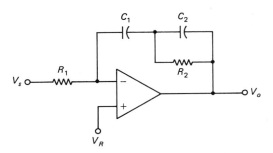

Figure P. 9.11 Another integrator and lead network.

THE STATE-SPACE MODEL

10.1 TIME AND FREQUENCY RESPONSES

In Sec. 2.4 you were introduced to state variables as a way of developing a system description as a set of first-order differential equations. If there is no direct feed-through (constant transfer) from the input to the output, a plant's state equations can be summarized in vector-matrix form as

$$\dot{\mathbf{x}} = \mathbf{A}\mathbf{x} + \mathbf{B}\mathbf{u}$$

$$\mathbf{y} = \mathbf{C}\mathbf{x} \tag{10.1}$$

in which $\mathbf{u}$ is the input signal set, an m-element column vector, $\mathbf{x}$ is the state vector, an n-element column and $\mathbf{y}$ is the output signal set, an r-element column vector. $\mathbf{B}$ is called the input matrix, $\mathbf{A}$ the state matrix, and $\mathbf{C}$ the output matrix, and their elements are constants for our linear, continuous-time, time-invariant systems. Representing quantities in vector form and operating on them with matrices has been a topic of study in mathematics since Coyley presented matrices in 1858. By analogy and extension with physical space, each element in the column matrix is thought of as the value along a dimension, so that the vector represents a point (or state) in an n-dimensional space. As you may know from field theory, navigation, or many other practical experiences, there is nothing unique about a physical coordinate system. Likewise, there is nothing unique about the choice of representing a point in n space, and any coordinate system can be transformed into any other by a suitable matrix multiplication. This is the mathematical equivalent of our knowledge that the choice of variables used to represent the dynamics of a system is not unique. Since the state variables are not unique, the matrices are not unique either. Only the input and output variables, which are chosen by design, are unique in a given problem. This fact is very useful, because it means that if one can derive conditions under which a state matrix with a special form has some property, the result can often be extended to all state-space descriptions.

Multiple-input multiple-output (MIMO) systems are, by nature, quite complicated in their physical form, so that scalar-level signal models of them are also complex. Figure 10.1 shows a plant and two diagrams representing mathematical models. Figure 10.1a is just a block that represents the plant itself, with its input and output signals. It may very well be that these signals are the only access the control engineer has to the plant. In some cases, the control engineer can participate in the plant design to ensure that signals useful to the control operation are brought out. So the plant may be a "black box" or a "gray box," terms commonly used to describe our knowledge or access. In Fig. 10.1b the diagram represents the state equations, not the physics of the plant. It is quite compact because of the use of vectors to collect the variables. I have included the initial state as an input, even though previously we have generally assumed it to be zero.

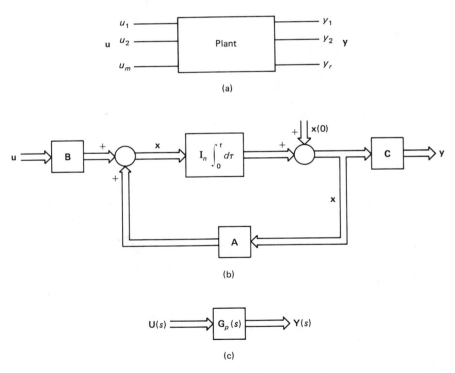

Figure 10.1 Block diagrams for (a) the plant (b) its state equation model, and (c) its transfer function model.

It turns out to be useful in developing our understanding and some tools for design, as you will see shortly. Figure 10.1c is the frequency domain model. It is even more compact than the state-space diagram because the complexity of the plant is in the matrix elements, each one of which is a transfer function from one of the inputs to one of the outputs.

Things get very complicated very quickly when one looks at the models on an individual (scalar) signal basis, as shown in Figs. 10.2 and 10.3. These are very simple examples. If Fig. 10.2 had to show 3 states instead of 2, there would have to be 10 more blocks, plus more summing junctions and many more wires. Figure 10.3 doesn't grow with the internal complexity represented by the states, again because the internals are largely contained in the individual transfer functions.

It is possible to extend transfer function analysis and design to the MIMO case. As you might imagine, the algebra gets to be quite a chore and, until fairly recently, the state-space methods I was aware of also required a lot of algebra for high-order systems. Now that personal computers are generally available, operations with matrices whose elements are constants are easy to do. This gives

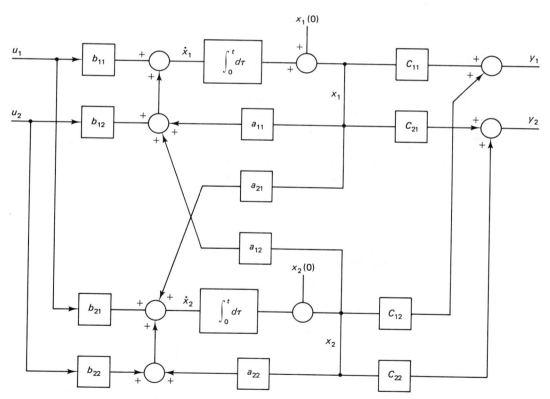

Figure 10.2 A scalar version of Fig. 10.1b for $m = n = r = 2$.

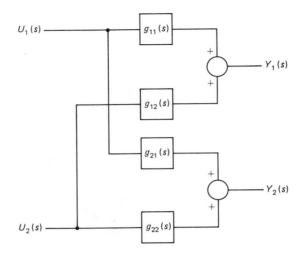

Figure 10.3 Expansion of Fig. 10.1c for $m = r = 2$.

state-space methods based on these operations a tremendous advantage over methods requiring transform algebra. For this reason, all the subsequent material will be on state-space models and design methods that use constant-element matrices. Transform notions are still useful for analysis and, after decoupling, for design.

Let's look at the homogeneous equation and solve it:

$$\dot{\mathbf{x}} = \mathbf{A}\mathbf{x} \qquad \mathbf{x}(0) = \mathbf{x}_0 \tag{10.2}$$

If this were a scalar equation, we would know that the solution is an exponential function which can also be represented by a power series:

$$x(t) = x_0 e^{at} = x_0 \sum_{k=0}^{\infty} \frac{(at)^k}{k!} \tag{10.3}$$

The validity of the power series form of the solution doesn't depend on the nature of the constant a, so it is likely that an analog to the power series is a solution to (10.2). Let's try

$$\mathbf{x}(t) = \left[\sum_{k=0}^{\infty} \frac{(\mathbf{A}t)^k}{k!} \right] \mathbf{x}_0 \tag{10.4}$$

Differentiating, we have

$$\dot{\mathbf{x}}(t) = \left[\sum_{k=1}^{\infty} \frac{\mathbf{A}^k t^{k-1}}{(k-1)!} \right] \mathbf{x}_0 = \left[\mathbf{A} \sum_{k=0}^{\infty} \frac{\mathbf{A}^k t^k}{k!} \right] \mathbf{x}_0 = \mathbf{A}\mathbf{x}(t)$$

The power series works for the vector-matrix equation as well. Again because of the analogy, this power series is represented by an exponential function. The power series, rather than the antilog, is the definition of the exponential notation in this case:

$$e^{\mathbf{A}t} = \sum_{k=0}^{\infty} \frac{(\mathbf{A}t)^k}{k!} \tag{10.5}$$

It is possible to derive the solution to the complete system entirely in the time domain, but it is more straightforward to use our experience with Laplace transforms. To this end, let's first transform the homogeneous equation and look at its solution.

$$s\mathbf{I}_n\mathbf{X}(s) - \mathbf{x}_0 = \mathbf{A}\mathbf{X}(s) \tag{10.6}$$

where $\mathbf{I}_n$ is the $n \times n$ identity matrix.

$$(s\mathbf{I}_n - \mathbf{A})\mathbf{X}(s) = \mathbf{x}_0$$
$$\mathbf{X}(s) = [s\mathbf{I}_n - \mathbf{A}]^{-1}\mathbf{x}_0 \tag{10.7}$$

Evidently, $[s\mathbf{I}_n - \mathbf{A}]^{-1}$ is the Laplace transform of $e^{\mathbf{A}t}$. These functions have an honored place in system theory and are called *transition matrix* and its transform. They are usually denoted as $\boldsymbol{\phi}(t)$ for the time version and $\boldsymbol{\Phi}(s)$ for its transform. This name comes from the fact that, given any state $\mathbf{x}(t_1)$ and the absence of a forcing function, any future state is given by $\mathbf{x}(t_2) = \boldsymbol{\phi}(t_2 - t_1)\mathbf{x}(t_1)$. Now, let's solve (10.1). Transforming,

$$s\mathbf{I}_n\mathbf{X}(s) - \mathbf{x}_0 = \mathbf{A}\mathbf{X}(s) + \mathbf{B}\mathbf{U}(s) \qquad \mathbf{Y}(s) = \mathbf{C}\mathbf{X}(s)$$
$$\mathbf{X}(s) = [s\mathbf{I}_n - \mathbf{A}]^{-1}[\mathbf{x}_0 + \mathbf{B}\mathbf{U}(s)] = \boldsymbol{\Phi}(s)\mathbf{x}_0 + \boldsymbol{\Phi}(s)\mathbf{B}\mathbf{U}(s) \tag{10.8}$$

The transfer function from input to state is

$$\mathbf{G}_s(s) = \boldsymbol{\Phi}(s)\mathbf{B} \tag{10.9}$$

and from input to output the transfer function is

$$\mathbf{G}_p(s) = \mathbf{C}\boldsymbol{\Phi}(s)\mathbf{B} \tag{10.10}$$

The roots of $|s\mathbf{I}_n - \mathbf{A}| = 0$ (also called the eigenvalues of $\mathbf{A}$) are the poles of $\boldsymbol{\Phi}(s)$ and the system transfer matrix.

The time domain solution for the state is the inverse Laplace transform of (10.8). The term from the forcing function is the product of two transforms, and we know that the corresponding time function is the convolution of the individual time functions. Thus,

$$\mathbf{x}(t) = \boldsymbol{\phi}(t)\mathbf{x}_0 + \int_0^t \boldsymbol{\phi}(t - \tau)\mathbf{B}\mathbf{u}(\tau) \, d\tau \qquad (10.11)$$

$$\mathbf{y}(t) = \mathbf{C}\mathbf{x}(t) = \mathbf{C}\boldsymbol{\phi}(t)\mathbf{x}_0 + \mathbf{C} \int_0^t \boldsymbol{\phi}(t - \tau)\mathbf{B}\mathbf{u}(\tau) \, d\tau \qquad (10.12)$$

As with the SISO system, the responses to two special input functions are of interest and general use. The impulse and the step function, respectively, are the basic forcing functions for analysis and specification. In the present case, we have the possibility of many inputs, so let's define a matrix of impulse functions to stimulate the system. Let $\mathbf{u}(t) = \mathbf{d} \, \text{imp}(t)$, $\mathbf{d} = [d_i]$, where the d_i are constants of our choice and $\text{imp}(t)$ is the unit impulse function defined in (5.1). With this input, the integral in (10.11) evaluates to its argument at $\tau = 0$:

$$\mathbf{x}(t) = \boldsymbol{\phi}(t)\mathbf{x}_0 + \boldsymbol{\phi}(t)\mathbf{B}\mathbf{d} \qquad (10.13)$$

$$\mathbf{y}(t) = \mathbf{C}\boldsymbol{\phi}(t)\mathbf{x}_0 + \mathbf{C}\boldsymbol{\phi}(t)\mathbf{B}\mathbf{d} \qquad (10.14)$$

We see that $\mathbf{C}\boldsymbol{\phi}(t)\mathbf{B}$ is the MIMO generalization of the impulse response or weighting function from SISO analysis. Also, this state-space result makes it clear that applying impulses to the inputs is equivalent to setting initial conditions on those states which are directly accessible from the inputs through the matrix $\mathbf{B}$.

Next, let's define an input vector of step functions. Let $\mathbf{u}(t) = \mathbf{u}_s u(t)$, $\mathbf{u}_s = [u_i]$, where the u_i are constants of our choice and $u(t)$ is the scalar unit step function of (5.2). Using this forcing function in (10.11) reduces the response calculation to the problem of the integral of the transition matrix:

$$\mathbf{x}(t) = \boldsymbol{\phi}(t)\mathbf{x}_0 + \int_0^t \boldsymbol{\phi}(t - \tau)\mathbf{B}\mathbf{u}_s \, d\tau$$

Let $\mu = t - \tau$, $d\mu = -d\tau$. Then the integral becomes

$$-\int_t^0 \boldsymbol{\phi}(\mu) \, d\mu = \int_0^t \boldsymbol{\phi}(\mu) \, d\mu$$

so that

$$\mathbf{x}(t) = \boldsymbol{\phi}(t)\mathbf{x}_0 + \int_0^t \boldsymbol{\phi}(\mu) \, d\mu \mathbf{B}\mathbf{u}_s \qquad (10.15)$$

$$\mathbf{y}(t) = \mathbf{C}\boldsymbol{\phi}(t)\mathbf{x}_0 + \mathbf{C} \int_0^t \boldsymbol{\phi}(\mu) \, d\mu \mathbf{B}\mathbf{u}_s \qquad (10.16)$$

The response to the step input turns out to be the integral of the impulse response, as it is with SISO systems.

In the next section I will give some properties of the transition matrix and some useful algorithms.

10.2   THE TRANSITION MATRIX

Since the time origin is arbitrary for our kinds of systems, the unforced transition from one state to the next must be a function of the time difference. If t_1 and t_2 are two time intervals, then $\mathbf{x}(t_1 + t_2) = \boldsymbol{\phi}(t_1 + t_2)\mathbf{x}(0) = \boldsymbol{\phi}(t_2)\mathbf{x}(t_1) = \boldsymbol{\phi}(t_2)\boldsymbol{\phi}(t_1)\mathbf{x}(0)$. This implies that $\boldsymbol{\phi}(t_1 + t_2) = \boldsymbol{\phi}(t_1)\boldsymbol{\phi}(t_2)$. From the series $\boldsymbol{\phi}(0) = \mathbf{I}_n$, which also makes sense physically. It follows from the last two items that $\boldsymbol{\phi}(t)\boldsymbol{\phi}(-t) = \mathbf{I}_n$, or $\boldsymbol{\phi}(t)^{-1} = \boldsymbol{\phi}(-t)$.

Hand calculation of the transition matrix isn't easy, as I mentioned before. For low-order systems, probably the most accurate route is to find $\boldsymbol{\Phi}(s)$, do a partial-fraction expansion on each element, and invert by recognition or table. The following algorithm does two things: Starting from the state matrix in any form, it gives the coefficients of the system polynomial, and it gives a sequence of matrices from which adj $\boldsymbol{\Phi}^{-1}(s)$ can be found. One may apply any convenient root-finding routine to the system polynomial to get the poles.

Let $\boldsymbol{\Phi}(s)$ be written in the form

$$\boldsymbol{\Phi}(s) = \frac{\mathbf{I}_n s^{n-1} + \mathbf{H}_1 s^{n-2} + \mathbf{H}_2 s^{n-3} + \cdots + \mathbf{H}_{n-1}}{D(s)} \tag{10.17}$$

with $D(s) = s^n + a_1 s^{n-1} + \cdots + a_n$ the system polynomial. Why can I write the leading coefficients in these polynomials as unity?

Algorithm 10.1   Transition Matrix Transform and System Polynomial Coefficients

Define $\mathbf{H}_0 = \mathbf{I}_n$. Then for $k = 1$ to n,

$$a_k = -k^{-1} \text{trace}(\mathbf{AH}_{k-1}) \qquad \mathbf{H}_k = \mathbf{AH}_{k-1} + a_k \mathbf{I}_n \tag{10.18}$$
Done.

Note that you don't need to calculate $\mathbf{H}_n$, but if you do, it should be zero. I have written this algorithm in APL to find the system polynomial and follow it in the function by a Laguerre root finder. It is not only simple to implement, but it is faster than the commercial eigenvalue finder, based on the standard QR algorithm, that came with my APL interpreter. This algorithm also gives the determinant of $\mathbf{A}$, since

$$|s\mathbf{I}_n - \mathbf{A}|_{s=0} = |-\mathbf{A}| = a_n$$
$$|\mathbf{A}| = a_n(-1)^n \tag{10.19}$$

For $n > 3$, it isn't a good use of time these days to hand-calculate $\boldsymbol{\phi}(t)$ by any method. The easiest method to set up for machine computation is truncation of the series expression (10.5). Each term is calculated, added, and tested to see if the addition is negligibly small by some measure. (A word about notation may be needed here. A measure of a matrix, usually called a *norm*, is commonly symbolized by double magnitude signs: $\| \mathbf{M} \|$ = norm of $\mathbf{M}$. When an operation applies to all values of an array index, an asterisk (*) will be used in the position of that index. Thus $\mathbf{x}[1, *] = 0$ means that the first row of $\mathbf{x}$ is made zero.) If the time interval is small compared to the dynamics of the system, convergence will be considerably more rapid than if a large time interval is chosen. I use the sum of the element magnitudes as a measure of the matrix size, since this is easy to find in APL.

Algorithm 10.2   Transition Matrix

Let $\mathbf{M} = \mathbf{S} = \mathbf{I}_n$, $k = 1$, and ϵ be a small positive number. Then

Loop: $\mathbf{M} = \mathbf{MA}t/k$, $\mathbf{S} = \mathbf{S} + \mathbf{M}$

If $\| \mathbf{M} \| < \epsilon \| \mathbf{S} \|$ stop

$k = k + 1$

Go to Loop.

Presumably, one wishes to use the transition matrix to find the time response for a significant amount of time. Using this algorithm for small t and then multiplying the result for each increment in time is faster than calculating the transition matrix for the longer period. It is also accurate enough. I have tested the step response for borderline-stable systems, as indicated by their Nyquist plots, and gotten a stable oscillation as expected. An algorithm for calculating the sequence of values of ϕ for a set of time increments spanning the transient response might go as follows.

Algorithm 10.3   Transition Matrix Sequence

Let TS be a small time step and N be the number of time steps to reach the desired time $N \times TS$. Let F be an $(N + 1) \times n \times n$ array to hold the result and $\mathbf{FS}$ be a matrix for the current value of ϕ in the sequence:

$\mathbf{FS} = \phi(TS)$, $\mathbf{F}[1, *, *] = \mathbf{FC} = \mathbf{I}_n$.

For $k = 2$ to $N + 1$, $\mathbf{FC} = \mathbf{FC} \cdot \mathbf{FS})$, $\mathbf{F}[k, *, *] = \mathbf{FC}$.

Done.

If one is not interested in the transition matrix as such, but only in the state response, a lot of storage can be saved by not keeping the sequence values of the matrix but instead by calculating and saving only the state at each time step as follows.

Algorithm 10.4   State Impulse and Initial Condition Response

Let TS be a small time step, N the number of time steps to reach the total time $TS \times N$, $\mathbf{X}$ the $n \times (N + 1)$ sequence of state values, and $\mathbf{XI}$ the current state.

$$FS = \phi(TS), \quad \mathbf{X}[*, 1] = \mathbf{x}_0 + 0.5\mathbf{Bd}, \quad \mathbf{XI} = \mathbf{x}_0 + \mathbf{Bd}.$$

For $k = 2$ to N, $\mathbf{XI} = FS \; \mathbf{XI} \; \mathbf{x}[*, k] = FS \; \mathbf{XI}.$

Done.

For a single-input system, once the impulse response sequence is obtained, the step response can be found by direct integration, and the response to any input function can be found by convolution. A multiple-input system with general inputs requires convolution with the transition matrix sequence. However, for a multiple-step input the response is the integral of the impulse response with corresponding weights. For these purposes, $\mathbf{x}_0$ must be set to zero in Algorithm 10.4. An easy method to use for the step response integral is the rectangular approximation expressed in the next algorithm.

Algorithm 10.5   State Step Response Sequence

Let TS and N be the time step and number of steps, as before. Let $\mathbf{XI}$ be the state impulse sequence for the desired imput set, $\mathbf{d} = \mathbf{u}_s$, with $\mathbf{x}_0 = 0$. Let $\mathbf{XS}$ be the required step response sequence, $n \times (N + 1)$. Then

$$\mathbf{XS}[*, 1] = 0.$$

For $k = 2$ to $N + 1$, $\mathbf{XS}[*, k] = \mathbf{XS}[*, k - 1] + \mathbf{XI}[*, k - 1].$

$$\mathbf{XS} = TS \; \mathbf{XS}.$$

Done.

Notice that the step response sequence is one time step longer than the impulse response sequence. This is because the first element is 0, assuming it to be the response at $t = 0$, and the integral increment is the value of the integrand at the left of the time step interval. More accurate methods are certainly available, but this is the simplest to code. In APL it is XS ← ((n, 1) ρ0), TSx + \XI.

10.3   STATE EQUATIONS FROM TRANSFER FUNCTIONS

Sometimes we have the model of a subsystem or component already in transfer function form and we want to incorporate it into a state-space model. Sometimes the model of a MIMO system is available as a transfer function matrix and we want the state variable model. For these practical reasons, and because the methods and results are interesting in themselves, I will present a few ways to translate from transfer functions to state equations.

Consider the following transfer function, which might be for a subsystem,

$$\frac{Y(s)}{U(s)} = \frac{50(1 + s/5)}{s(1 + s/2)(1 + s/50)} \tag{10.20}$$

$$= \frac{1000(s + 5)}{s(s + 2)(s + 50)} \tag{10.21}$$

$$= \frac{1000s + 5000}{s^3 + 52s^2 + 100s} \tag{10.22}$$

$$= \frac{50}{s} - \frac{31.25}{s + 2} - \frac{18.75}{s + 50} \tag{10.23}$$

The plethora of forms for this transfer function reflects the variety of analysis tools we use and the nonuniqueness of a mathematical model. Equation (10.20) is closest to the physical model and most suitable for Bode plot analysis, (10.21) is the best form for root locus analysis, and (10.22) and (10.23) lead to different state-space models. Consider (10.20) first. If it weren't for the s term in the numerator, it would be in a form we used in Chap. 2. We could cross-multiply and convert, literally or conceptually, to a time domain differential equation and assign state variables to the derivatives of $y(t)$. Let's start this process and see how it goes when we use an intermediate variable to carry it through:

$$Y(s) = \frac{1000s + 5000}{s^3 + 52s^2 + 100s} U(s)$$

Let

$$Z(s) = \frac{U(s)}{s^3 + 52s^2 + 100s} \tag{10.24}$$

so that

$$Y(s) = (1000s + 5000)Z(s) \tag{10.25}$$

Equation (10.24) is in the all-pole form we've used before. We can go to the time domain next:

$$\dddot{z} + 52\ddot{z} + 100\dot{z} = u(t) \tag{10.26}$$

This is a third-order system if we want $z(t)$ itself, and we do because we need it to form $y(t)$. We can, as in Chap. 2, assign the states as

$$x_1 = z \qquad x_2 = \dot{z} = \dot{x}_1 \qquad x_3 = \ddot{z} = \ddot{x}_2$$

$$\dot{x}_3 = -100\dot{z} - 52\ddot{z} + u(t) = -100x_2 - 52x_3 + u(t)$$

This yields state and input matrices:

$$A = \begin{bmatrix} 0 & 1 & 0 \\ 0 & 0 & 1 \\ 0 & -100 & -52 \end{bmatrix} \quad B = \begin{bmatrix} 0 \\ 0 \\ 1 \end{bmatrix} \tag{10.27}$$

Now we need an output equation. Converting (10.25) to the time domain gives $y(t) = 1000\dot{z} + 5000z = 5000x_1 + 1000x_2$. The output matrix is

$$C = [5000 \quad 1000 \quad 0] \tag{10.28}$$

The forms in (10.27) are examples of a general form, sometimes called the *controllable canonical form*, in which the polynomial coefficients appear in one row of the state matrix, the state for that row is the only one driven by the input, and 1s appear to the right of the main diagonal. The row with the coefficients doesn't have to be the bottom. If we had numbered the derivatives of z in reverse order, $x_3 = z$, $x_2 = \dot{z}$, and $x_1 = \ddot{z}$, the coefficients would be in the top row, and x_1 would be the directly driven state. This corresponds to multiplying the first set of states by a matrix

$$T = \begin{bmatrix} 0 & 0 & 1 \\ 0 & 1 & 0 \\ 1 & 0 & 0 \end{bmatrix} \tag{10.29}$$

Can you find a transformation that puts the coefficients into column 1? What does it do to the input and output matrices?

Now let's look at the partial-fraction expansion form in (10.23). In this form the output is the sum of three simple first-order processes, each of which can be described by one state variable. We have two direct options here—we can make the relation between state and output simple or that between state and input simple. Taking the first path, let

$$y(t) = x_1(t) + x_2(t) + x_3(t) \tag{10.30}$$

where $X_1(s) = 50U(s)/s$, $X_2(s) = -31.25U(s)/(s + 2)$, $X_3(s) = -18.75U(s)/(s + 50)$. Transforming each of these to the time domain yields

$$\dot{x}_1 = 50u(t) \qquad \dot{x}_2 = -2x_2 - 31.25u(t) \qquad \dot{x}_3 = -50x_3 - 18.75u(t)$$

These equations are represented by the matrix set

$$A = \begin{bmatrix} 0 & 0 & 0 \\ 0 & -2 & 0 \\ 0 & 0 & -50 \end{bmatrix} \quad B = \begin{bmatrix} 50 \\ -31.25 \\ -18.75 \end{bmatrix}$$

$$C = [1 \quad 1 \quad 1] \tag{10.31}$$

The other path begins by writing $y(t) = 50x_1 - 31.25x_2 - 18.75x_3$. This will lead, as you may verify, to the same **A** and new **B** and **C** matrices that are the transpose of the old **C** and **B** matrices, respectively. Each of these is an example of the diagonalized form of the state equations in which the state matrix has the system poles on its diagonal and zeros elsewhere, and each state is directly driven from the input.

Suppose the transfer function has a repeated root. For example,

$$\frac{Y(s)}{U(s)} = \frac{100(1 + s/10)}{(1 + s/2)^2 (1 + s/100)} \tag{10.32}$$

$$= \frac{4000(s + 10)}{(s + 2)^2 (s + 100)}$$

$$= \frac{-37.484}{s + 100} + \frac{-317.1}{s + 2} + \frac{326.53}{(s + 2)^2} \tag{10.33}$$

Again, we can write the output as the sum of three states. To simplify the internal representation, I will put the constants in the output matrix:

$$y(t) = -37.484x_1 - 317.1x_2 + 326.53x_3 \tag{10.34}$$

$$X_1(s) = \frac{U(s)}{s + 100} \qquad X_2(s) = \frac{U(s)}{s + 2}$$

$$X_3(s) = \frac{X_2(s)}{s + 2} \tag{10.35}$$

You see that the squared pole is taken care of by relating it to the state for the first-degree pole. If the highest power of a pole is k, there will be a sequence of k terms of increasing power in the partial-fraction expansion and a chain of k states related to each other in a way similar to the derivative sequence in the controllable canonical form. Transforming (10.35) to the time domain yields

$$\dot{x}_1 = -100x_1 + u(t) \qquad \dot{x}_2 = -2x_2 + u(t) \qquad \dot{x}_3 = x_2 - 2x_3 \tag{10.36}$$

The system matrices are

$$\mathbf{A} = \begin{bmatrix} -100 & 0 & 0 \\ 0 & -2 & 0 \\ 0 & 1 & -2 \end{bmatrix} \qquad \mathbf{B} = \begin{bmatrix} 1 \\ 1 \\ 0 \end{bmatrix}$$

$$\mathbf{C} = [-37.484 \quad -317.1 \quad 326.53] \tag{10.37}$$

As in the simple pole case, the poles appear on the diagonal of the state matrix, but this time there is an off-diagonal 1 to represent the coupling between the state variables for the multiple pole. In this form, the input directly drives the state variable for the simple pole and the first state variable in the chain for the multiple pole.

If x_2 and x_3 are interchanged, so that x_3 is the state variable for the single pole at -2 and x_2 is that for the double pole, the state and input matrices will be

$$\mathbf{J} = \begin{bmatrix} -100 & 0 & 0 \\ 0 & -2 & 1 \\ 0 & 0 & -2 \end{bmatrix} \qquad \mathbf{B} = \begin{bmatrix} 1 \\ 0 \\ 1 \end{bmatrix} \qquad (10.38)$$

This is known as *Jordan normal form*, and the square with the repeated pole is a *Jordan block*. Again, observe that only the first state in the chain is directly driven by the input.

To summarize and generalize, some of the easier ways to convert transfer functions to state equations are

1. Conversion to the controllable canonic form, in which the denominator coefficients appear in a row of the state matrix and the numerator coefficients appear in the output matrix. If

$$\frac{Y(s)}{U(s)} = \frac{b_1 s^{n-1} + \cdots + b_n}{s^n + a_1 s^{n-1} + \cdots + a_n} \qquad (10.39)$$

   Then we can have

$$\mathbf{A} = \begin{bmatrix} 0 & 1 & 0 & \cdots & 0 \\ 0 & 0 & 1 & \cdots & 0 \\ & \cdot & \cdot & \cdots & \\ 0 & 0 & 0 & \cdots & 1 \\ -a_n & -a_{n-1} & \cdot & \cdots & -a_1 \end{bmatrix} \qquad \mathbf{B} = \begin{bmatrix} 0 \\ \cdot \\ \cdot \\ 0 \\ 1 \end{bmatrix}$$

$$\mathbf{C} = [b_{n-1} \quad \cdots \quad b_1] \qquad (10.40)$$

2. Partial-fraction expansion to explicit-pole forms.

   (*a*) If the poles are all simple, the state matrix will be a diagonal matrix with the poles on its diagonal. The expansion coefficients can appear in either the input or output matrices. If

$$\frac{Y(s)}{U(s)} = \sum_{k=1}^{n} \frac{c_k}{s - p_k} \qquad (10.41)$$

   then we can have

$$
A = \begin{bmatrix} p_1 & 0 & \cdots & 0 \\ 0 & p_2 & 0 & \cdot & 0 \\ \cdot & & \cdot & & \cdot \\ \cdot & & & \cdot & \cdot \\ 0 & \cdot & \cdot & 0 & p_n \end{bmatrix} \qquad B = \begin{bmatrix} 1 \\ \cdot \\ \cdot \\ \cdot \\ 1 \end{bmatrix}
$$

$$
C = [c_1 \quad c_2 \quad \cdots \quad c_n] \tag{10.42}
$$

(b) If some multiple poles are present, the partial-fraction sequences they generate can be represented by chains of coupled states, with none but the first variable in each chain directly driven by the input. Each chain will then produce a Jordan block in the state matrix. Let $Y_m(s)$ be a repeated-pole expansion sequence portion of $Y(s)$ with multiplicity m. Then

$$
\frac{Y_m(s)}{U(s)} = \sum_{k=1}^{m} \frac{c_k}{(s - p)^k} \tag{10.43}
$$

This gives an m-row portion of the state and input matrices as

$$
A = \begin{bmatrix} \cdot & \cdot & \cdot & & & 0 & & \\ \cdot & p & 1 & 0 & 0 & \cdots & 0 & \cdot \\ \cdot & 0 & p & 1 & 0 & \cdots & 0 & \cdot & 0 \\ 0 & \cdot & 0 & 0 & p & 1 & \cdots & 0 & \cdot \\ \cdot & \cdot & \cdot & \cdot & \cdot & \cdots & \cdot & \cdot \\ \cdot & 0 & 0 & 0 & 0 & \cdots & p & \cdot \\ & & & 0 & & & & \cdot & \cdot \end{bmatrix} \qquad B = \begin{bmatrix} \cdot \\ 0 \\ \cdot \\ 0 \\ 1 \\ \cdot \end{bmatrix}
$$

$$
C = [\cdots \quad c_1 \quad c_2 \quad \cdots \quad c_m \quad \cdots] \tag{10.44}
$$

The examples and results given above are all for SISO cases. Let's look at a MIMO case in which the plant is described by a transfer function matrix. The following description of a boiler plant was given by Mayr in 1973:

$$
Y(s) = \begin{bmatrix} \dfrac{1}{1 + 4s} & \dfrac{0.7}{1 + 5s} & \dfrac{0.3}{1 + 5s} & \dfrac{0.2}{1 + 5s} \\[2mm] \dfrac{0.6}{1 + 5s} & \dfrac{1}{1 + 4s} & \dfrac{0.4}{1 + 5s} & \dfrac{0.35}{1 + 5s} \\[2mm] \dfrac{0.35}{1 + 5s} & \dfrac{0.4}{1 + 5s} & \dfrac{1}{1 + 4s} & \dfrac{0.6}{1 + 5s} \\[2mm] \dfrac{0.2}{1 + 5s} & \dfrac{0.3}{1 + 5s} & \dfrac{0.7}{1 + 5s} & \dfrac{1}{1 + 4s} \end{bmatrix} U(s) \tag{10.45}
$$

Looking at this, you might think that there are only two different poles, so there ought to be just two state variables required. If you try it, you will quickly discover that there isn't enough flexibility to meet all the cross-couplings, nor a way to account for the four different inputs. What is required is a way to account for the transfer from each input to all the outputs. This is done by looking at the corresponding column and deciding how many states are represented in it. For example, input u_1 drives all the outputs through column 1. There are two different poles in that column, so if I define two states, I will be able to account for all the transfers from input 1. It so happens that the same is true of the other three columns, so that one needs a total of eight states to represent this system. A set of equations for doing this is

$$\dot{x}_1 = -0.25x_1 + u_1$$
$$\dot{x}_2 = -0.2x_2 + u_1$$
$$\dot{x}_3 = -0.25x_3 + u_2$$
$$\dot{x}_4 = -0.2x_4 + u_2$$
$$\dot{x}_5 = -0.25x_5 + u_3$$
$$\dot{x}_6 = -0.2x_6 + u_3$$
$$\dot{x}_7 = -0.25x_7 + u_4$$
$$\dot{x}_8 = -0.2x_8 + u_4$$

(10.46)

$$y_1 = x_1 + 0.7x_4 + 0.3x_6 + 0.2x_8$$
$$y_2 = 0.6x_2 + x_3 + 0.4x_6 + 0.35x_8$$
$$y_3 = 0.35x_2 + 0.4x_4 + x_5 + 0.6x_8$$
$$y_4 = 0.2x_2 + 0.3x_4 + 0.7x_6 + x_7$$

The system matrices are

$$A = \begin{bmatrix} -0.25 & 0 & 0 & 0 & 0 & 0 & 0 & 0 \\ 0 & -0.2 & 0 & 0 & 0 & 0 & 0 & 0 \\ 0 & 0 & -0.25 & 0 & 0 & 0 & 0 & 0 \\ 0 & 0 & 0 & -0.2 & 0 & 0 & 0 & 0 \\ 0 & 0 & 0 & 0 & -0.25 & 0 & 0 & 0 \\ 0 & 0 & 0 & 0 & 0 & -0.2 & 0 & 0 \\ 0 & 0 & 0 & 0 & 0 & 0 & -0.25 & 0 \\ 0 & 0 & 0 & 0 & 0 & 0 & 0 & -0.2 \end{bmatrix}$$

$$\mathbf{B} = \begin{bmatrix} 1 & 0 & 0 & 0 \\ 1 & 0 & 0 & 0 \\ 0 & 1 & 0 & 0 \\ 0 & 1 & 0 & 0 \\ 0 & 0 & 1 & 0 \\ 0 & 0 & 1 & 0 \\ 0 & 0 & 0 & 1 \\ 0 & 0 & 0 & 1 \end{bmatrix} \qquad (10.47)$$

$$\mathbf{C} = \begin{bmatrix} 1 & 0 & 0 & 0.7 & 0 & 0.3 & 0 & 0.2 \\ 0 & 0.6 & 1 & 0 & 0 & 0.4 & 0 & 0.35 \\ 0 & 0.35 & 0 & 0.4 & 1 & 0 & 0 & 0.6 \\ 0 & 0.2 & 0 & 0.3 & 0 & 0.7 & 1 & 0 \end{bmatrix}$$

Each input drives two states, and each output is a combination of four states, one from each of the pairs driven by an input. In general, if there are r inputs, and the jth column of the transfer function matrix has n_j poles that must be used in a state submatrix, this approach will lead to a total number of states

$$n = \sum_{j=1}^{r} n_j \qquad (10.48)$$

The state matrix will be block diagonal, which is useful for determining controllability by inspection, as you will see in the next section. The submatrices of $\mathbf{A}$ don't have to be in explicit-pole form, but it is likely to be a good idea to factor all the polynomials of the transfer function matrix to find the poles so as not to have unnecessary states.

10.4   CONTROLLABILITY AND OBSERVABILITY

The systems we examined up through Chap. 9 were all designed to make the output response some desired function, either a constant or a replica of the input signal. So long as the behavior was stable and within linear operating ranges, we were not too concerned about the internal variables. In fact, in some cases we attempted to cover a pole with a zero. This has the effect of making the pole transparent at the output. The effective transfer function is determined by the zeros and poles left that don't cancel. Although cascade compensation of this type is possible for large systems, it is more difficult to design than the state feedback method which will be presented in the next chapter. Some other aspects of pole cancelation are explored in the problems at the end of this chapter. To use state feedback, it is necessary that the state variables be drivable from the inputs and that they be measurable or findable from the outputs. While this seems intuitively obvious, a formal definition of these qualities has been very useful, both as a means of determining whether or not a given system can be designed with state feedback and in providing a mechanism for the design itself.

10.4.1 Controllability

A system is *controllable* if and only if it can be driven from any given initial state to any given final state in a finite time. No limit is placed on the input amplitude for this purpose. Although the necessary and sufficient conditions for a system to satisfy this definition can be derived in a few mathematical lines, I feel it is interesting and instructive to try a few simple cases and work toward the general result. A better physical insight and some extra information will be your reward for this approach. Consider first a single-state-variable system.

$$\dot{x} = ax + bu \qquad (10.49)$$

Since we are only concerned with the input and state, no output definition is needed. We know that the motion of this system is described by

$$x(t) = x_0 e^{at} + \int_0^t e^{a(t-\tau)} bu(\tau) \, d\tau \qquad (10.50)$$

The starting state x_0 is arbitrary, and we can define a final state $x_f = x(t_f)$, also arbitrary. With x and t defined, it appears sufficient to let $u(t)$ be a constant u_1 over the time interval. Making all these substitutions in (10.49),

$$x_f = x_0 e^{at_f} + \int_0^{t_f} e^{a(t_f-\tau)} bu_1 \, d\tau$$

$$x_f - x_0 e^{at_f} = bu_1 \int_0^{t_f} e^{a(t_f-\tau)} \, d\tau$$

Doing the integral and solving for u_1 yields

$$u_1 = \frac{a(x_f - x_0 e^{at_f})}{b(e^{at_f} - 1)} \qquad (10.51)$$

Evidently, a constant input is good enough. The only condition the system description has to meet for this solution is that b isn't zero. There is no stability requirement. Whether a is positive or negative, a suitable input still exists. Notice that if $|at_f|$ becomes small compared to 1, u_1 becomes inverse to the time interval. That is, the shorter the time allowed to get between the states, the harder the system must be driven. Once the final state is achieved, it can be held by another constant input, $u_2 = -ax_f/b$, which you can find by setting $\dot{x} = 0$ in (10.49).

Next, I will try some second-order systems. First, let's look at an explicit-pole state matrix with distinct poles:

$$\dot{x}_1 = p_1 x_1 + b_{11} u_1(t) + b_{12} u_2(t)$$

$$\dot{x}_2 = p_2 x_2 + b_{21} u_1(t) + b_{22} u_2(t) \tag{10.52}$$

If rank $(\mathbf{B}) = 2$, $\mathbf{B}^{-1}$ exists and the controls can be separated by a network relating new inputs $\mathbf{v}$ to $\mathbf{u}$ by $\mathbf{u} = \mathbf{B}^{-1}\mathbf{v}$. Then x_1 is driven by v_1, and x_2 is driven by v_2. This gives us two independent first-order systems, and the solutions are the same as (10.51), with appropriate subscripts. In fact, if we wanted to test the solution it wouldn't be necessary to construct the network, only to calculate $\mathbf{u}$ after finding $\mathbf{v}$ and apply $\mathbf{u}$ to the original system. Also, in this case, it isn't even necessary that p_1 be different from p_2.

If rank $(\mathbf{B}) = 1$, the only other sensible choice, then one column of $\mathbf{B}$ is a multiple of the other. This means that $u_1(t)$ will affect the two states in the same proportion as $u_2(t)$. For example, if $u_1(t)$ drives x_1 twice as hard as it drives x_2, then $u_2(t)$ also drives x_1 twice as hard as it drives x_2. This means that one input is redundant and can be made zero. The system is then reducible to a single-input two-state system:

$$\dot{x}_1 = p_1 x_1 + b_1 u(t)$$

$$\dot{x}_2 = p_2 x_2 + b_2 u(t) \tag{10.53}$$

The solution to these equations is

$$x_1(t) = x_{01} e^{p_1 t} + \int_0^t e^{p_1(t-\tau)} b_1 u(\tau)\, d\tau$$

$$x_2(t) = x_{02} e^{p_2 t} + \int_0^t e^{p_2(t-\tau)} b_2 u(\tau)\, d\tau \tag{10.54}$$

If u is a constant, an arbitrary final state cannot be satisfied, since (10.54) will give us two algebraic equations in one unknown. Since we don't have diversity in space (two independent inputs) we can try diversity in time. Suppose u is one value for half the time interval and another value for the other half. Let

$$u(t) = u_1 \qquad \text{for } 0 \le t < t_f/2$$

$$\quad = u_2 \qquad \text{for } t_f/2 < t \le t_f \tag{10.55}$$

where u_1 and u_2 are constants. The convolution integral breaks up into segments for which the input is constant. The basic integral is

$$\int_a^b e^{p_i(t_f - \tau)} \, d\tau = e^{p_i t_f} \int_a^b e^{-p_i \tau} \, d\tau$$

$$= e^{p_i t_f} \frac{(e^{-p_i a} - e^{-p_i b})}{p_i} \tag{10.56}$$

Putting $t = t_f$, $x_{fi} = x_i(t_f)$, (10.55), and (10.56) into (10.54) produces

$$x_{fi} - x_{0i} e^{p_i t_f} = \frac{b_i}{p_i} e^{p_i t_f} [(1 - e^{-p_i t_f/2}) u_1$$

$$+ (e^{-p_i t_f/2} - e^{-p_i t_f}) u_2] \qquad i = 1, 2 \tag{10.57}$$

In order to have the coefficients of u_j in one equation independent of those in the other, p_1 must be different from p_2. In order to drive each state, b_1 and b_2 must be nonzero, but they can be equal to each other since they don't affect the linear independence of the equations.

If $| p_i t_f | \ll 1$, the coefficients of the u_j are small, and u_j will be inverse to t_f, as in the single-state-variable case. With two states having a diagonal state matrix, we see that the system is controllable if either rank $(\mathbf{B}) = 2$ or the poles are distinct. In the first case, it takes one step to reach the final state, and in the second case it takes two steps. Again, stability is not a requirement. What about holding the state? The constant-value (zero-derivative) condition is

$$0 = \mathbf{A}\mathbf{x}_f + \mathbf{B}\mathbf{u}_h \tag{10.58}$$

If we have two independent inputs, then

$$\mathbf{u}_h = -\mathbf{B}^{-1}\mathbf{A}\mathbf{x}_f \tag{10.59}$$

Otherwise, the final state can't be held unless it is zero. Of course, if the system is unstable, noise will drive it off equilibrium and corrective action (feedback) will have to be taken to put it back.

This process generalizes for larger systems with diagonal state matrices. For an nth-order system, if rank $(\mathbf{B}) = n$, we have essentially n independent first-order systems and can drive the system between any two states in one step. If rank $(\mathbf{B}) = 1$, we essentially have a single-input system so that we can divide the time interval into n sections, generating n equations for the u_j which are independent and solvable providing the poles are distinct, and we can drive the system between any two states in n steps. For intermediate rank values, we can expect that segmenting the interval will work with fewer segments required. Equation (10.59) gives the input set that holds the final state if rank $(\mathbf{B}) = n$.

In Sec. 10.3, you saw that a system with repeated poles in its characteristic equation has an explicit-pole state equation form with Jordan blocks instead of being simply diagonal. What is required for such a subsystem to meet the controllability definition? To see the dynamics let's look at another second-order example. Let

$$\dot{x}_1 = px_1 + x_2$$

$$\dot{x}_2 = px_2 + bu(t) \tag{10.60}$$

In this system, $u(t)$ is the forcing function for x_2, and x_2 is the forcing function for x_1. The general solution is

$$x_2(t) = x_{02}e^{pt} + \int_0^t e^{p(t-\tau)} bu(\tau)\, d\tau$$

$$x_1(t) = x_{01}e^{pt} + \int_0^t e^{p(t-\tau)} bx_2(\tau)\, d\tau \tag{10.61}$$

As before, if u is made constant, no value can be found which will satisfy the two scalar conditions of (10.61). Let's try the two-step function (10.55) again. In order to use x_{f1}, we need the whole time function for $x_2(t)$, which will be in parts corresponding to the two input steps:

$$x_2(t) = x_{02}e^{pt} + \frac{b}{p} e^{pt} (1 - e^{-pt})u_1 \qquad 0 \le t < t_f/2$$

$$= x_{02}e^{pt} + \frac{b}{p} e^{pt}[(1 - e^{-pt_f/2})u_1$$

$$+ (e^{-pt_f/2} - e^{-pt})u_2] \qquad t_f/2 < t \le t_f \tag{10.62}$$

The response of x_1 due to the initial value of x_2 is

$$x_{1u}(t_f) = x_{02}t_f e^{pt_f} \tag{10.63}$$

To find the response of x_1 to the forcing function, replace t by t_f in (10.61) and t by τ in (10.62) before substituting into (10.61). Since $x_2(t)$ changes form at $t_f/2$, the integral has to be split. Call that part of $x_2(t)$ due to the forcing function $x_{2d}(t)$. Then

$$\int_0^{t_f/2} e^{p(t_f-\tau)} x_{2d}(\tau)\, d\tau = \frac{b}{p} e^{pt_f}\left[\frac{t_f}{2} + \frac{1}{p}(e^{pt_f/2} - 1)\right]u_1$$

$$\int_{t_f/2}^{t_f} e^{p(t_f-\tau)} x_{2d}(\tau)\, d\tau = \frac{b}{p} e^{pt_f}$$

$$\times \left\{ (1 - e^{-pt_f/2}) \frac{t_f}{2} u_1 + e^{-pt_f/2} \left[\frac{t_f}{2} + \frac{1}{p} (e^{-pt_f/2} - 1) \right] u_2 \right\}$$

The final state equations are

$$\frac{p}{b} (x_{f1} e^{-pt_f} - x_{01} - x_{02} t_f) = \left[t_f - \frac{1}{p} + \left(\frac{1}{p} - \frac{t_f}{2} \right) e^{-pt_f/2} \right] u_1$$

$$+ e^{-pt_f/2} \left(\frac{t_f}{2} - \frac{1}{p} + \frac{1}{p} e^{-pt_f/2} \right) u_2$$

$$\frac{P}{b} (x_{f2} e^{-pt_f} - x_{02}) = (1 - e^{-pt_f/2}) u_1 + e^{-pt_f/2} (1 - e^{-pt_f/2}) u_2 \quad (10.64)$$

Equation (10.64) could be simplified, but the objective here is to show that the two-step control works for this case. The coefficients of the u_j in the two equations are independent, and the only requirement for controllability is that b be nonzero. Also, one can see the trend for the general case. Since each state variable derivative is driven by the next state in the chain, each equation for the final state value will have integrals of the time solution of the next lower state variable in the chain, introducing higher powers of t_f which will guarantee linear independence of the coefficients from the other equations. All that is necessary is to use n steps in $u(t)$. An arbitrary final state can't be held because the input only drives the bottom state in the chain.

Now it's time to summarize what we've seen so far. For the explicit-pole state representations, the state can be moved between two conditions in at most n steps, providing the following conditions are met.

Case	Controllable if
$n = 1$	b nonzero.
$n = 2$	rank $(\mathbf{B}) = 2$ or $\mathbf{B}$ has no zeros and $\mathbf{A}$ has distinct poles or the system is a Jordan block.
$n > 2$	rank $(\mathbf{B}) = n$, or if rank $(\mathbf{B}) = q < n$, which means that q of the state variables can be driven directly, the poles for the remaining $n - q$ state variables must be distinct from each other and the pole(s) of the directly driven variables, or be part of Jordon blocks. $\mathbf{B}$ can have all-zero rows only in Jordan blocks.

The basic mathematical property we're looking for is linear independence of n equations so we can solve for n input constants whether they are spread over several inputs or over time or both. Linear independence is classically tested by calculating a determinant. It would be good to have a determinant that reflects the conditions given above rather than requiring explicit calculation of the equation coefficients. Consider the distinct-pole case. For $n = 2$,

$$\begin{bmatrix} b_{11} & b_{12} & p_1 b_{11} & p_1 b_{12} \\ b_{21} & b_{22} & p_2 b_{21} & p_2 b_{22} \end{bmatrix}$$

has rank 2 if either rank $(\mathbf{B}) = 2$ (columns 1 and 2) or if the poles are distinct (columns 2 and 3, for instance). A matrix for testing a third-order system can be constructed by using the square of the state matrix to generate columns of $p_i^2 b_{jk}$ which will be distinct from the others if the poles are distinct. This leads to the general form

$$[\mathbf{B}_p \quad \mathbf{PB}_p \quad \mathbf{P}^2\mathbf{B}_p \quad \cdots \quad \mathbf{P}^{n-1}\mathbf{B}_p]$$

where $\mathbf{P}$ is the explicit-pole state matrix and $\mathbf{B}_p$ is its input matrix, as a test matrix. You may verify that this works for the Jordan block as well.

It isn't necessary to convert an original state description to the explicit-pole form in order to use the controllability test. As I said earlier, a given state matrix can be converted to this form by a suitable matrix multiplication (coordinate transformation). Suppose the given state description is

$$\dot{\mathbf{x}} = \mathbf{Ax} + \mathbf{Bu}(t)$$

and that a matrix $\mathbf{S}$ relates $\mathbf{x}$ to the explicit-pole state by

$$\mathbf{x} = \mathbf{Sx}_p$$

Then $\mathbf{S\dot{x}}_p = \mathbf{ASx}_p + \mathbf{Bu}(t)$,

$$\dot{\mathbf{x}}_p = \mathbf{S}^{-1}\mathbf{ASx}_p + \mathbf{S}^{-1}\mathbf{Bu}(t)$$

Since $\mathbf{x}_p$ is the explicit-pole state vector,

$$\mathbf{P} = \mathbf{S}^{-1}\mathbf{AS} \quad \text{and} \quad \mathbf{B}_p = \mathbf{S}^{-1}\mathbf{B}$$

The controllability matrix general term is

$$\mathbf{P}^k\mathbf{B}_p = (\mathbf{S}^{-1}\mathbf{AS})^k\mathbf{S}^{-1}\mathbf{B} = \mathbf{S}^{-1}\mathbf{ASS}^{-1}\mathbf{AS}\cdots\mathbf{S}^{-1}\mathbf{ASS}^{-1}\mathbf{B} = \mathbf{S}^{-1}\mathbf{A}^k\mathbf{B}$$

Since $\mathbf{S}^{-1}$ is a common factor to every term and it has rank n, it may be removed from the test matrix without affecting the rank. In general then, the controllability matrix may be defined as

$$\mathbf{C}_t = [\mathbf{B} \quad \mathbf{AB} \quad \mathbf{A}^2\mathbf{B} \quad \cdots \quad \mathbf{A}^{n-1}\mathbf{B}] \qquad (10.65a)$$

Constructing the controllability matrix is an iterative process similar to the series construction for the transition matrix. In this case, one knows how many matrix multiplications are to be done, and the dimensions of the results are determined by the dimension of **A** and **B**.

Algorithm 10.6   *The Controllability Matrix*

Let **A** be the $n \times n$ state matrix and **B** be the $n \times r$ input matrix. Let **Ct** be the controllability matrix, $n \times nr$, and **Bp** be an intermediate product matrix. Then either

$$\mathbf{Ct} = \mathbf{Bp} = \mathbf{B}.$$

For $k = 2$ to n, $\mathbf{Bp} = \mathbf{ABp}.$

Attach **Bp** to the right of **Ct**.

Next k.

or

$$\mathbf{Ct}\ [1 \text{ to } n, 1 \text{ to } nr] = 0.$$

$\mathbf{Ct}[*, 1 \text{ to r}] = \mathbf{Bp} = \mathbf{B}.$

For $k = 2$ to n, $\mathbf{Bp} = \mathbf{ABp}.$

$\mathbf{Ct}[*, 1 + (k - 1)\text{ r to } kr] = \mathbf{Bp}$

Next k.

Once the controllability matrix is formed, one needs to find its rank, and sometimes it is useful to know which columns (or rows) of a matrix are linearly independent. Although the rank of an $n \times n$ matrix can be between 0 and n, I am interested only in the question of n or not. To find the answer, I can select n columns at a time and find their determinant from Algorithm 10.1 and (10.19). If the determinant is zero, I make another selection and try again. I keep this up until either a nonzero determinant appears or all possible selections are exhausted. In mechanizing this process, one needs a systematic way to select the columns. I chose to think of the column positions as entries in a vector of binary values. A selected column is represented by a 1 and a not-selected column by a 0. The selection process steps through all the binary numbers, checking each number to see if the total number of 1s is n and, if so, finding the determinant of the selected columns. These ideas are embodied in the following two algorithms.

Algorithm 10.7   *Decimal Integer to Binary Vector Conversion*

This is an adaptation of the standard divide-by-2 method which can be found in any logic circuits text. Let the function name be BIN, with input K and output Z.

Let M = round-up ($\log_2 K$).

Let **Z** be an M vector of zeros; $T = K, j = M$.

Loop: $T = T/2$. $Z[j]$ = round-up(T − round-down (T))

T = round-down (T); $j = j - 1$.

If $j > 0$ go to Loop.

Done.

Algorithm 10.8   Rank (Independent Columns) of a Matrix

Let **A** be an $M \times N$ matrix, with $M \le N$. Assume a function which returns the determinant of a square matrix, D = DENT **F**, D the output and **F** the input.

Let **C** be a vector of integers from 1 to N. Let $K = 2^N, J = 1$.

Test: **B** = BIN J. If sum(B) $<>$ M go to NXTJ.

R = **B** followed by N − length (**B**) zero elements.

R = elements (**C**) with the same index as nonzero elements of **R**.

If | DENT $A[*, \mathbf{R}]$ | > 0 (or some small number) then return **R**.

NXTJ: $J = J + 1$. If $K > J$ go to Test.

R = 0.

10.4.2   Observability

A system is *observable* if its initial state can be determined from output measurements over a finite time. Presumably, if the input isn't zero, we can calculate its effect on the output and subtract it to get the initial-condition response. Then we might as well set the input to zero and consider just the unforced system:

$$\dot{\mathbf{x}} = \mathbf{Ax} \qquad \mathbf{y} = \mathbf{Cx} \qquad (10.65b)$$

Again, although a few lines of matrix calculus will derive the conditions on **A** and **C** for observability, I will work through a few low-order cases to bring out the dynamics of the measurement. Besides, they're simpler than the corresponding controllability problems.

Consider the scalar case:

$$\dot{x} = ax \qquad y = cx \qquad (10.66)$$

The response is

$$x(t) = x_0 e^{at} \qquad y(t) = cx_0 e^{at} \qquad (10.67)$$

so that one measurement at a time t_1 will nail down the value of x_0. The only requirement of the system is that c be nonzero.

Next, consider a diagonal system:

$$\dot{x}_1 = p_1 x_1 \qquad \dot{x}_2 = p_2 x_2$$

$$y_1 = c_{11} x_1 + c_{12} x_2$$

$$y_2 = c_{21} x_1 + c_{22} x_2 \tag{10.68}$$

The responses for the state variables are

$$x_1 = x_{01} e^{p_1 t} \qquad x_2 = x_{02} e^{p_2 t}$$

so that

$$y_1 = c_{11} x_{01} e^{p_1 t} + c_{12} x_{02} e^{p_2 t}$$

$$y_2 = c_{21} x_{01} e^{p_1 t} + c_{22} x_{02} e^{p_2 t} \tag{10.69}$$

If the rank of C is 2, one measurement of the output vector at t_1 will be enough to solve (10.69) for the initial state. If the rank of C is 1, there is really only one independent output, so two measurements at different times will be needed:

$$y(t_1) = c_1 x_{01} e^{p_1 t_1} + c_2 x_{02} e^{p_2 t_1}$$

$$y(t_2) = c_1 x_{01} e^{p_1 t_2} + c_2 x_{02} e^{p_2 t_2} \tag{10.70}$$

You will find by writing the determinant that a solution is possible only if the c_i are both nonzero and the poles are distinct. For a diagonal system, it is important that no column of C be zeros, since that means the corresponding state could not affect the output. In general, for a diagonal system with distinct poles $n + 1 -$ rank (C) vector measurements will be enough to determine the initial state.

An unforced second-order Jordan block has the state equations

$$\dot{x}_1 = p x_1 + x_2 \qquad \dot{x}_2 = p x_2 \tag{10.71}$$

From the calculation for the forced case, (10.61) and (10.63), the response is

$$x_2 = x_{02} e^{pt} \qquad x_1 = x_{01} e^{pt} + x_{02} t e^{pt} \tag{10.72}$$

Since the coefficient of x_{02} in the x_1 response includes t as well as the exponential, it can be made distinct by taking measurements at two different times. Therefore, even if C has rank 1, and measures only x_1 directly, one can still distinguish the two initial state values. For the extreme case,

$$y(t_1) = c_1 e^{pt_1}(x_{01} + t_1 x_{02})$$

$$y(t_2) = c_1 e^{pt_2}(x_{01} + t_2 x_{02}) \tag{10.73}$$

The determinant of these equations is just $t_2 - t_1$, which substantiates the discussion. For a Jordan block, the minimum conditions for state feedback to work are that the input drive the lowest (highest-numbered) state variable in the chain and that the output be connected to the highest (lowest-numbered) variable in the chain. This is complete state controllability and observability.

The conditions on the output and state matrices turn out to be the same as those for the controllability problems. The same test matrix form can be used, although the commonest form for the observability test matrix is generated by using C, basically a row matrix, rather than C^τ. The standard form is

$$\mathbf{O}_t = \begin{bmatrix} \mathbf{C} \\ \mathbf{CA} \\ \mathbf{CA}^2 \\ \cdot \\ \cdot \\ \cdot \\ \mathbf{CA}^{n-1} \end{bmatrix} \tag{10.74}$$

Since the rank of a matrix is unchanged by matrix transposition, the algorithms for test matrix formation and independent column selection can also be used to generate $\mathbf{O}_t$ and select independent rows.

10.5   SECOND- AND THIRD-ORDER EXAMPLES

I have delayed doing examples until the end of this chapter so that I could display all the properties and functions we've studied for each example system. I hope you haven't waited too long.

■ Example 10.1

This is a second-order system described by the matrices

$$\mathbf{A} = \begin{bmatrix} -2 & 1 \\ -10 & -5 \end{bmatrix} \quad \mathbf{B} = \begin{bmatrix} 0 \\ 10 \end{bmatrix}$$

$$\mathbf{C} = \begin{bmatrix} 1 & 0 \end{bmatrix} \tag{10.75}$$

The state matrix is full and has a nonzero determinant. The input drives only the second state variable, and the output measures only the first state variable.

$$s\mathbf{I}_2 - \mathbf{A} = \begin{bmatrix} s+2 & -1 \\ 10 & s+5 \end{bmatrix}$$

$$D(s) = | s\mathbf{I}_2 - \mathbf{A} | = s^2 + 7s + 20 \qquad (10.76)$$

The poles are at $-3.5 \pm j2.784$. It's easy enough to calculate $\mathbf{\Phi}(s) = [s\mathbf{I} - \mathbf{A}]^{-1}$ directly, but for illustration I will use Algorithm 10.1.

$$\mathbf{H}_0 = \mathbf{I}_2 = \begin{bmatrix} 1 & 0 \\ 0 & 1 \end{bmatrix} \qquad a_1 = -\text{trace}(\mathbf{A}) = 7$$

$$\mathbf{H}_1 = \mathbf{A}\mathbf{H}_0 + a_1\mathbf{I}_2 = \begin{bmatrix} -2 & 1 \\ -10 & -5 \end{bmatrix} + \begin{bmatrix} 7 & 0 \\ 0 & 7 \end{bmatrix} = \begin{bmatrix} 5 & 1 \\ -10 & 2 \end{bmatrix}$$

$$a_2 = \frac{-\text{trace}(\mathbf{A}\mathbf{H}_1)}{2} = \frac{-\text{trace}\begin{bmatrix} -20 & 0 \\ 0 & -20 \end{bmatrix}}{2} = 20$$

$$\mathbf{\Phi}(s) = \frac{s\mathbf{H}_0 + \mathbf{H}_1}{D(s)} = \frac{\begin{bmatrix} s+5 & 1 \\ -10 & s+2 \end{bmatrix}}{s^2 + 7s + 20} \qquad (10.77)$$

The transfer from the input to the state is

$$\mathbf{H}_s(s) = \mathbf{\Phi}(s)\mathbf{B} = \frac{1}{D(s)} \begin{bmatrix} 10 \\ 10(s+2) \end{bmatrix} = \frac{10}{D(s)} \begin{bmatrix} 1 \\ s+2 \end{bmatrix} \qquad (10.78)$$

Notice the zero at -2 in the transfer function to x_2. The transfer function from input to output is

$$\mathbf{G}_p(s) = \mathbf{C}\mathbf{\Phi}(s)\mathbf{B} = \frac{1}{D(s)} 10 \qquad (10.79)$$

To find $\mathbf{\phi}(t)$ and the various time responses, I will use the following Laplace transform pairs taken from a table:

$$\frac{b}{s^2 + 2as + a^2 + b^2} \leftrightarrow e^{-at} \sin bt \qquad (10.80)$$

$$\frac{s + a}{s^2 + 2as + a^2 + b^2} \leftrightarrow e^{-at} \cos bt \qquad (10.81)$$

For our problem, $a = 3.5$ and $b = \sqrt{7.75} = 2.784$. To inverse-transform the diagonal terms of $\mathbf{\Phi}(s)$, I put $s + 5 = s + 3.5 + 1.5$ and $s + 2 = s + 3.5 - 1.5$. With these rearrangements to fit the transforms,

$\phi(t) =$

$$\begin{bmatrix} e^{-3.5t}[\cos{(2.78t)} + 0.528 \sin{(2.78t)}] & 0.3521e^{-3.5t} \sin{(2.78t)} \\ -3.521e^{-3.5t} \sin{(2.78t)} & e^{-3.5t}[\cos{(2.78t)} - 0.528 \sin{(2.78t)}] \end{bmatrix} \qquad (10.82)$$

Likewise, the state impulse response matrix is

$$\mathbf{H}_s(t) = \begin{bmatrix} 3.521e^{-3.5t} \sin{(2.78t)} \\ 10e^{-3.5t}[\cos{(2.78t)} - 0.528 \sin{(2.78t)}] \end{bmatrix} \qquad (10.83)$$

To obtain the system step response matrix, I need the following integrals:

$$\int e^{ax} \sin{(bx)} \, dx = e^{ax} \frac{a \sin{(bx)} - b \cos{(bx)}}{a^2 + b^2} \qquad (10.84a)$$

$$\int e^{ax} \cos{(bx)} \, dx = e^{ax} \frac{a \cos{(bx)} + b \sin{(bx)}}{a^2 + b^2} \qquad (10.84b)$$

Integrating $\mathbf{H}_s$ from 0 to t gives the state step response matrix:

$$\mathbf{IH}_s(t) = \begin{bmatrix} 0.5\{1 - e^{-3.5t}[\cos{(2.78t)} + 1.257 \sin{(2.78t)}]\} \\ 1 + e^{-3.5t}[2.335 \sin{(2.78t)} - \cos{(2.78t)}] \end{bmatrix} \qquad (10.85)$$

A lot of algebra has been used, and one should always look for checks. Does the steady-state value of the step response satisfy the original state equation when $\dot{\mathbf{x}} = 0$?

$$0 = \mathbf{A}\mathbf{x}_{ss} + \mathbf{B}\mathbf{u}_s \qquad (10.86)$$

I wrote APL functions to calculate the impulse and step responses from (10.83) and (10.85) with more accurate values than those shown for the constants. They are plotted in Figs. 10.4 and 10.5. I then used the transition matrix algorithm to generate the step response shown in Fig. 10.6. Both approaches used a 0.01 s time interval. The series for the transition matrix is tested with the sum-of-element-magnitudes norm and $\epsilon = 10^{-4}$. The series stopped after the t^3 term. It gave

$$\phi(0.01) = \begin{bmatrix} 0.9797136667 & 0.00965483333 \\ -0.0965483333 & 0.9507491667 \end{bmatrix}$$

should you wish to check it. Can you find any difference between Figs. 10.5 and 10.6?

Since this is a second-order system, the controllability matrix is only two terms, [**B**   **AB**]:

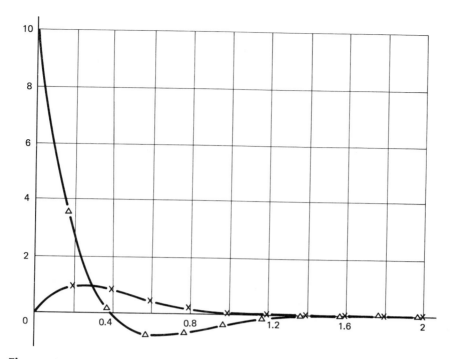

Figure 10.4 State impulse response for Example 10.1. $\times$, X_1; $\triangle$, X_2.

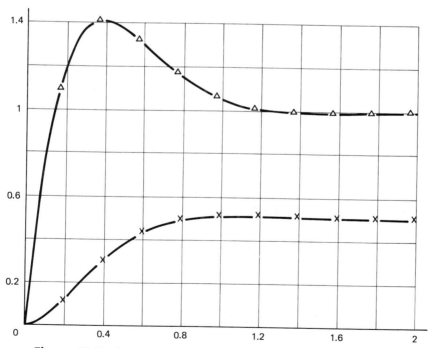

Figure 10.5 Step response for Example 10.1. $\times$, X_1; $\triangle$, X_2.

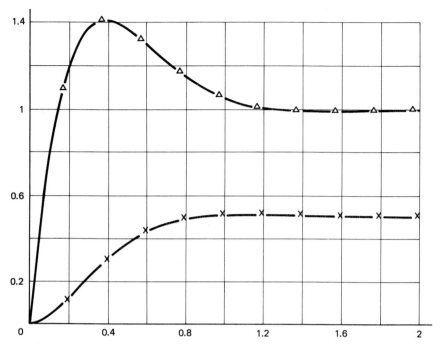

Figure 10.6   Step response for Example 10.1 calculated using a truncated-series transition matrix. $\times$, X_1; $\triangle$, X_2.

$$\mathbf{C}_t = \begin{bmatrix} 0 & 10 \\ 10 & -50 \end{bmatrix}$$ $|\,C_t\,|$ is not zero, so the system is controllable.

Likewise, the observability matrix is just

$$\mathbf{O}_t = \begin{bmatrix} \mathbf{C} \\ \mathbf{CA} \end{bmatrix} = \begin{bmatrix} 1 & 0 \\ -2 & 1 \end{bmatrix}$$

and has a nonzero determinant. The system is completely controllable and observable, and so it is suitable for state feedback.   ■

■ Example 10.2 Rosenbrock's Problem

This system has two inputs, two outputs, and three states. The matrices are

$$\mathbf{A} = \begin{bmatrix} -1 & 1 & 0 \\ 0 & -1 & 0 \\ 0 & 0 & -1 \end{bmatrix} \qquad \mathbf{B} = \begin{bmatrix} -\frac{1}{6} & 0 \\ \frac{2}{3} & 1 \\ 0 & \frac{1}{2} \end{bmatrix}$$

$$\mathbf{C} = \begin{bmatrix} 3 & -\frac{3}{4} & -\frac{1}{2} \\ 2 & -1 & 0 \end{bmatrix} \tag{10.87}$$

The state matrix shows a Jordan block for the first two states, and the third state is isolated from the other two. The third state is driven by the second input only and is measured by the first output only. The structure shows that all the poles are at -1. The input matrix and the output matrix are both rank 2, using any pair of rows from **B** and any pair of columns from **C**. This being so, the states will be controllable and observable. The state matrix is rank 3 and so invertible, which is handy for checking the steady-state response values. From our work with the examples in Sec. 10.4 on responses, I can write down the transition matrix:

$$\boldsymbol{\phi}(t) = \begin{bmatrix} e^{-t} & te^{-t} & 0 \\ 0 & e^{-t} & 0 \\ 0 & 0 & e^{-t} \end{bmatrix} \tag{10.88}$$

Nevertheless, let us go through the frequency response computation both for the instruction and because it's not too hard. Following Algorithm 10.1,

$$\mathbf{H}_0 = \begin{bmatrix} 1 & 0 & 0 \\ 0 & 1 & 0 \\ 0 & 0 & 1 \end{bmatrix} \qquad a_1 = -\text{trace}(\mathbf{A}) = 3$$

$$\mathbf{H}_1 = \mathbf{A}\mathbf{H}_0 + a_1 \mathbf{I}_3 = \begin{bmatrix} 2 & 1 & 0 \\ 0 & 2 & 0 \\ 0 & 0 & 2 \end{bmatrix}$$

$$a_2 = -0.5\,\text{trace}(\mathbf{A}\mathbf{H}_1) = 3$$

$$\mathbf{H}_2 = \mathbf{A}\mathbf{H}_1 + a_2 \mathbf{I}_3 = \begin{bmatrix} 1 & 1 & 0 \\ 0 & 1 & 0 \\ 0 & 0 & 1 \end{bmatrix}$$

$$a_3 = \frac{-\text{trace}(AH_2)}{3} = 1$$

$$D(s) = s^3 + 3s^2 + 3s + 1 = (s + 1)^3 \tag{10.89}$$

$$\boldsymbol{\Phi}(s) = \frac{s^2\mathbf{H}_0 + s\mathbf{H}_1 + \mathbf{H}_2}{D(s)}$$

$$= \begin{bmatrix} (s+1)^{-1} & (s+1)^{-2} & 0 \\ 0 & (s+1)^{-1} & 0 \\ 0 & 0 & (s+1)^{-1} \end{bmatrix} \tag{10.90}$$

You may recognize that (10.90) is the Laplace transform of (10.88).
The state impulse response matrix is

$$\mathbf{H}_s(t) = \boldsymbol{\phi}(t)\mathbf{B} = \begin{bmatrix} e^{-t}(4t - 1)/6 & te^{-t} \\ 2e^{-t}/3 & e^{-t} \\ 0 & 0.5e^{-t} \end{bmatrix} \tag{10.91}$$

The output impulse response matrix is

$$\mathbf{G}_p(t) = \mathbf{C}\boldsymbol{\phi}(t)\mathbf{B} = \begin{bmatrix} (2t - 1)e^{-t} & (3t - 1)e^{-t} \\ e^{-t}(4t - 3)/3 & (2t - 1)e^{-t} \end{bmatrix} \qquad (10.92)$$

To construct the step response matrices, we need two integrals, the first of which should be very familiar to you:

$$\int_0^t e^{-\tau} \, d\tau = 1 - e^{-t} \qquad (10.93)$$

$$\int_0^t \tau e^{-\tau} \, d\tau = 1 - (t + 1)e^{-t} \qquad (10.94)$$

Applying these integrals to (10.91) gives us the state step response matrix:

$$\mathbf{IH}_s(t) = \begin{bmatrix} [3 - (3 + 4t)e^{-t}]/6 & 1 - (t + 1)e^{-t} \\ 2(1 - e^{-t})/3 & 1 - e^{-t} \\ 0 & 0.5(1 - e^{-t}) \end{bmatrix} \qquad (10.95)$$

and the output step response matrix is

$$\mathbf{IG}_p(t) = \begin{bmatrix} 1 - (2t + 1)e^{-t} & 2 - (2 + 3t)e^{-t} \\ [1 - (1 + 4t)e^{-t}]/3 & 1 - (1 + 2t)e^{-t} \end{bmatrix} \qquad (10.96)$$

The steady-state state is

$$\mathbf{x}_{ss} = -\mathbf{A}^{-1}\mathbf{B}\mathbf{u}_s$$

$$\mathbf{x}_{ss} = \begin{bmatrix} 0.5 & 1 \\ \frac{2}{3} & 1 \\ 0 & 0.5 \end{bmatrix} \mathbf{u}_s \qquad (10.97)$$

and the steady-state output is $\mathbf{y}_{ss} = \mathbf{C}\mathbf{x}_{ss}$,

$$\mathbf{y}_{ss} = \begin{bmatrix} 1 & 2 \\ \frac{1}{3} & 1 \end{bmatrix} \mathbf{u}_s \qquad (10.98)$$

These values check with the functions when $t \to \infty$.

Again I wrote APL functions to plot the impulse and step responses from each input to the state and output variables. The plots are shown in Figs. 10.7–10.12. This is not a very exciting system, although those t terms from the Jordan block give the responses a little curl at the beginning. The series for the transition matrix stopped after the t^4 term. The time step is 0.1 s, which is a larger fraction of the dominant time constant than was the case in Example 10.1. The matrix is

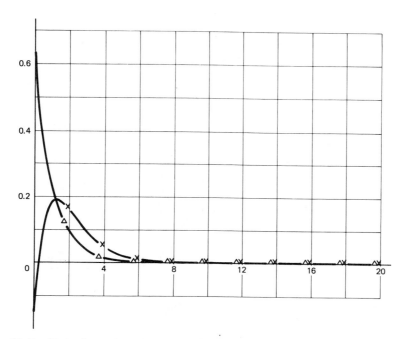

Figure 10.7   State impulse response for Example 10.2, $u_1 = 1$ and $u_2 = 0$. x, X_1; $\triangle$, X_2; $X_3 = 0$.

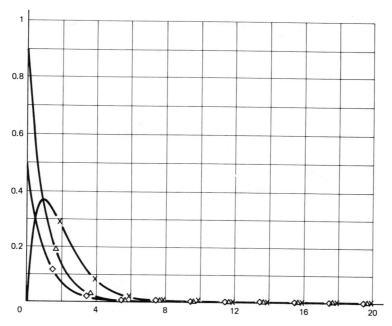

Figure 10.8   State impulse response for Example 10.2, $u_1 = 0$ and $u_2 = 1$. x, X1; $\triangle$, X2; $\diamond$, X3.

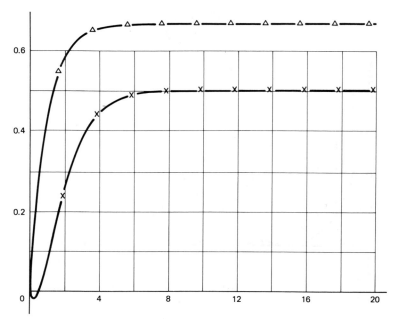

Figure 10.9 State step response for Example 10.2, $u_1 = 1$ and $u_2 = 0$. x, X1; △, X_2; $X_3 = 0$.

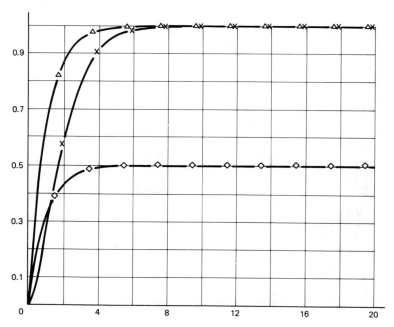

Figure 10.10 State step response for Example 10.2, $u_1 = 0$ and $u_2 = 1$. x, X_1; △, X2; ◇, X3.

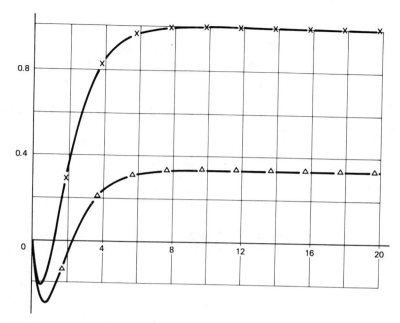

Figure 10.11 Output step response for Example 10.2, $u_1 = 1$ and $u_2 = 0$. x, Y1; △, Y2.

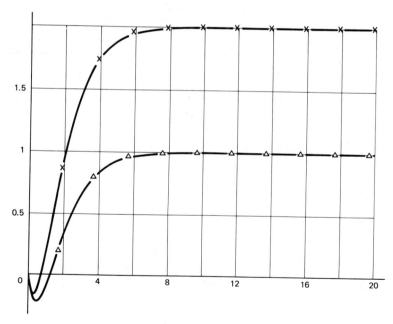

Figure 10.12 Output step response for Example 10.2, $u_1 = 0$ and $u_2 = 1$. x, Y1; △, Y2.

$$\varphi(0.1) = \begin{bmatrix} 0.9048375 & 0.09048333333 & 0 \\ 0 & 0.9048375 & 0 \\ 0 & 0 & 0.9048375 \end{bmatrix}$$

Figures 10.13–10.16 give the state and output step responses computed using the truncated transition matrix.

The controllability matrix has three terms this time, $[\mathbf{B} \quad \mathbf{AB} \quad \mathbf{A^2B}]$.

$$\mathbf{C}_t = \begin{bmatrix} -\frac{1}{6} & 0 & \frac{5}{6} & 1 & -1.5 & -2 \\ \frac{2}{3} & 1 & -\frac{2}{3} & -1 & \frac{2}{3} & 1 \\ 0 & 0.5 & 0 & -0.5 & 0 & 0.5 \end{bmatrix} \tag{10.99}$$

The rank algorithm says the first three columns are linearly independent. Their determinant, as given by Algorithm 10.1, is $\frac{2}{9}$. The observability matrix is

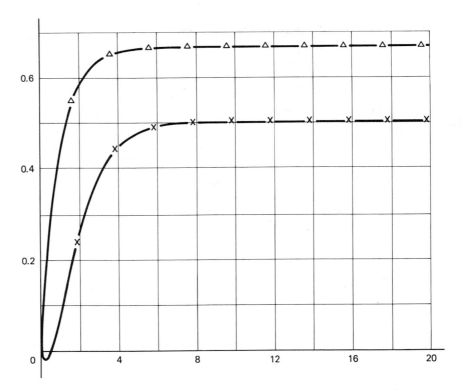

Figure 10.13   State step response for Example 10.2, $u_1 = 1$ and $u_2 = 0$, calculated with a truncated-series transition matrix. x, X_1; △, X_2; $X_3 = 0$.

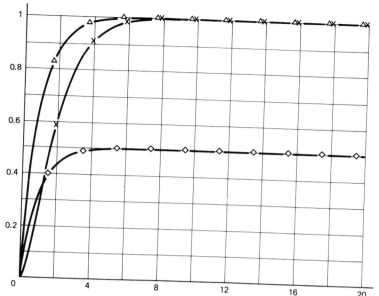

Figure 10.14   State step response for Example 10.2, $u_1 = 0$ and $u_2 = 1$, calculated using a truncated-series transition matrix. x, X1; △, X2; ◇, X3.

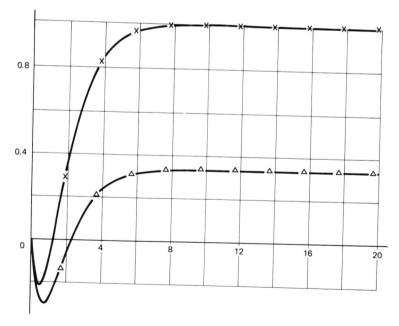

Figure 10.15   Output step response for Example 10.2, $u_1 = 1$ and $u_2 = 0$, calculated using a truncated-series transition matrix. x, Y_1; △, Y_2.

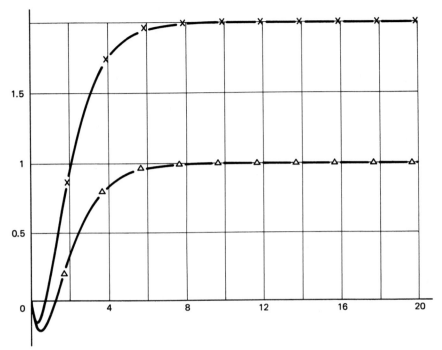

Figure 10.16 Output step response for Example 10.2, $u_1 = 0$ and $u_2 = 1$, calculated using a truncated-series transition matrix. x, Y1, $\triangle$, Y2.

$$
\mathbf{O}_t = \begin{bmatrix}
3 & -\frac{3}{4} & -0.5 \\
2 & -1 & 0 \\
-3 & 3.75 & 0.5 \\
-2 & 3 & 0 \\
3 & -6.75 & -0.5 \\
2 & -5 & 0
\end{bmatrix}
$$

The rank algorithm says the first three rows of $\mathbf{O}_t$ are linearly independent. Their determinant is -3. This confirms my preliminary assessment based on our work in Sec. 10.4. ■

FURTHER READING

An excellent treatment of state-space analysis and design can be found in the text and example problems of Chapters 5 and 6 of Ogata's book.

Ogata, Katsuhiko, *Discrete-Time Control Systems,* Englewood Cliffs, N.J., Prentice-Hall, 1987.

Some of the systems used in this chapter came from the following papers.

Davison, E. J. and Ferguson, I. J., "The Design of Controllers for the Multivariable Robust Servomechanism Problem Using Parameter Optimization Methods", *IEEE Trans.* vol. AC-26 no. 1, February 1981.

Hammarstrom, L. G. *et. al.*, "On Modeling Accuracy for Multivariable Distillation Control", *Chem. Eng'g. Commun.* vol 19 no. 1–3 pp 77–91, 1982.

PROBLEMS

Section 10.1

10.1   Make scalar signal diagrams for systems described by the following sets of matrices:

(a)
$$\mathbf{A} = \begin{bmatrix} -2 & 1 \\ 1 & -8 \end{bmatrix} \qquad \mathbf{B} = \begin{bmatrix} 1 & 0.1 \\ 0.2 & 5 \end{bmatrix}$$

$$\mathbf{C} = \begin{bmatrix} 1 & 0.1 \\ 0.1 & 1 \end{bmatrix}$$

(b)
$$\mathbf{A} = \begin{bmatrix} 0 & 1 & 0 \\ 0 & 0 & 0 \\ 0 & 0 & -1 \end{bmatrix} \qquad \mathbf{B} = \begin{bmatrix} 0 \\ 1 \\ 2 \end{bmatrix}$$

$$\mathbf{C} = \begin{bmatrix} 1 & 0 & 10 \\ 10 & 2 & 0 \end{bmatrix}$$

(c) $-\mathbf{A} = \mathbf{B} = \mathbf{C} = \mathbf{I}_3$

10.2   Suppose the input to a system is a set of ramp functions, $\mathbf{u}(t) = [r_i] \cdot \text{ramp}(t)$. Find the general form of the state response. Integration by parts may help.

10.3   Let $\mathbf{H}_s(t)$ be the matrix which gives the state response to a set of impulse inputs. Let $\mathbf{IH}(t)$ be the matrix which gives the state response to a set of step inputs. Suppose $u_1(t) = 10\,\text{imp}(t)$ and $u_2(t) = -0.5u(t)$ and the number of inputs is >2. Write an expression for the total state response. This is an example of superposition formulated for vector signals.

Section 10.2

10.4   Suppose the state matrix is

$$\mathbf{A} = \begin{bmatrix} 0 & 1 & 0 \\ 0 & 0 & 1 \\ -2 & -5 & -20 \end{bmatrix}$$

Form the series for the transition matrix up through the t^3 term. Use the largest element magnitude as the matrix norm. If $\epsilon = 10^{-4}$, how large can

t be to use just the first two terms? If the first three terms are used, how large can t be? How do these times compare to the system time constants?

10.5   Use Algorithm 10.1 to find the transition matrix transform for the state matrix in Prob. 10.4. If $\mathbf{B} = [0 \quad 0 \quad 1]^\tau$ and $\mathbf{C} = [1 \quad 0 \quad 0]$, what is the input-output transfer function?

10.6   Write a program (or use the function TRANSITION in the LSP) from Algorithm 10.2 to study the relationship between time interval, system time constants, and the number of terms at which the series for the transition matrix terminates. Specifically, let the state matrix be

$$\mathbf{A} = \begin{bmatrix} 0 & 1 \\ -ab & -(a+b) \end{bmatrix}$$

where $-a$ and $-b$ are the poles of the system transfer function. Choose $a = 1$, $b < a$, and plot the number of terms used against the time interval, up to 50 terms. For fixed b/a, is there a dependence on a? How small does b/a have to be for the choice of b to have a negligible effect?

10.7   Write programs based on Algorithms 10.2 and 10.4 to generate the unforced state response to an initial condition. If you use a language that requires it, set the state matrix maximum dimension at 10×10. Include user prompts and input statements to request the initial state, time step, and total time or number of steps. Save the resulting state sequence in a variable and prompt the user for a screen plot or printed output.

Section 10.3

10.8   Make state-space models of both types for each of the following transfer functions:

(a)
$$\frac{Y(s)}{U(s)} = \frac{25(1 + s/0.7)}{s(1 + s/0.05)(1 + s/5)}$$

(b)
$$\frac{Y(s)}{U(s)} = \frac{89.1}{s^2 + 6s + 8.91}$$

(c)
$$\frac{Y(s)}{U(s)} = \frac{22(s^2 + 2s + 1)}{(s^2 + 2s + 11.25)(s^2 + 6s + 58)}$$

10.9   Although complex poles are a small problem for hand calculations, they are more so for computer programs. If one has to provide for the possibility of complex matrix entries, it doubles the storage allocation and requires subroutines to handle complex matrix arithmetic. One should, therefore, keep to real numbers for the state-space model. If a transfer function has complex poles, real numbers can be obtained by avoiding the explicit-pole form, as you may have seen in the last problem. If you want the explicit-pole form as much as possible, the transfer function can be treated as a set of parallel blocks, one containing the complex pair and one for each of the real poles. This will lead to a block diagonal form. Model the transfer function below in this way.

$$\frac{Y(s)}{U(s)} = \frac{36}{(s^2 + 2s + 1.25)(s + 3.6)(s + 8)}$$

10.10 In an application which doesn't call for a special form for the state matrix, an easy approach for converting a transfer function, and incorporating it into a larger model, is to treat it as a cascade of blocks (possibly driven by a state from the larger model) whose output may be another state driving the larger model. The transfer function for Prob. 10.9 is shown in Fig. P10.10 labeled with this idea in mind. Write the state equations for this portion of the model.

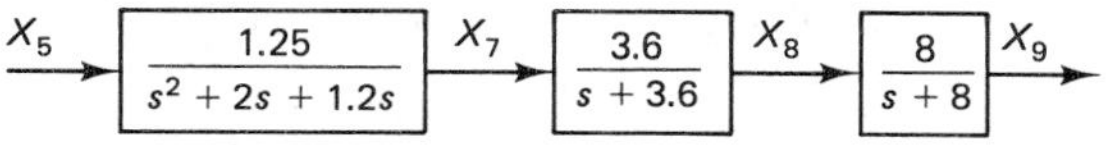

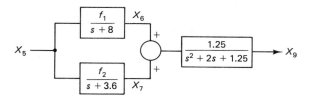

Figure P10.10

10.11 Figure P10.11 shows another block diagram arrangement for the transfer function in Prob. 10.9. Determine the constants f_1 and f_2. Write the state equations for this portion of the model.

Figure P10.11

10.12 A transfer function matrix model for a distillation column was given by Hammarstrom et al. in 1982 as

$$\begin{bmatrix} X_D \\ X_B \end{bmatrix} = \begin{bmatrix} \dfrac{12.8\exp(-s)}{16.7s + 1} & \dfrac{-18.9\exp(-3s)}{21s + 1} \\[2em] \dfrac{6.6\exp(-7s)}{10.9s + 1} & \dfrac{-19.5\exp(-3s)}{14.4s + 1} \end{bmatrix} \begin{bmatrix} U_1(s) \\ U_2(s) \end{bmatrix}$$

X_D is the top product composition, X_B is the bottom product composition, U_1 is the reflux flow rate, U_2 is the boil-up rate, and the time unit is minutes.

Model the exponentials by single-section all-pass transfer functions $(1 - as)/(1 + as)$. Choose a in each case so that a 90° phase lag occurs at the same frequency for the all-pass as it does for the exponential it replaces. Then develop a state-space model.

Section 10.4

10.13 Equation (10.51) gives the input value for a single-state system needed to drive it between two states.

(a) Derive an approximate expression for $|\,at_f\,| \ll 1$.

(b) Derive an approximate expression for $|\,at_f\,| \gg 1$, $x_f = 0$

(c) Let $x_f = 0$ and $x_0 = b = 1$. Plot u_1 against t_f for $0.1 \le t_f \le 10$ for the two cases $a = 1$ and $a = -1$.

10.14 Given the state description

$$\dot{x}_1 = -0.1x_1 + 10u_1 - 5u_2$$

$$\dot{x}_2 = -0.12x_2 - 3u_1 + 7u_2$$

(a) Decouple the inputs to give a new system which has states independently driven by new inputs v_1 and v_2.

(b) Write the expressions for the constant inputs needed to drive the system between two arbitrary states.

(c) If $t_f = 20$ and $\mathbf{x}_0 = [10 \quad -10]^\tau$, find v_1 and v_2 and then u_1 and u_2 needed to drive the system to the zero state.

(d) If $|\,u_j\,| \le 1$ and $t_f = 20$, what are the possible initial states that can be driven to the origin?

10.15 For the single-input two-state-variable case, what happens if the time interval is divided unevenly? Let $u(t) = u_1$, $0 \le t < t_r$, $= u_2$, $t_r < t \le t_f$. Substitute this input into (10.54) and solve for the input sequence that will drive the system between two arbitrary states. Are the conditions on solvability any different?

10.16 Given the system

$$\dot{x}_1 = -0.1x_1 + 10u(t)$$

$$\dot{x}_2 = -0.12x_2 + 7u(t)$$

(a) For $t_f = 20$ and $\mathbf{x}_0 = [10 \quad -10]$ find a two-step input sequence that will drive the system to the zero state.

(b) For a bounded input, $|\,u\,| \le 1$, $t_f = 20$, and equal time intervals, what is the set of initial conditions from which the system can be driven to the zero state?

(c) Find, for both unbounded and bounded inputs, the set of initial conditions that can be driven to the origin by one input value in 20 s or less.

10.17 Simplify (10.57) so that no common factors remain in the coefficients of the u_j. Solve the equations for u_1 and u_2.

(a) Write an approximate expression for $|\,p_i t_f\,| \ll 1$.

(b) Write an approximate expression for $|\,p_i t_f\,| \gg 1$, $x_f = 0$.

10.18 Simplify (10.64) by removing common factors from the coefficients of u_1 and u_2. Multiply through by p so that all the t_f terms are converted to pt_f.
   (a) What is the determinant for these equations?
   (b) Write approximate expressions for $|pt_f| \ll 1$.
   (c) Write approximate expressions for $|pt_f| \gg 1$, $x_f = 0$.

10.19 Construct the controllability test matrix for general second- and third-order Jordon blocks. What are the points that make the columns independent?

10.20 Construct the controllability test matrix for the general controllable canonical form single-input system. What is its determinant? What condition in the state matrix will make the determinant of the controllability test matrix zero? What is the physical meaning of this condition?

10.21 A system is described by the matrices

$$\mathbf{A} = \begin{bmatrix} -0.6 & 1 & 2 \\ 0 & -0.4 & 3 \\ 0 & 0 & -0.2 \end{bmatrix} \quad \mathbf{B} = \begin{bmatrix} 1 & 0 \\ 0.5 & -0.5 \\ 0 & 1 \end{bmatrix}$$

$$\mathbf{C} = \begin{bmatrix} 1 & 0 & -1 \\ 0 & -1 & 1 \end{bmatrix}$$

Construct the observability and controllability test matrices. Find sets of independent columns in $\mathbf{C}_t$ and rows in $\mathbf{O}_t$. Work from both ends of each matrix to see if there is more than one set of linearly independent vectors of the rank of the matrix.

10.22 Solve (10.69) for the initial state.

10.23 Solve (10.70) for the initial state.

10.24 Solve (10.74) for the initial state.

10.25 If the system poles are in the left half s plane, the response to an initial state will decay to zero as time goes by. Measurement devices have limited resolution, so the estimate of the initial state will become less accurate if the measurement is delayed too long. On the other hand, if the measurement is made too soon, it seems that the states might not be sufficiently distinct for an accurate determination. Let each output measurement have a fixed uncertainty δy. Let the uncertainties in the initial state variables for a two-state Jordan block be δx_1 and δx_2. Find measurement times t_1 and t_2 that minimize some convenient measure of the total state uncertainty, such as $(\delta x_1)^2 + (\delta x_2)^2$. Assume p is negative real.

10.26 This problem looks at controllability from a transfer function description viewpoint. Figure P10.26 shows a cascade of blocks with state variable labels on two of them. Write the state equations for this system. Write the

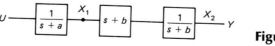

Figure P10.26

controllability and observability test matrices and find their ranks. From the diagram it is physically obvious that the pole at $-b$ is measurable. The cancelation in the overall transfer function by the zero causes lack of controllability.

10.27 Figure P10.27 is similar in purpose to Fig. P10.26. This time the canceled pole is driven by the input so that it is obviously controllable. Write the state equations using the labeled state variables. Construct the controllability and observability test matrices and determine their ranks. The zero should cause observability to fail.

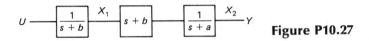

Figure P10.27

Chapter 11

STATE-SPACE DESIGN

11.1   POLE LOCATION BY STATE FEEDBACK

Figure 11.1 shows the general arrangement for state feedback. Conceptually, signals representing the n state variables are brought to a network in which each signal is, possibly, multiplied by r values and distributed to the r plant inputs. Each plant input is the sum of n contributions from the feedback network, labeled **K**, and a new control input labeled **v**. The network is represented by a matrix **K** with $n \times r$ elements. The element k_{ij} multiplies the ith state, and the result is added to the jth input. Since there are only n poles in an nth-order system, and we have $n \times r$ gain values at our disposal, we have more freedom than we need for pole adjustment. The extra flexibility is sometimes useful because we can generate different solutions and base a choice on some measure of cost-effectiveness, or we can choose to simplify the network so that only n gains need to be found. We will explore some of this later in this section. First we will look at the single-input case because it has just n gains to be set.

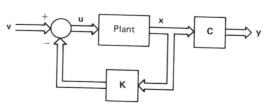

Figure 11.1   State feedback. **K** is a matrix representing a network of constant gains.

11.1.1   The Single-Input Case

Figure 11.2 shows a single-input plant with state feedback. Each state variable is multiplied by a single gain; then these signals are added up and subtracted from the new control signal v to form the plant input u. The matrix $\mathbf{K}$ is a row matrix, which is necessary for $\mathbf{Kx}$ to form a scalar signal. $\mathbf{B}$ is a column matrix which distributes the single input to drive the various states. The system equations are now

$$\dot{\mathbf{x}} = \mathbf{Ax} + \mathbf{Bu} \qquad u = v - \mathbf{Kx}$$
$$\dot{\mathbf{x}} = (\mathbf{A} - \mathbf{BK})\mathbf{x} + \mathbf{Bv} \tag{11.1}$$

The matrix $\mathbf{BK}$ is sometimes called an outer product of the vectors $\mathbf{B}$ and $\mathbf{K}$.

To make the discussion more concrete, let's consider a third-order system in controllable canonical form:

$$\mathbf{A} = \begin{bmatrix} 0 & 1 & 0 \\ 0 & 0 & 1 \\ -a_3 & -a_2 & -a_1 \end{bmatrix} \quad \mathbf{B} = \begin{bmatrix} 0 \\ 0 \\ 1 \end{bmatrix} \quad \mathbf{K} = [k_1 \quad k_2 \quad k_3] \tag{11.2}$$

I haven't specified an output matrix because we are not concerned with the output. We are assuming, for the time being, that all the state variables are directly measurable. The $\mathbf{BK}$ product is all zeros except for the bottom row:

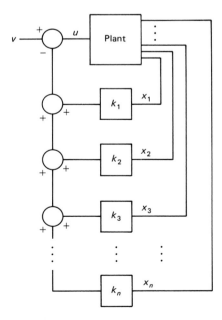

Figure 11.2   Single-input state feedback system.

$$\mathbf{BK} = \begin{bmatrix} 0 & 0 & 0 \\ 0 & 0 & 0 \\ k_1 & k_2 & k_3 \end{bmatrix} \tag{11.3}$$

The modified state matrix is

$$\mathbf{A} - \mathbf{BK} = \begin{bmatrix} 0 & 1 & 0 \\ 0 & 0 & 1 \\ -a_3 - k_1 & -a_2 - k_2 & -a_1 - k_3 \end{bmatrix} \tag{11.4}$$

The system polynomial is now

$$|s\mathbf{I}_3 - \mathbf{A} + \mathbf{BK}| = \begin{vmatrix} s & -1 & 0 \\ 0 & s & -1 \\ a_3 + k_1 & a_2 + k_2 & s + a_1 + k_3 \end{vmatrix}$$

$$= s^3 + (a_1 + k_3)s^2 + (a_2 + k_2)s + a_3 + k_1 \tag{11.5}$$

The desired poles can be formed into factors and multiplied out to give the desired system equation, $s^3 + d_1 s^2 + d_2 s + d_3$. Equating coefficients of the two polynomials will give the required k values:

$$d_1 = a_1 + k_3 \qquad d_2 = a_2 + k_2 \qquad d_3 = a_3 + k_1$$

$$k_1 = d_3 - a_3 \qquad k_2 = d_2 - a_2 \qquad k_3 = d_1 - a_1 \tag{11.6}$$

This is a beautifully simple solution and certainly justifies the name of the canonical form. You can see that the process generalizes to higher-order systems. The next question is, Does a transformation exist for adapting this solution to an arbitrary single-input state description?

Of course, you say. More important someone's even discovered it. Before I present the transformation though, let me show how it affects the solution. Call the transformation matrix $\mathbf{T}$. Let the plant be described by the standard symbols

$$\dot{\mathbf{x}} = \mathbf{Ax} + \mathbf{Bu} \qquad \mathbf{u} = \mathbf{v} - \mathbf{Kx}$$

$$\dot{\mathbf{x}} = (\mathbf{A} - \mathbf{BK})\mathbf{x} + \mathbf{Bv}$$

as before, where $\mathbf{x}$ is the set of state variables to which we actually have access. Let the controllable canonic state description be $\mathbf{x}_c = \mathbf{Tx}$. Then $\mathbf{x} = \mathbf{T}^{-1}\mathbf{x}_c$,

$$\mathbf{T}^{-1}\dot{\mathbf{x}}_c = (\mathbf{A} - \mathbf{BK})\mathbf{T}^{-1}\mathbf{x}_c + \mathbf{Bv}$$

$$\dot{\mathbf{x}}_c = \mathbf{T}(\mathbf{A} - \mathbf{BK})\mathbf{T}^{-1}\mathbf{x}_c + \mathbf{TBv}$$

$$= [\mathbf{TAT}^{-1} - (\mathbf{TB})(\mathbf{KT}^{-1})]\mathbf{x}_c + (\mathbf{TB})\mathbf{v} \tag{11.7}$$

Evidently the canonical form matrices are

$$\mathbf{A}_c = \mathbf{TAT}^{-1} \qquad \mathbf{B}_c = \mathbf{TB} \qquad \mathbf{K}_c = \mathbf{KT}^{-1} \qquad (11.8)$$

Define a row matrix of the desired polynomial's coefficients:

$$\mathbf{d} = [d_n \quad d_{n-1} \quad \cdots \quad d_1] \qquad (11.9)$$

and a similar row matrix for the original system polynomial:

$$\mathbf{a} = [a_n \quad a_{n-1} \quad \cdots \quad a_1] \qquad (11.10)$$

Then the solution for the controllable canonical system is

$$\mathbf{K}_c = [\mathbf{d} - \mathbf{a}] \qquad (11.11)$$

and that for the original system is

$$\mathbf{K} = [\mathbf{d} - \mathbf{a}]\mathbf{T} \qquad (11.12)$$

That certainly looks tidy doesn't it? Notation sometimes hides a lot. In this case, **T** is defined by its inverse, which is the product of two matrices. One is constructed from the original system polynomial coefficients, and the other is the controllability matrix $\mathbf{C}_t$. Define

$$\mathbf{W} = \begin{bmatrix} a_{n-1} & a_{n-2} & \cdots & a_1 & 1 \\ a_{n-2} & \cdots & & a_1 & 1 & 0 \\ \cdots & \cdots & & \cdots & \cdots & \cdot \\ a_1 & 1 & 0 & \cdots & 0 & 0 \\ 1 & 0 & 0 & \cdots & 0 & 0 \end{bmatrix} \qquad (11.13)$$

Then $\mathbf{T} = (\mathbf{C}_t\mathbf{W})^{-1}$.

The matrix **W** is not hard to build. The following algorithm is a description which can be followed easily in APL, and with a lot of looping in the more primitive languages.

Algorithm 11.1   The W Matrix

Let **W** be *a* row of n zeros followed by the coefficients a_{n-1} down to a_0 ($a_0 = 1$).

Duplicate the row n times to form an $n \times 2n$ matrix.
For $i = 1$ to n, shift row i to the left $i - 1$ positions.

Drop the first n columns.
The result is **W**.

Another solution for **K** is Ackermann's formula. Let $P(s)$ be the desired system polynomial. If one substitutes a matrix for s, one then has a polynomial matrix function whose result is a matrix. Denote this by **P(A)**. Then Ackermann's formula is

$$\mathbf{K} = \mathbf{B}_c{}^\tau \mathbf{C}_t^{-1} \mathbf{P(A)} \qquad (11.15)$$

$\mathbf{B}_c{}^\tau$ is a row of $n - 1$ zeros followed by a 1. An apparent advantage of Ackermann's formula is that it isn't necessary to find the original system polynomial. However, finding the plant's poles is a necessary element in understanding the system, so one is likely to do it even if it isn't necessary for finding the feedback gains.

11.1.2   An Extension to the Multiple-Input Case

A simple approach to the multiple-input system is to make it look like a single-input system as in Fig. 11.3. The state signals are weighted and summed into one signal as before, and this signal is then distributed to the inputs by an arbitrary set of gains. Although the minus signs are shown subtracting the feedback signals from the new input signals, they can be moved back to follow the k_{pi} gains. This

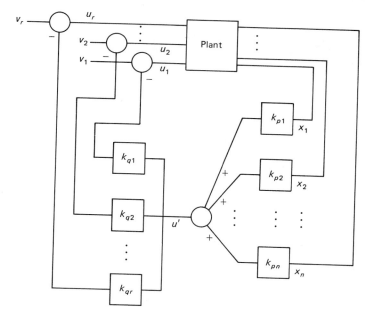

Figure 11.3 Multiple-input system with single-input style state feedback.

makes u' a virtual input line. The requirement for this scheme to work is that the plant seen from u' be controllable. A virtual input matrix $\mathbf{B}' = \mathbf{B}\mathbf{K}_q$ has to be tested with $\mathbf{A}$ for controllability. The original plant has to be controllable before this or any scheme is applied. Boiling the feedback down to one signal isn't going to help if the plant isn't controllable in the first place.

■ Example 11.1

$$\mathbf{A} = \begin{bmatrix} -1 & 1 & 0 \\ 0 & -1 & 1 \\ 0 & 0 & -1 \end{bmatrix} \quad \mathbf{B} = \begin{bmatrix} -\frac{1}{6} & 0 \\ \frac{2}{3} & 1 \\ 0 & \frac{1}{2} \end{bmatrix} \tag{11.16}$$

The state matrix is a single Jordan block, but the input matrix is not so simple. However, we can expect that the Jordan block is controllable from u_2. The controllability test matrix is

$$\mathbf{C}_t = \begin{bmatrix} -\frac{1}{6} & 0 & \frac{5}{6} & 1 & -1.5 & -1.5 \\ \frac{2}{3} & 1 & -\frac{2}{3} & -0.5 & \frac{2}{3} & 0 \\ 0 & 0.5 & 0 & -0.5 & 0 & 0.5 \end{bmatrix} \tag{11.17}$$

The determinant of the first three columns is $-\frac{2}{9}$, so the system is controllable. You may recognize that every other column, starting from the first, is due to a product formed from the first column in $\mathbf{B}$. Since the first column in $\mathbf{B}$ has the gains through which u_1 drives the state, I speak of this column as belonging to u_1. In this manner, columns 1, 3, and 5 in $\mathbf{C}_t$ belong to u_1 and measure the controllability of the system through that input. Likewise, columns 2, 4, and 6 measure the controllability from u_2. For this system, the determinant of columns 1, 3, and 5 is zero, so the system is not controllable from u_1 alone. The determinant of columns 2, 4, and 6 is $-\frac{1}{8}$, so the system can be controlled from u_2 alone. This being so, let us first try $\mathbf{K}_q = \begin{bmatrix} 0 & 1 \end{bmatrix}^\mathsf{T}$.

I choose as our objective locating the system poles at -1, -1, and -10. This will make the system look more like a critically damped second-order plant. The desired system polynomial is

$$P(s) = (s + 1)^2 (s + 10) = s^3 + 12s^2 + 21s + 10 \tag{11.18}$$

The virtual input matrix is $\mathbf{B}\mathbf{K}_q$, the second column in $\mathbf{B}$. I used Ackermann's formula to find $\mathbf{K}_p = \begin{bmatrix} 0 & 0 & 18 \end{bmatrix}$. The modified state matrix is

$$\mathbf{A} - \mathbf{B}\mathbf{K} = \begin{bmatrix} -1 & 1 & 0 \\ 0 & -1 & -17 \\ 0 & 0 & -10 \end{bmatrix} \tag{11.19}$$

Since the matrix is triangular, the poles are on the main diagonal and have the required values.

Figures 11.4 and 11.5 show the responses of the original plant to steps in u_1 and u_2, respectively. Since neither the other states nor u_1 drive x_3, the system has a second-order-type response to that input. Notice it takes about 4 s to come close to the final values. The response to u_2 is a bit quicker, and all three states take about the same time to close on the steady state. Figures 11.6 and 11.7 show the step responses for the modified system. The response to u_1 is the same, as it should be. Since x_3 isn't driven, it stays at zero, and the relationship between u_1 and the other states is the same as it was before. The response to u_2 is quite different. x_3 is speeded up because it has the higher-frequency pole, and the other two are slowed down to look like the second-order response. The result is that the response to each input is similar in its time course to the critically damped quadratic, while the third state is speeded up so that on the time scale of the slower states it appears to be nearly constant. Another important effect is that the steady-state values are different for the u_2 response with and without the state

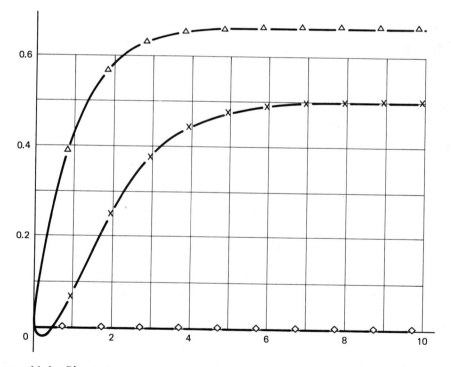

Figure 11.4 Plant step response for Example 11.1, $u_1 = 1$, $u_2 = 0$. ×, x_1; △, x_2; ◇, x_3.

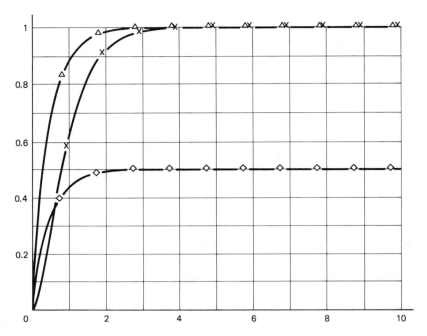

Figure 11.5 Step response for plant in Example 11.1, $u_1 = 0$, $u_2 = 1$. x, x_1; $\triangle$, x_2; $\lozenge$, x_3.

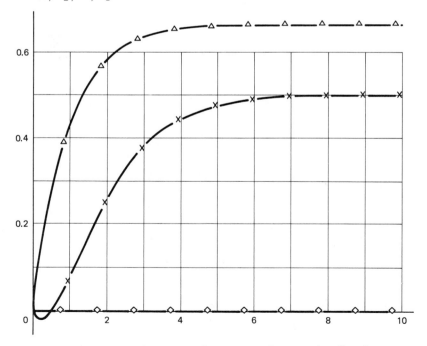

Figure 11.6 Step response for Example 11.1, with state feedback to a single input; $u_1 = 1$, $u_2 = 0$. x, x_1; $\triangle$, x_2; $\lozenge$, x_3.

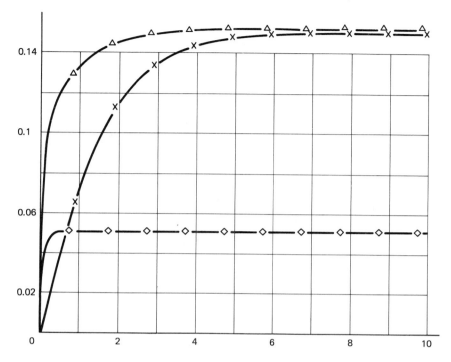

Figure 11.7   Step response for Example 11.1, with state feedback to a single input. $u_1 = 0$, $u_2 = 1$. x, x_1; $\triangle$, x_2; $\diamond$, x_3.

feedback. This shows that state feedback is a form of minor-loop compensation, and we have some more work to do before we achieve both specified closed-loop poles and a specified steady-state transfer.

Suppose we try driving both inputs with state feedback. That is, let's choose $\mathbf{K}_q = [1 \quad 1]^\tau$. Now the virtual input matrix is the sum across the rows of $\mathbf{B}$, $\mathbf{BK}_q = [-\frac{1}{6} \quad \frac{5}{3} \quad 0.5]^\tau$. It turns out that $\mathbf{K}_p = [0 \quad 0 \quad 18]$ again. It happens that in this case the same poles are formed whether the feedback is to one or both inputs. On reflection, this is sensible because only x_3 is being shifted, so that only x_3 is fed back. Feeding to one input or two only changed the coupling between x_3 and the other states. If the plant were not in explicit-pole form, the results would be quite different. The modified state matrix is now

$$\mathbf{A} - \mathbf{BK} = \begin{bmatrix} -1 & 0 & 3 \\ 0 & -1 & -29 \\ 0 & 0 & -10 \end{bmatrix} \quad (11.20)$$

Figures 11.8 and 11.9 show the responses to steps in u_1 and u_2 for this case. Again the response to u_1 is the same. The time course for the response to u_2 is similar in shape to that for the previous case, but the steady-state values are

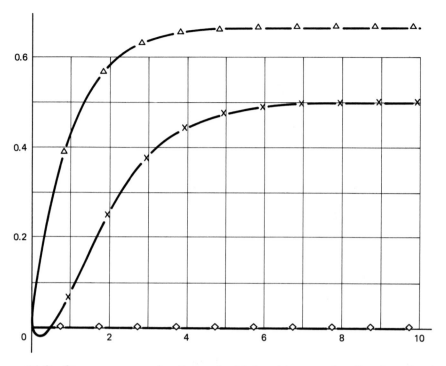

Figure 11.8   Step response for Example 11.1, with state feedback to both inputs. $u_1 = 1$, $u_2 = 0$. x, x_1; $\triangle$, x_2; $\diamond$, x_3.

different from those for the previous case and for the original plant because of the increased coupling between x_3 and the other two state variables. ∎

■ Example 11.2

You may have noticed that the matrices in Example 11.1 are very similar to those in Example 10.2. I simply added a 1 to complete the coupling between the state variables. Now I present the original state matrix for your consideration. Let

$$\mathbf{A} = \begin{bmatrix} -1 & 1 & 0 \\ 0 & -1 & 0 \\ 0 & 0 & -1 \end{bmatrix} \quad \mathbf{B} = \begin{bmatrix} -\frac{1}{6} & 0 \\ \frac{2}{3} & 1 \\ 0 & \frac{1}{2} \end{bmatrix} \quad (11.21)$$

The controllability matrix is

$$\mathbf{C}_t = \begin{bmatrix} -\frac{1}{6} & 0 & \frac{5}{6} & 1 & -1.5 & -2 \\ \frac{2}{3} & 1 & -\frac{2}{3} & -1 & \frac{2}{3} & 1 \\ 0 & 0.5 & 0 & -0.5 & 0 & 0.5 \end{bmatrix} \quad (11.22)$$

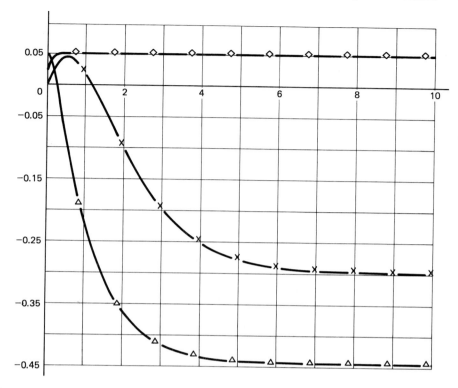

Figure 11.9 Step response for Example 11.1, with state feedback to both inputs. $u_1 = 0$, $u_2 = 1$. x, x_1; $\triangle$, x_2; $\diamond$, x_3.

The determinant of the first three columns is $\frac{2}{9}$, so the plant is controllable. The determinants for each group of columns belonging to a single input are both zero, which means the plant isn't controllable from either input. A single-input feedback won't work. How about a single line fed to both inputs? The independent pole, for x_3, isn't distinct from the Jordan block; consequently it violates one of the conditions needed for single-input controllability. No single-column virtual input matrix will form a controllable pair with **A**. A direct solution to this problem is to use a preliminary state feedback network to provide enough cross-coupling to make the system single-input-controllable. Figure 11.10 illustrates this approach. A cross-coupling network is chosen arbitrarily, and the composite system is tested for single-input controllability. If the composite plant has it, then the same single-input procedure as before is used to shift the poles. I will show several cases in this example. Try

$$\mathbf{K}_c = \begin{bmatrix} 1 & 0 & 1 \\ 0 & 1 & 0 \end{bmatrix} \tag{11.23}$$

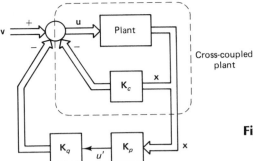

Figure 11.10 Applying a cross-coupling network $\mathbf{K}_c$ to produce single-input controllability.

The cross-coupled state matrix is

$$\mathbf{A}_c = \mathbf{A} - \mathbf{B}\mathbf{K}_c = \begin{bmatrix} -\frac{5}{6} & 1 & \frac{1}{6} \\ -\frac{2}{3} & -2 & -\frac{2}{3} \\ 0 & -0.5 & -1 \end{bmatrix} \tag{11.24}$$

The controllability matrix for the pair $\mathbf{A}_c,\mathbf{B}$ is

$$\mathbf{C}_c = \begin{bmatrix} -\frac{1}{6} & 0 & 0.8055 & 1.0333 & -1.949 & -3.402777 \\ \frac{2}{3} & 1 & -\frac{11}{9} & -\frac{7}{3} & 2.1296 & 4.6111 \\ 0 & 0.5 & -\frac{1}{3} & -1 & 0.9444 & 2.16667 \end{bmatrix} \tag{11.25}$$

The determinant of the first three columns is $\frac{2}{3}$. The determinant of columns 1, 3, and 5, belonging to u_1, is zero. The determinant of the columns belonging to u_2 is -0.41667, so the new plant is drivable from u_2. As a matter of interest, the poles of the new plant are -1, $-\frac{4}{3}$, and $-\frac{3}{2}$. If we stopped here, we'd have a stable plant. Now set $\mathbf{K}_q = [0 \quad 1]^\tau$ to represent a single feedback line to u_2. The virtual input matrix is the second column in $\mathbf{B}$. This gives $\mathbf{K}_p = [9.4, 19.8, 9.4]/3$. The total feedback is

$$\mathbf{K} = \mathbf{K}_c + \mathbf{K}_q\mathbf{K}_p = \begin{bmatrix} 1 & 0 & 1 \\ \frac{9.4}{3} & 6.6 & \frac{9.4}{3} \end{bmatrix} \tag{11.26}$$

The state matrix with all the feedback is

$$\mathbf{A} - \mathbf{B}\mathbf{K} = \begin{bmatrix} -\frac{5}{6} & 1 & \frac{1}{6} \\ -3.8 & -8.6 & -3.8 \\ -\frac{4.7}{3} & -3.8 & -\frac{7.7}{3} \end{bmatrix} \tag{11.27}$$

The poles, to five figures, are -1, -1, and -10, as required. The cross-coupling caused by the feedback quickly obscures the structure of the original plant description by making the state matrix dense.

The initial **K** I used above is not the only one that will make the plant single-input controllable.

$$\mathbf{K}_c = \begin{bmatrix} 0 & 1 & 0 \\ 1 & 0 & 1 \end{bmatrix} \tag{11.28}$$

also works. The cross-coupled state matrix is

$$\mathbf{A}_c = \begin{bmatrix} -1 & \frac{7}{6} & 0 \\ -1 & -\frac{5}{3} & -1 \\ -0.5 & 0 & -1.5 \end{bmatrix} \tag{11.29}$$

The controllability test matrix for the pair $\mathbf{A}_c, \mathbf{B}$ is

$$\mathbf{C}_c = \begin{bmatrix} -\frac{1}{6} & 0 & \frac{4.9}{9} & \frac{7}{6} & -2.0462963 & -3.69444 \\ \frac{2}{3} & 1 & -\frac{4}{9} & -\frac{13}{6} & 0.5462963 & 3.19444 \\ 0 & 0.5 & \frac{1}{12} & -\frac{3}{4} & -0.597222 & 0.541667 \end{bmatrix} \tag{11.30}$$

Now the determinant of the columns belonging to u_1 is 0.1759, and the columns belonging to u_2 have a determinant of zero. Choosing $\mathbf{K}_q = [1 \quad 0]^\tau$ to drive the u_1 input gives a $\mathbf{K}_p = [-9.316 \quad 9.421 \quad -9.316]$. The total feedback is

$$\mathbf{K} = \begin{bmatrix} -9.316 & 10.421 & -9.3116 \\ 1 & 0 & 1 \end{bmatrix}$$

The net state matrix is now

$$\mathbf{A} - \mathbf{BK} = \begin{bmatrix} -2.553 & 2.737 & -1.553 \\ 5.21 & -7.947 & 5.21 \\ -0.5 & 0 & -1.5 \end{bmatrix} \tag{11.31}$$

The poles are again -1, -1, and -10.

What happens if I choose all the elements of $\mathbf{K}_c$ to be unity? The cross-coupled state matrix is

$$\mathbf{A}_c = \begin{bmatrix} -\frac{5}{6} & \frac{7}{6} & \frac{1}{6} \\ -\frac{5}{3} & -\frac{8}{3} & -\frac{5}{3} \\ -0.5 & -0.5 & -1.5 \end{bmatrix} \tag{11.32}$$

The controllability test matrix for the pair $\mathbf{A}_c, \mathbf{B}$ is

$$\mathbf{C}_c = \begin{bmatrix} -\frac{1}{6} & 0 & \frac{5.5}{6} & 1.25 & -\frac{23}{9} & -\frac{16}{3} \\ \frac{2}{3} & 1 & -1.5 & -3.5 & \frac{26}{9} & \frac{28}{3} \\ 0 & 0.5 & -\frac{1}{4} & -\frac{5}{4} & \frac{2}{3} & 3 \end{bmatrix} \tag{11.33}$$

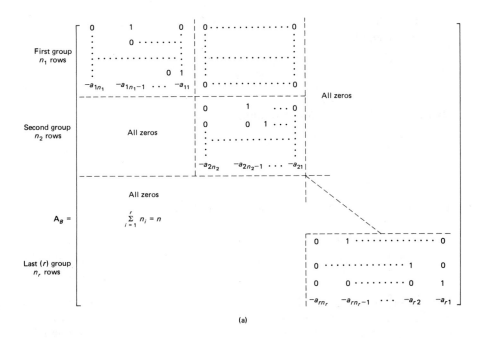

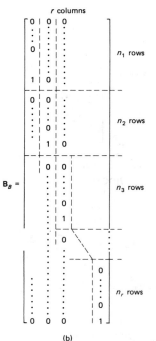

Figure 11.11 (a) Block-controllable state matrix.

The determinants are, for columns 1, 2, and 3: $\frac{2}{9}$; for columns 1, 3, and 5: 0.06481; and for columns 2, 4, and 6: -0.583333. We now have a plant which is controllable from either or both inputs. Suppose I choose to drive both inputs through $\mathbf{K}_q$. If I choose $\mathbf{K}_q = [1 \quad 1]^\tau$, then the controllability test through $\mathbf{B}'$ fails. If I choose $\mathbf{K}_q = [1 \quad -1]^\tau$, $\mathbf{A}_c, \mathbf{B}'$ are a controllable pair, so the design of $\mathbf{K}_p$ can proceed. $\mathbf{K}_p = [-5.8 \quad -9.4 \quad -5.8]$. The total feedback is

$$\mathbf{K} = \mathbf{K}_c + \mathbf{K}_q \mathbf{K}_p = \begin{bmatrix} -4.8 & -8.4 & -4.8 \\ 6.8 & 10.4 & 6.8 \end{bmatrix} \tag{11.34}$$

The net state matrix is

$$\mathbf{A} - \mathbf{BK} = \begin{bmatrix} -1.8 & -0.4 & -0.8 \\ -3.6 & -5.8 & -3.6 \\ -3.4 & -5.2 & -4.4 \end{bmatrix} \tag{11.35}$$

The poles are as required.

In this example, we have three different $\mathbf{K}$s that will give the same closed-loop pole values. In later sections, we will try to make use of the flexibility inherent in multiple-input state feedback. ■

11.1.3 Block-Controllable Form

A more reliable but complex procedure for reassigning the plant poles is to convert the model to block-controllable form. In this form, the state matrix is partitioned into blocks, each one of which represents a subsystem in controllable canonical form. These blocks are arranged along the main diagonal, with zeros everywhere else in the matrix. For a system with r inputs, there are r blocks. The input matrix has r columns, each of which has a 1 in the row appropriate to drive the lowest state in its block. Figure 11.11 shows the general forms of the state and input matrices. Once the model is transformed into this form, the feedback matrix is nearly as easy to find as in the single-input case. The hard part is obtaining the two transformation matrices.

I will not derive the transformations, but I will try to lead you through their construction and describe their effect in use. To begin, one must first construct the controllability test matrix $\mathbf{C}_t$ for the pair $\mathbf{A}, \mathbf{B}$, the original system description. The construction procedure assumes that $\mathbf{C}_t$ has rank n and that $\mathbf{B}$ has rank r. If $\mathbf{B}$ has a rank less than r, then there are some redundant inputs which can be made zero and the size of $\mathbf{B}$ reduced to its independent inputs. One must then construct a matrix from n independent columns in $\mathbf{C}_t$. The objective here is to choose the n columns from each of the r groups belonging to each input and to do it so that an approximately equal number of columns is in each group. The groups of columns are arranged together, and all the columns belonging to a particular input are put together in a submatrix. Figure 11.12 illustrates the sorting procedure.

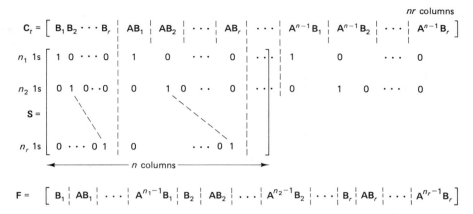

$$\mathbf{F} = \left[\begin{array}{c|c|c|c|c|c|c|c|c} \mathbf{B}_1 & \mathbf{AB}_1 & \cdots & \mathbf{A}^{n_1-1}\mathbf{B}_1 & \mathbf{B}_2 & \mathbf{AB}_2 & \cdots & \mathbf{A}^{n_2-1}\mathbf{B}_2 & \cdots & \mathbf{B}_r & \mathbf{AB}_r & \cdots & \mathbf{A}^{n_r-1}\mathbf{B}_r \end{array}\right]$$

Figure 11.12 Grouping n columns from the controllability test matrix.

First, generate a selection matrix $\mathbf{S}$ by constructing an $r \times r$ identity matrix and repeating it until the column count exceeds n. Line $\mathbf{S}$ up under $\mathbf{C}_t$, column for column. Mark off the first n columns of these matrices. The 1s in the first row in $\mathbf{S}$ mark the columns in $\mathbf{C}_t$ to form the first group in the result matrix $\mathbf{F}$. These are the first (approximately) n/r columns associated with u_1. Next, the second row in $\mathbf{S}$ marks the columns belonging to u_2 to be placed in the second group in

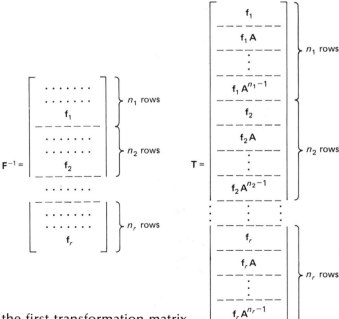

Figure 11.13 Formation of the first transformation matrix.

F. Notice that there is one and only one 1 in each column in **S**, and so there will be r groups in **F** with a total of n columns. The number in each group is equal to the number of 1s in the corresponding row in **S**, and this number is used later several times to sort and select, so it must be saved.

F is inverted, and its rows marked off into r groups, starting at the top and using the same numbers that marked the column groups in **F** starting at the left. The bottom row in each group is selected and used as the basis for generating a set of rows in a new matrix **T**, **T** is the first transformation matrix. Figure 11.13 shows the row selection from $\mathbf{F}^{-1}$ and the generation of **T**. Again, one uses the group numbers. The first row selected in $\mathbf{F}^{-1}$, $\mathbf{f}_1$, is n_1 rows down, and it is used to generate n_1 rows in **T**. The next row selected is $\mathbf{f}_2$, $m_2 = n_1 + n_2$ rows down in $\mathbf{F}^{-1}$, and it is used to generate n_2 rows in **T**.

Figure 11.14 shows the transforming action of **T** on **A** and **B**. The results are very similar to the forms we are looking for. The only difference is the possible

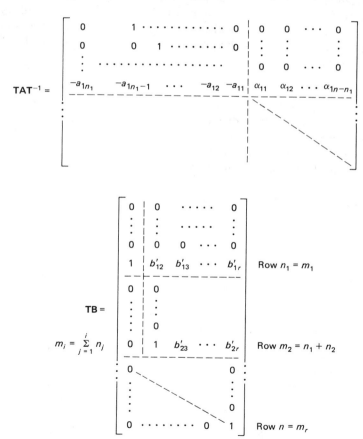

Figure 11.14 Action of **T** on the original plant matrices.

$$\mathbf{B}_T = \begin{bmatrix} 1 & b'_{12} & b'_{13} & \cdots & b'_{1r} \\ 0 & 1 & b'_{23} & \cdots & b'_{2r} \\ 0 & 0 & 1 & \cdots & b'_{3r} \\ & \cdots & \cdots & \cdots & \\ 0 & \cdots & \cdots & 0 & 1 \end{bmatrix}^{-1}$$

Figure 11.15 The second transformation matrix.

presence in the m_i rows of nonzero constants where there should be zeros. If they appear, these defects can be removed in two more operations. First, the nonzero rows in **TB** are collected and put into a matrix and inverted. The result is the second transformation matrix $\mathbf{B}_T$. This process is shown in Fig. 11.15. Using $\mathbf{B}_T$ to postmultiply **TB** cleans out the unwanted entries, as illustrated in Fig. 11.16. Each zero row in **TB** generates a new zero row in the same position when it is used to multipy $\mathbf{B}_T$. The m_ith row in **TB** is the ith row in $\mathbf{B}_T^{-1}$, so it produces zeros in the m_ith row of the product, except for a 1 in the ith column of that row because that entry is the product with the ith column of $\mathbf{B}_T$. Thus

$$\mathbf{B}_B = \mathbf{TBB}_T \tag{11.36}$$

The product of $\mathbf{B}_B$ and the transformed feedback matrix $\mathbf{K}_B$ is shown in Fig. 11.17. The m_ith row of $\mathbf{B}_B$, which corresponds to u_i, has a 1 in the ith position which picks out the ith row in $\mathbf{K}_B$ and places it in the m_ith row of the product. This row lines up with the proper row in $\mathbf{A}_B$ to give the controllable form for the ith block. The entries in the result read

$$\begin{aligned}(\mathbf{A}_B - \mathbf{B}_B\mathbf{K}_B)\,[i, *] = [&-a_{i,ni} - k'_{i1},\ -a_{i,ni-1} - k'_{i2},\ \ldots,\ -a_{i1} - k'_{in_i}, \\ &\alpha_{i1} - k'_{i,ni+1},\ \alpha_{i2} - k'_{i,ni+2},\ \ldots,\ \alpha_{i,n-ni} - k'_{in}]\end{aligned} \tag{11.37}$$

$$\underbrace{\begin{array}{r} \\ \text{Row } z_j \\ \\ \text{Row } n_1 \\ \\ \text{Row } m_i \\ \\ \\ \end{array}\begin{bmatrix} & \cdots & \\ 0 & \cdots\cdots & 0 \\ & \cdots & \\ 1 & \cdots\cdots & b'_{1r} \\ & \cdots & \\ 0 & \cdots\ 1\ \cdots & b'_{ir} \\ & \cdots & \\ 0 & \cdots\cdots & 1 \end{bmatrix}}_{\mathbf{TB}} \underbrace{\begin{bmatrix} 1 & \cdots\cdots & b'_{1r} \\ & \cdots & \\ 0 & \cdots\ 1\ \cdots & b'_{ir} \\ & \cdots & \end{bmatrix}^{-1}}_{\mathbf{B}_T} = \underbrace{\begin{bmatrix} 0 & \cdots\cdots & 0 \\ 1 & 0\ \cdots & 0 \\ & \cdots & \\ 0 & \cdots\ 1\ \cdots & 0 \\ & \cdots & \\ 0 & \cdots\cdots & 1 \end{bmatrix}}_{\mathbf{B}_B} \begin{array}{l} (\text{Row } z_j) \times \mathbf{B}_T \\ (\text{Row } n_1) \times \mathbf{B}_T \\ \\ (\text{Row } m_i) \times \mathbf{B}_T \\ \\ \\ \end{array}$$

Figure 11.16 Action of $\mathbf{B}_T$ on **TB**.

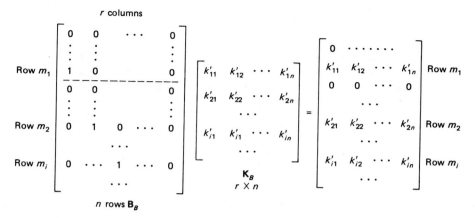

Figure 11.17 Product $\mathbf{B}_B \mathbf{K}_B$.

Evidently, we can remove the unwanted constants with the k' entries which aren't needed for pole shifting. The equations for the k' values are

$$k'_{ij} = d_{i,n_i+1-j} - a_{i,n_i+1-j} \qquad 1 \le j \le n_i \tag{11.38}$$

$$k'_{ij} = \alpha_{i,j-n_i} \qquad n_i + 1 \le j \le n \tag{11.39}$$

where d_{ij} is a coefficient of the desired polynomial for the ith block. In assigning the desired poles to the various blocks, it seems reasonable to factor the polynomials for the various subsystems and assign poles to the subsystems as close as possible to those already there. This is in line with the philosophy of making minimal alterations to the system.

At this point in the procedure, one has

$$\mathbf{A}_B - \mathbf{B}_B \mathbf{K}_B = \mathbf{TAT}^{-1} - \mathbf{TBB}_T \mathbf{K}_B$$

$$= \mathbf{T}(\mathbf{A} - \mathbf{B}(\mathbf{B}_T \mathbf{K}_B \mathbf{T}))\mathbf{T}^{-1} \tag{11.40}$$

If one chooses the actual feedback matrix to be

$$\mathbf{K} = \mathbf{B}_T \mathbf{K}_B \mathbf{T} \tag{11.41}$$

the actual system will be a coordinate transformation of the block-controllable description with the same poles.

■ Example 11.3

Let's look again at the system in Example 11.2. The selection matrix is

$$S = \begin{bmatrix} 1 & 0 & 1 \\ 0 & 1 & 0 \end{bmatrix} \tag{11.42}$$

Since there are two 1s in the first row, $n_1 = 2$, and there is only one 1 in the second row, so $n_2 = 1$. The row indexes are $m_1 = n_1 = 2$ and $m_2 = n_1 + n_2 = 3$. The matrix **F** will be columns 1 and 3 and then column 2 from C_t, arranged left to right.

$$F = \begin{bmatrix} -\frac{1}{6} & \frac{5}{6} & 0 \\ \frac{2}{3} & -\frac{2}{3} & 1 \\ 0 & 0 & 0.5 \end{bmatrix} \tag{11.43}$$

$$F^{-1} = \begin{bmatrix} 1.5 & 1.875 & -3.75 \\ 1.5 & 0.375 & -0.75 \\ 0 & 0 & 2 \end{bmatrix} \tag{11.44}$$

Selecting the rows for **T**,

$$\mathbf{f}_1 = \mathbf{F}^{-1}[m_1, *] = [1.5 \quad 0.375 \quad -0.75]$$

$$\mathbf{f}_2 = \mathbf{F}^{-1}[m_2, *] = [0 \quad 0 \quad 2]$$

$$\mathbf{T} = \begin{bmatrix} \mathbf{f}_1 \\ \mathbf{f}_1 \mathbf{A} \\ \mathbf{f}_2 \end{bmatrix}$$

$$= \begin{bmatrix} 1.5 & 0.375 & -0.75 \\ -1.5 & 1.125 & 0.75 \\ 0 & 0 & 2 \end{bmatrix} \tag{11.45}$$

$$\mathbf{T}^{-1} = \begin{bmatrix} 0.5 & -\frac{11}{6} & 0.25 \\ \frac{2}{3} & \frac{2}{3} & 0 \\ 0 & 0 & 0.5 \end{bmatrix} \tag{11.46}$$

Transforming,

$$\mathbf{TAT}^{-1} = \begin{bmatrix} 0 & 1 & 0 \\ -1 & -2 & 0 \\ 0 & 0 & -1 \end{bmatrix} \tag{11.47a}$$

$$\mathbf{TB} = \begin{bmatrix} 0 & 0 \\ 1 & 1.5 \\ 0 & 1 \end{bmatrix} \tag{11.47b}$$

The second and third rows of TB are the inverse of the second transformation matrix, B_T.

$$\mathbf{B}_T = \begin{bmatrix} 1 & -1.5 \\ 0 & 1 \end{bmatrix} \tag{11.48}$$

By the selection numbers, the first input will drive the upper left 2×2 part of the state matrix. This is in controllable form with coefficients for $(s + 1)^2$. Choosing the same desired poles as in the previous examples, this row needs no alteration. The second input drives the third row, which is a scalar system. I want to shift the pole from -1 to -10, which means I need to add -9 in that location. Therefore, the only nonzero entry in $\mathbf{K}_B$ is a 9 in the bottom right-hand corner:

$$\mathbf{K}_B = \begin{bmatrix} 0 & 0 & 0 \\ 0 & 0 & 9 \end{bmatrix} \tag{11.49}$$

The actual feedback network required is represented by

$$\mathbf{K} = \mathbf{B}_T \mathbf{K}_B \mathbf{T}$$

$$= \begin{bmatrix} 0 & 0 & -27 \\ 0 & 0 & 18 \end{bmatrix} \tag{11.50}$$

The net state matrix is

$$\mathbf{A} - \mathbf{BK} = \begin{bmatrix} -1 & 1 & -4.5 \\ 0 & -1 & 0 \\ 0 & 0 & -10 \end{bmatrix} \tag{11.51}$$

This is certainly a much cleaner result than the ones obtained by using preliminary cross-coupling. The net state matrix is still in explicit-pole form instead of obscured-pole (non-) form. ∎

11.2   REGULATOR DESIGN

The purpose of a regulator system is to maintain a set of outputs at specified constant value even in the presence of external disturbances to the system. Figure 11.18 shows a block diagram for a single-input single-output system with a disturbance acting at the output. An integrator is used to drive the plant to zero steady-state error against a constant disturbance, as we saw in Chap. 5. If $d(t) = d_0$ is a constant, then zero output error means that the signal to the left of the disturbance must be $r_0 - d_0$. If either the disturbance or the reference inputs change, a transient will occur whose features are determined by the transfer function of the process, but the output will settle to zero error again. Figure 11.19 shows the MIMO analogy to the SISO regulator. In this system both state feedback and cascade gain networks are used to help assign the closed-loop poles of the

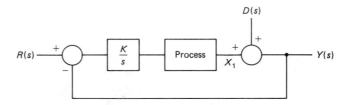

Figure 11.18 Disturbance rejection in a SISO regulator.

system. For the design procedure I will present, the number of plant inputs and outputs must be the same, and the plant must be controllable. Since there are m integrators, the regulator has $n + m$ states and $n + m$ closed-loop poles. Starting at the input, the basic equations that describe this regulator are

$$\dot{\mathbf{v}} = \mathbf{r} - \mathbf{y} = \mathbf{r} - \mathbf{Cx} - \mathbf{d} \qquad \mathbf{u} = \mathbf{K_1v} - \mathbf{K_2x}$$

$$\dot{\mathbf{x}} = \mathbf{Ax} + \mathbf{Bu} \qquad \mathbf{y} = \mathbf{Cx} + \mathbf{d} \qquad (11.52)$$

One can choose to augment the plant states $\mathbf{x}$ by either the integrator outputs $\mathbf{v}$ or the plant inputs $\mathbf{u}$. The state equations formulated for each of these choices are

$$\dot{\mathbf{x}} = (\mathbf{A} - \mathbf{BK_2})\mathbf{x} + \mathbf{BK_1v}$$

$$\dot{\mathbf{v}} = -\mathbf{Cx} + \mathbf{r} - \mathbf{d} \qquad (11.53)$$

and

$$\dot{\mathbf{x}} = \mathbf{Ax} + \mathbf{Bu}$$

$$\dot{\mathbf{u}} = -(\mathbf{K_1C} + \mathbf{K_2A})\mathbf{x} - \mathbf{K_2Bu} + \mathbf{K_1r} - \mathbf{K_1d} \qquad (11.54)$$

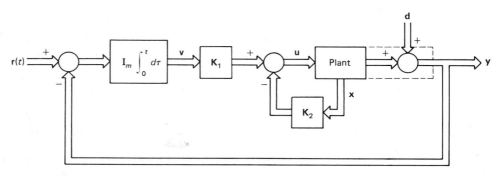

Figure 11.19 MIMO regulator.

Although (11.53) looks simpler, its solution leads to a generalized inverse, which is not as convenient as a square matrix inversion. I will present the solution of (11.54) and let you do (11.53) as a problem. A solution must accomplish two things: It must place the closed-loop poles in desired locations, and it must give values for the $\mathbf{K}_i$ matrices. To do the pole-placement part of the problem we want to force the total state matrix to look like a combination of state, input, and feedback matrices as in the last section. First, group (11.54) into one vector-matrix equation.

$$\begin{bmatrix} \dot{\mathbf{x}} \\ \cdots \\ \dot{\mathbf{u}} \end{bmatrix} = \begin{bmatrix} \overset{n \times n}{\mathbf{A}} & \overset{n \times m}{\mathbf{B}} \\ -(\mathbf{K}_1\mathbf{C} + \mathbf{K}_2\mathbf{A}) & -\mathbf{K}_2\mathbf{B} \\ m \times n & m \times m \end{bmatrix} \begin{bmatrix} \mathbf{x} \\ \cdots \\ \mathbf{u} \end{bmatrix} + \begin{bmatrix} \overset{n \times m}{\mathbf{0}} \\ \cdots \\ \mathbf{K}_1 \end{bmatrix} [\mathbf{r} - \mathbf{d}] \quad (11.55)$$

To get the $\mathbf{A} - \mathbf{B}\mathbf{K}$ form from the state matrix above, separate the rows into two matrices as follows.

$$\begin{bmatrix} \mathbf{A} & \mathbf{B} \\ \mathbf{0} & \mathbf{0} \end{bmatrix} - \begin{bmatrix} \mathbf{0} & \mathbf{0} \\ \mathbf{K}_1\mathbf{C} + \mathbf{K}_2\mathbf{A} & \mathbf{K}_2\mathbf{B} \end{bmatrix}$$

The second of these matrices can be represented as the outer product

$$\begin{bmatrix} \mathbf{0} \\ \cdots \\ \mathbf{I}_m \end{bmatrix} [\mathbf{K}_1\mathbf{C} + \mathbf{K}_2\mathbf{A} \cdots \mathbf{K}_2\mathbf{B}]$$

Now I can define an $n + m$ dimensional system with state feedback as

$$\mathbf{A}' = \begin{bmatrix} \mathbf{A} & \mathbf{B} \\ \mathbf{0} & \mathbf{0} \end{bmatrix} \quad \mathbf{B}' = \begin{bmatrix} \mathbf{0} \\ \mathbf{I}_m \end{bmatrix}$$

$$\mathbf{K}' = [\mathbf{K}_1\mathbf{C} + \mathbf{K}_2\mathbf{A} \quad \mathbf{K}_2\mathbf{B}] \quad (11.56)$$

The pair $\mathbf{A}',\mathbf{B}'$ are controllable if $\mathbf{A},\mathbf{B}$ are controllable. Any of the methods of the last section, if suitable, can be used to find $\mathbf{K}'$ for the desired set of closed-loop poles. The next step is finding $\mathbf{K}_1$ and $\mathbf{K}_2$ for a given $\mathbf{K}'$. The $\mathbf{K}_i$ can be factored out of $\mathbf{K}'$ as

$$\mathbf{K}' = [\mathbf{K}_1 \quad \mathbf{K}_2] \begin{bmatrix} \mathbf{C} & \mathbf{0} \\ \mathbf{A} & \mathbf{B} \end{bmatrix} \quad (11.57)$$

from which

$$[\mathbf{K}_1 \quad \mathbf{K}_2] = \mathbf{K}' \begin{bmatrix} \mathbf{C} & \mathbf{0} \\ \mathbf{A} & \mathbf{B} \end{bmatrix}^{-1} \quad (11.58)$$

The design procedure is

1. Form the state and input matrices of (11.56).
2. Find a $\mathbf{K}'$ that places the poles where you want them.
3. Use (11.58) to find the $\mathbf{K}_i$.

Notice that $\mathbf{C}$ has to have full rank for the last step to work.

■ Example 11.4

I will use the system in Example 11.2 as the basis for a regulator design and follow the steps just listed.

1. The state and input matrices are

$$\mathbf{A}' = \begin{bmatrix} -1 & 1 & 0 & -\frac{1}{6} & 0 \\ 0 & -1 & 0 & \frac{2}{3} & 1 \\ 0 & 0 & -1 & 0 & 0.5 \\ 0 & 0 & 0 & 0 & 0 \\ 0 & 0 & 0 & 0 & 0 \end{bmatrix} \qquad \mathbf{B}' = \begin{bmatrix} 0 & 0 \\ 0 & 0 \\ 0 & 0 \\ 1 & 0 \\ 0 & 1 \end{bmatrix} \quad (11.59)$$

2. I will use the block-controllable transformations. The transformed state matrix is

$$\mathbf{A}_B = \begin{bmatrix} 0 & 1 & 0 & 0 & 0 \\ 0 & 0 & 1 & 0 & 0 \\ 0 & -1 & -2 & 0 & 0 \\ 0 & 0 & 0 & 0 & 1 \\ 0 & 0 & 0 & 0 & -1 \end{bmatrix} \quad (11.60)$$

The first block has poles at 0, -1, and -1, and the second block has poles at 0 and -1. Since the plant's poles are at -1, I choose to shift each block to the ITAE set centered at -1. The coefficients are 1, 1.75, 2.15, 1 and 1, 1.4, 1, respectively. The feedback matrix is

$$\mathbf{K}_B = \begin{bmatrix} 1 & 1.15 & -0.25 & 0 & 0 \\ 0 & 0 & 0 & 1 & 0.4 \end{bmatrix} \quad (11.61)$$

Transforming back,

$$\mathbf{K}' = \begin{bmatrix} -0.6 & 2.325 & -1.5 & -0.25 & -0.975 \\ 0 & 0 & 1.2 & 0 & 0.4 \end{bmatrix} \quad (11.62)$$

3. To save looking all the way back to Example 10.2, the output matrix is

$$\mathbf{C} = \begin{bmatrix} 3 & -0.75 & -0.5 \\ 2 & -1 & 0 \end{bmatrix}$$

$$[\mathbf{K}_1 \quad \mathbf{K}_2] = \begin{bmatrix} 3 & -6 & -2.4 & -0.975 & 0 \\ -1 & 3 & 3 & 0.75 & -0.7 \end{bmatrix} \quad (11.63)$$

$\mathbf{K}_1$ is 2 × 2, and $\mathbf{K}_2$ is 2 × 3. Although the values given have been rounded to four figures for clarity, the arithmetic was double-precision.

I used (11.53), with $\mathbf{y} = [\mathbf{C} \quad 0] (\mathbf{x}^\tau, \mathbf{v}^\tau)^\tau + \mathbf{d}$, to form a five-variable state description of the regulator and ran the step responses. These are shown in Figs. 11.20–11.23. Observe that the steady-state values of the outputs match the inputs, but both outputs have a transient response to a step in a single input. Likewise, a disturbance step in either channel causes a transient in both channels. The shape of the response to a step in d_i is the same as that for r_i; that is, it is reversed in sign and shifted so that the disturbance response starts at one and ends at zero in the driven channel. Coupled transient response is adequate for regulator design in many cases, but not for tracking systems in which each output must follow one, and only one, of a set of changing inputs. Tracking system design is the subject of the next section. The signs and swing amplitudes in the transient response are wilder than one would expect from a SISO all-pole ITAE system, so I suspect we're seeing the effects of some zeros in the transfer functions. ■

Suppose we need a regulator which isn't unity gain, probably because we're using transducers to produce voltage representations of the outputs. Figure 11.24

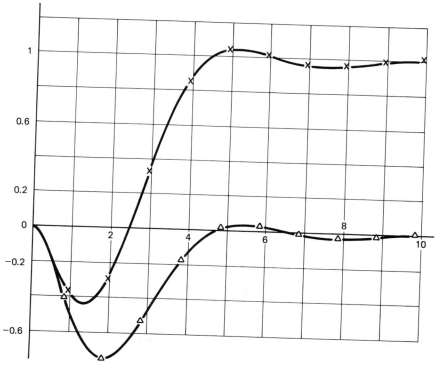

Figure 11.20 Step response for the regulator in Example 11.4. $r_1 = 1$, $r_2 = 0$. x, y_1; △, y_2.

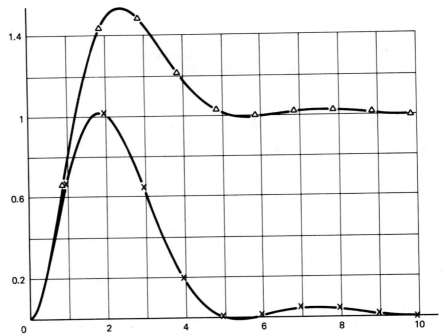

Figure 11.21 Step response for the regulator in Example 11.4. $r_1 = 0$, $r_2 = 1$. x, y_1, △, y_2.

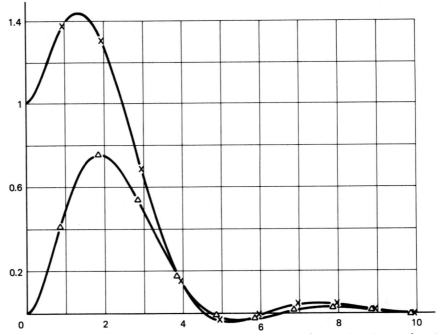

Figure 11.22 Disturbance response for the regulator in Example 11.4; $d_1 = 1$, $d_2 = 0$. x, y1; △, y2.

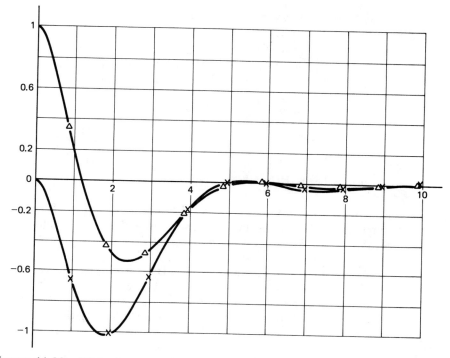

Figure 11.23 Disturbance response for the regulator in Example 11.4; $d_1 = 0$, $d_2 = 1$. x, y_1; △, y_2.

shows this option with the transducer transfers included in a matrix **H**. For the present purpose, I assume **H** to be constant and diagonal. Diagonality isn't mathematically necessary, but it's physically sensible. The basic equations, starting at the input junction, are

$$\dot{\mathbf{v}} = \mathbf{r} - \mathbf{Hy} = \mathbf{r} - \mathbf{HCx} - \mathbf{Hd} \qquad \mathbf{u} = \mathbf{K}_1\mathbf{v} - \mathbf{K}_2\mathbf{x}$$

$$\dot{\mathbf{x}} = \mathbf{Ax} + \mathbf{Bu} \qquad \mathbf{y} = \mathbf{Cx} + \mathbf{d}$$

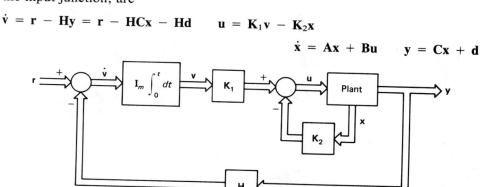

Figure 11.24 Nonunity-gain regulator.

Again, I will write the system in terms of $\mathbf{x}$ and $\mathbf{v}$ for verification and $\mathbf{x}$ and $\mathbf{u}$ for design.

$$\begin{bmatrix} \dot{\mathbf{x}} \\ \dot{\mathbf{v}} \end{bmatrix} = \begin{bmatrix} \mathbf{A} - \mathbf{BK}_2 & \mathbf{BK}_2 \\ -\mathbf{HC} & \mathbf{0} \end{bmatrix} \begin{bmatrix} \mathbf{x} \\ \mathbf{v} \end{bmatrix} + \begin{bmatrix} \mathbf{0} \\ \mathbf{I}_m \end{bmatrix} [\mathbf{r} - \mathbf{Hd}]$$

$$y = [\mathbf{C} \quad \mathbf{0}] [x^\tau \quad v^\tau]^T + \mathbf{d} \tag{11.64}$$

and

$$\begin{bmatrix} \dot{\mathbf{x}} \\ \dot{\mathbf{u}} \end{bmatrix} = \begin{bmatrix} \mathbf{A} & \mathbf{B} \\ -\mathbf{K}_1\mathbf{HC} - \mathbf{K}_2\mathbf{A} & -\mathbf{K}_2\mathbf{B} \end{bmatrix} \begin{bmatrix} \mathbf{x} \\ \mathbf{u} \end{bmatrix} + \begin{bmatrix} \mathbf{0} \\ \mathbf{K}_1 \end{bmatrix} [\mathbf{r} - \mathbf{Hd}] \tag{11.65}$$

The only change in the state matrix of (11.64) from that of (11.54) is the presence of $\mathbf{HC}$ instead of $\mathbf{C}$. This means the design procedure is the same until the last step. Equation (11.58) is modified to read

$$[\mathbf{K}_1 \quad \mathbf{K}_2] = \mathbf{K}' \begin{bmatrix} \mathbf{HC} & \mathbf{0} \\ \mathbf{A} & \mathbf{B} \end{bmatrix}^{-1} \tag{11.65}$$

Since $\mathbf{H}$ is full-rank, it doesn't change the invertibility of the coefficient matrix in (11.65).

■ Example 11.5

This structure and method strikes me as being a natural for a low-pass active filter. Suppose I have a chain of n integrators, each with gain a. I can take their outputs, each of which is a state variable, and feed them back to the beginning of the chain through weighting resistors to be determined. An input integrator followed by gain k_1 ensures that the dc gain will be b_0 as determined by the output

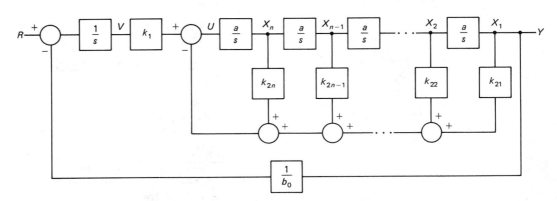

Figure 11.25 Active low-pass filter with dc gain b_0.

feedback. This structure is shown in block diagram form in Fig. 11.25. There are $n + 1$ states, so I can design an $n + 1$ pole filter. The state and input matrices for the integrator chain are

$$\mathbf{A} = a \begin{bmatrix} 0 & 1 & 0 & \cdots\cdots & 0 \\ 0 & 0 & 1 & \cdots\cdots & 0 \\ \multicolumn{5}{c}{\cdots\cdots\cdots\cdots\cdots} \\ 0 & \cdots\cdots & & 0 & 1 \\ 0 & \cdots\cdots & & 0 & 0 \end{bmatrix} \qquad (11.66)$$

The output matrix is

$$\mathbf{C} = \begin{bmatrix} 1 & 0 & \cdots & 0 \end{bmatrix} \qquad (11.67)$$

$$\mathbf{K}_1\mathbf{H}\mathbf{C} = \begin{bmatrix} k_1/b_0 & 0 & \cdots & 0 \end{bmatrix}$$

$$\mathbf{K}_2\mathbf{A} = a\begin{bmatrix} 0 & k_{21} & k_{22} & \cdots & k_{2n-1} \end{bmatrix} \qquad \mathbf{K}_2\mathbf{B} = k_{2n}a \qquad (11.68)$$

If I factor all the a out of the equations, the state matrix for the $n + 1$ system will be in controllable form with zero coefficients. The feedback vector is

$$\mathbf{K}' = \begin{bmatrix} k_1/(ab_0) & \mathbf{K}_2 \end{bmatrix} = \begin{bmatrix} d_n & d_{n-1} & \cdots & d_1 \end{bmatrix} \qquad (11.69)$$

where again the d_j are the desired polynomial coefficients.

As it stands, this design uses a minimum of one op amp per pole, while more conventional designs get at least two poles per op amp. However, the adjustment is very easy, and there is good flexibility in choosing component values. One can also use this approach for two-pole op-amp design as suggested in the problems. ∎

11.3   SERVO DESIGN

The MIMO servo is a system in which the number of outputs is equal to the number of inputs; each output responds to only one input and is the only output that responds to that input. When this is the case, the system is effectively m independent SISO systems or channels. If the process to be controlled is already in this condition, the design can proceed on a SISO basis. However, many processes and plants are internally cross-coupled, so that they must be externally decoupled, if possible, in order to channelize them. There are methods for doing this on a transfer function basis in which one finds a cascade transfer function matrix which diagonalizes the product of the compensator and the plant matrices. This involves adding either considerable network complexity or its equivalent— software complexity in a computer-controlled system. In 1967, Falb and Wolovich published a method which works in many cases and doesn't raise the order of the system. I will present the development of the method in the next subsection.

Once decoupling is achieved, it remains to design the channels to have the specified closed-loop response and verify the behavior of the entire system. These topics are taken up in the following subsection.

11.3.1 One Input for One Output; Decoupling

In the usual state variable description there is no direct connection between the input and the output; the output is a function of the state vector only. One can get a relation by differentiating the output and replacing $\dot{\mathbf{x}}$ by its equation. Thus

$$\dot{\mathbf{y}} = \mathbf{C}\dot{\mathbf{x}} = \mathbf{C}\mathbf{A}\mathbf{x} + \mathbf{C}\mathbf{B}\mathbf{u} \tag{11.70}$$

However, the input and output matrices are frequently sparse enough so that the **CB** product will have some empty rows, which means that some of the outputs won't be driven and some will. To form an arrangement in which each output has a direct connection to the input, each component of the output must be treated individually. So, write

$$y_i = \mathbf{C}_i\mathbf{x} \tag{11.71}$$

where $\mathbf{C}_i$ is row i in **C**. Now, on differentiating one has

$$\dot{y}_i = \mathbf{C}_i\mathbf{A}\mathbf{x} + \mathbf{C}_i\mathbf{B}\mathbf{u}$$

If the $\mathbf{C}_i\mathbf{B}$ product is all zeros, differentiate again and substitute for $\dot{\mathbf{x}}$ again.

$$\ddot{y}_i = \mathbf{C}_i\mathbf{A}^2\mathbf{x} + \mathbf{C}_i\mathbf{A}\mathbf{B}\mathbf{u}$$

There's no $\dot{\mathbf{u}}$ term because of the zero product of $\mathbf{C}_i\mathbf{B}$. If necessary, differentiate and substitute again, and keep it up until a nonzero relation appears between y_i and **u**. Call the number of differentiations at which the nonzero result appears d_i. Then

$$y_i{}^{[d_i]} = \mathbf{C}_i\mathbf{A}^{d_i}\mathbf{x} + \mathbf{C}_i\mathbf{A}^{d_i-1}\mathbf{B}\mathbf{u} \tag{11.72}$$

All the results for $i = 1$ to m are collected in a vector-matrix equation:

$$\begin{bmatrix} y_1{}^{[d_1]} \\ y_2{}^{[d_2]} \\ \cdots \\ \cdots \\ y_m{}^{[d_m]} \end{bmatrix} = \begin{bmatrix} \mathbf{C}_1\mathbf{A}^{d_1} \\ \mathbf{C}_2\mathbf{A}^{d_2} \\ \cdots \\ \cdots \\ \mathbf{C}_m\mathbf{A}^{d_m} \end{bmatrix}\mathbf{x} + \begin{bmatrix} \mathbf{C}_1\mathbf{A}^{d_1-1}\mathbf{B} \\ \mathbf{C}_2\mathbf{A}^{d_2-1}\mathbf{B} \\ \cdots \\ \cdots \\ \mathbf{C}_m\mathbf{A}^{d_m-1}\mathbf{B} \end{bmatrix}\mathbf{u} \tag{11.73}$$

or

$$\mathbf{y}' = \mathbf{M}\mathbf{x} + \mathbf{N}\mathbf{u} \tag{11.74}$$

Now here comes the slick trick. If **u** is replaced by $\mathbf{N}^{-1}\mathbf{v}$, where **v** is a set of inputs, there will be a component-to-component relation between **v** and **y'**. The dependency on **x** is removed by subtracting **Mx** from **v**. Setting

$$\mathbf{u} = \mathbf{N}^{-1}(\mathbf{v} - \mathbf{Mx}) \tag{11.75}$$

yields

$$\mathbf{y}' = \mathbf{Mx} + \mathbf{NN}^{-1}(\mathbf{v} - \mathbf{Mx}) = \mathbf{v} \tag{11.76}$$

Since the derivatives of the components of **y** are driven by the corresponding components of **v**, the y_i are integrals of the v_i. That is why this result is called *integral decoupling*. Equations (11.75) and (11.76) imply the block diagrams in Fig. 11.26.

The critical step in obtaining decoupling by this method is the inversion of **N**. There is no guarantee, and no way to test the system matrices beforehand, that $|\mathbf{N}|$ isn't zero. The system in Example 11.2 is a case in point. Although this system could be statically decoupled by the regulator design in Example 11.4, it can't be dynamically decoupled by the integral method. Fortunately, many systems can.

Another point is that the system has to be stable before the decoupling is designed and applied. Suppose there are three states and two inputs. After decoupling, the input-output relation may be an integrator in each channel—two

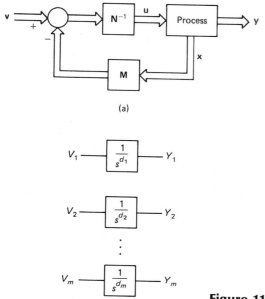

(a)

(b)

Figure 11.26 Decoupled plant. (a) Decoupling matrices. (b) Equivalent channels.

states. These poles can be shifted away from the origin by output feedback, but the third state isn't affected by such feedback. It's been hidden by the decoupling.

11.3.2 The Complete Design

Starting from the original process description,

$$\dot{x} = Ax + Bu \qquad y = Cx \tag{11.77}$$

one may need to add state feedback to give satisfactory pole locations. This will give a new state description:

$$\dot{x} = (A - BK)x + Bu \qquad y = Cx \tag{11.78}$$

Defining $A' = A - Bk$, the decoupling is designed, yielding matrices $Ni = N^{-1}$ and M. The state description is now

$$\dot{x} = (A - BK - BNiM)x + BNiv \qquad y = Cx \tag{11.79}$$

Using the channel equivalent transfer functions, cascade compensation may be provided and output feedback applied to place the required closed-loop poles and gains. The cascade compensator can be represented by its own state equations:

$$\dot{x}_c = A_c x_c + B_c(r - Hy) \qquad v = C_c x_c \tag{11.80}$$

This equation isn't used for design—that's done in the frequency domain. It's

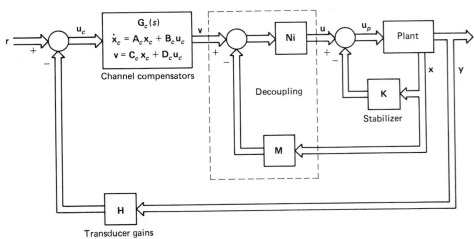

Figure 11.27 Complete servo system design. The design proceeds from right to left: plant stabilization, decoupling, channel design.

combined with (11.79) to form a total state description so that time domain performance can be verified.

$$
\begin{bmatrix} \dot{\mathbf{x}} \\ \cdots \\ \dot{\mathbf{x}}_c \end{bmatrix} =
\begin{bmatrix}
\overset{n \times n}{\mathbf{A} - \mathbf{BK} - \mathbf{BNiM}} & \cdot & \overset{n \times m_c}{\mathbf{BNiC}_c} \\
\cdots\cdots\cdots\cdots\cdots\cdots\cdots\cdots\cdots\cdots\cdots\cdots\cdots \\
\underset{m_c \times n}{-\mathbf{B}_c\mathbf{HC}} & \cdot & \underset{m_c \times m_c}{\mathbf{A}_c}
\end{bmatrix}
\begin{bmatrix} \mathbf{x} \\ \cdots \\ \mathbf{x}_c \end{bmatrix}
+
\begin{bmatrix} \overset{n \times m}{\mathbf{0}} \\ \cdots \\ \underset{m_c \times m}{\mathbf{B}_c} \end{bmatrix}
\mathbf{r} \quad (11.81)
$$

The steps in the design procedure are shown in Fig. 11.27. A rearrangement that combines the state feedbacks, forming a more compact system, is shown in Fig. 11.28.

■ Example 11.6

A model for a particular batch reactor is given as

$$
\mathbf{A} =
\begin{bmatrix}
1.38 & -0.2077 & 6.715 & -5.676 \\
-0.5814 & -4.29 & 0 & 0.675 \\
1.067 & 4.273 & -6.654 & 5.893 \\
0.048 & 4.273 & 1.343 & -2.104
\end{bmatrix}
$$

$$
\mathbf{B} =
\begin{bmatrix}
0 & 0 \\
5.679 & 0 \\
1.136 & -3.146 \\
1.136 & 0
\end{bmatrix}
$$

$$
\mathbf{C} =
\begin{bmatrix}
1 & 0 & 1 & -1 \\
0 & 1 & 0 & 0
\end{bmatrix}
\quad (11.82)
$$

The ranks of **B** and **C** are both 2. The poles of **A** are -8.67, -5.057, 0.0635, and 1.99. Since there are two right-plane poles, I will find a state feedback stabilizer. Transforming **A** and **B** to block-controllable form, I find

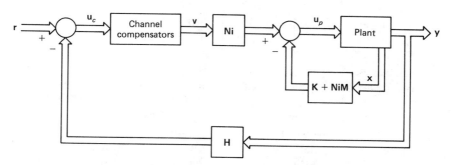

Figure 11.28 After the design, a compact implementation.

$$\mathbf{A}_B = \begin{bmatrix} 0 & 1 & 0 & 0 \\ -1.244 & -5.259 & 2.698 & 0.2498 \\ 0 & 0 & 0 & 1 \\ -11.062 & -1.383 & 19.54 & -6.409 \end{bmatrix} \qquad (11.83)$$

There are two controllable 2×2 blocks, and the one on the lower right has a positive entry at $\mathbf{A}_B[4,3]$ indicating a negative polynomial coefficient which, in turn, implies the positive pole. Factoring the quadratic whose coefficients are 1, $-\mathbf{A}_B[2,2]$, $-\mathbf{A}_B[2,1]$ I find the poles at -5.011 and -0.2483. Factoring the quadratic whose coefficients are 1, $-\mathbf{A}_B[4,4]$, $-\mathbf{A}_B[4,3]$ I find the poles at -8.664 and 2.255. These poles are not the same as the poles of the matrix because they don't include the effect of the out-of-block entries. I will leave the stable poles in the second row alone and use the feedback to zero out the out-of-block entries. I will shift the lower block poles to -8 and -2. This means I want a quadratic with coefficients 1, 10, 16. I find

$$\mathbf{K}_B = \begin{bmatrix} 0 & 0 & 2.698 & 0.2498 \\ -11.062 & -1.383 & 35.54 & 3.591 \end{bmatrix} \qquad (11.84)$$

Transforming back to the actual state description,

$$\mathbf{K} = \begin{bmatrix} -0.144 & 0.0024565 & -0.07939 & 0.06711 \\ -1.833 & -0.1404 & -1.141 & 0.6258 \end{bmatrix} \qquad (11.85)$$

To check the results at this point, I find the poles of the compensated state matrix $\mathbf{A} - \mathbf{B}\mathbf{K}$. They have the expected values.

Now for decoupling. $\mathbf{C}$ and $\mathbf{B}$ are full enough that only one differentiation is needed for each component of $\mathbf{y}$. Thus $\mathbf{M} = \mathbf{C}(\mathbf{A} - \mathbf{B}\mathbf{K})$ and $\mathbf{N} = \mathbf{C}\mathbf{B}$. $\mathbf{Ni} = \mathbf{N}^{-1}$ exists.

$$\mathbf{M} = \begin{bmatrix} -3.367 & -0.6495 & -4.973 & 4.29 \\ 0.2366 & -4.304 & 0.4508 & 0.2939 \end{bmatrix}$$

$$\mathbf{Ni} = \begin{bmatrix} 0 & 0.1761 \\ -0.3179 & 0 \end{bmatrix} \qquad (11.86)$$

The poles of $\mathbf{A} - \mathbf{B}(\mathbf{K} + \mathbf{Ni}\mathbf{M})$ are 0, 0, -5.039, and -1.192. Notice that I have two poles at the origin. These are the integrators that belong to the input-output channels. The other two poles are not the same as, but they are close to, pole values that existed before the decoupling was added.

I choose to use unity output feedback and make the forward-gain unity to place each channel's closed-loop pole at -1. Running the pole-finding function on $\mathbf{A} - \mathbf{B}(\mathbf{K} + \mathbf{Ni}\mathbf{M} + \mathbf{Ni}\mathbf{C})$ I see that the poles at the origin have moved to -1 and the other poles have stayed the same. Figures 11.29–11.32 show the responses

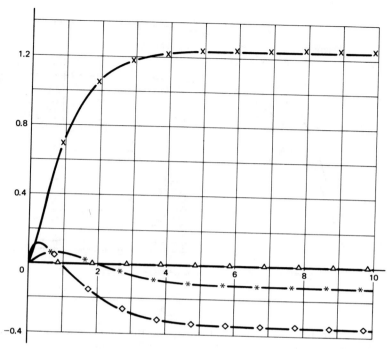

Figure 11.29 State response to a step in r1, Example 11.5. x, x_1; $\triangle$, x_2; $\diamond$, x_3, $\star$, x_4.

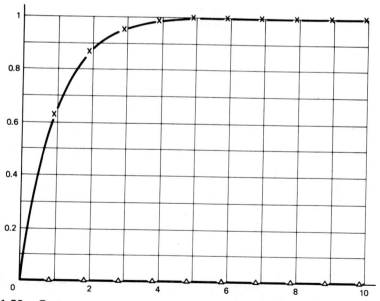

Figure 11.30 Output response to a step in r1, Example 11.5. x, Y_1; $\triangle$, Y_2.

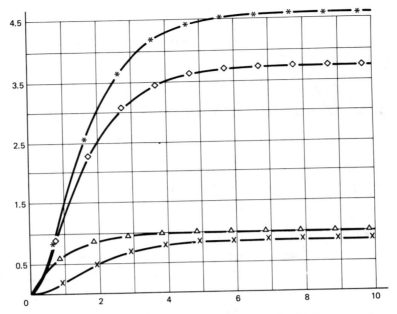

Figure 11.31 State response to a step in r_2, Example 11.5. x, x_1, △, x_2, ◇, x_3, ★, x_4.

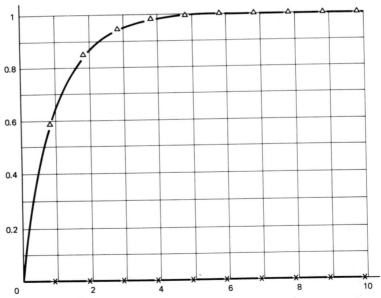

Figure 11.32 Output response to a step in r_2, Example 11.5. x, y_1; △, y_2.

to step inputs. They are very smooth and all look like single-pole responses with 1s time constants. While all the states respond to either input, the output decoupling is apparently perfect.

∎

11.4 ESTIMATING STATES: THE OBSERVER

In most real systems only some of the states are measurable. This means the rest have to be calculated or estimated for systems based on state feedback design. This problem was solved in 1966 by Luenberger, who introduced the idea of using a model of the real system and driving it with signals from the real system to generate the states. The signals available to us are the input and output, but not the initial state any more than the present state. Since the initial state isn't known and is possibly not zero on power-up, the output error is used to drive the model to match the output of the real system. This is where observability comes into play. Only if the system is observable will the output reflect the true state so that the model's state can be driven to match the real state. Figure 11.33 represents the modeling state estimator, frequently called the *observer*. You can see that the estimated state $\mathbf{x}'$ is driven by $\mathbf{u}$ and itself through known matrices, but we have to decide on how to treat the output error gain matrix $\mathbf{L}$.

The state equation for $\mathbf{x}'$ is

$$\dot{\mathbf{x}}' = \mathbf{A}\mathbf{x}' + \mathbf{B}\mathbf{u} + \mathbf{L}\mathbf{y} - \mathbf{L}\mathbf{y}'$$

$$= \mathbf{A}\mathbf{x}' + \mathbf{B}\mathbf{u} + \mathbf{L}\mathbf{C}(\mathbf{x} - \mathbf{x}') \tag{11.87}$$

If we form the error between states by subtracting $\mathbf{x}'$ from $\mathbf{x}$, we get

$$\dot{\mathbf{x}} - \dot{\mathbf{x}}' = \mathbf{A}(\mathbf{x} - \mathbf{x}') + \mathbf{L}\mathbf{C}(\mathbf{x} - \mathbf{x}')$$

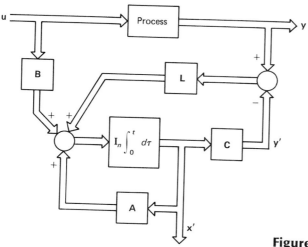

Figure 11.33 The state observer.

or, letting $\mathbf{e} = \mathbf{x} - \mathbf{x}'$,

$$\dot{\mathbf{e}} = (\mathbf{A} - \mathbf{LC})\mathbf{e} \tag{11.88}$$

The effect of the input disappears from the error because it is the same on both the real and estimated states, providing we know $\mathbf{B}$ exactly. Also, with perfect knowledge of $\mathbf{C}$, the error equation is autonomous, so the error will decay to zero regardless of its initial value if we assign the poles of $\mathbf{A} - \mathbf{LC}$ to be stable. Since transposition has no effect on the poles of a matrix, we may transpose $\mathbf{A} - \mathbf{LC}$ to $\mathbf{A}^\tau - \mathbf{C}^\tau \mathbf{L}^\tau$. This gives us exactly the same form as the state feedback problem, so we can use the same methods for its solution.

For testing the behavior of the system with state feedback and an observer, one should construct a total state description. Since it is likely that our values for the process matrices are in error, different labels should be used for the plant and observer elements. Figure 11.34 defines this situation with state feedback.

$$\begin{bmatrix} \dot{\mathbf{x}}_p \\ \dot{\mathbf{x}}_o \end{bmatrix} = \begin{bmatrix} \mathbf{A}_p & -\mathbf{B}_p\mathbf{K} \\ \mathbf{LC}_p & \mathbf{A}_o - \mathbf{LC}_o & -\mathbf{B}_o\mathbf{K} \end{bmatrix} \begin{bmatrix} \mathbf{x}_p \\ \mathbf{x}_0 \end{bmatrix} + \begin{bmatrix} \mathbf{B}_p \\ \mathbf{B}_o \end{bmatrix} \mathbf{v}$$

$$\mathbf{y} = [\mathbf{C}_p \quad \mathbf{0}] \begin{bmatrix} \mathbf{x}_p \\ \mathbf{x}_0 \end{bmatrix} \tag{11.89}$$

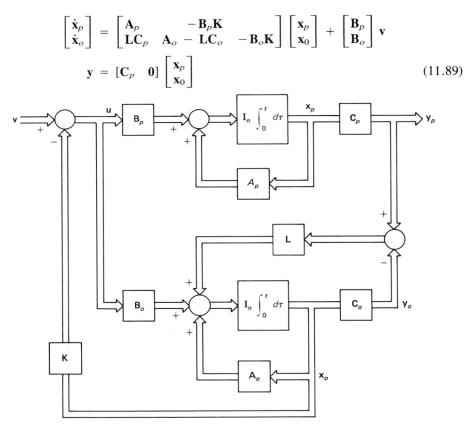

Figure 11.34 Model for process and observer, with state feedback. Subscripts *p* are for process elements, and *o* are for observer elements.

If one assumes that the plant and observer matrices are equal and replaces $\mathbf{x}_o$ by $\mathbf{x}_p - \mathbf{e}$, one obtains

$$\begin{bmatrix} \dot{\mathbf{x}}_p \\ \dot{\mathbf{e}} \end{bmatrix} = \begin{bmatrix} A - BK & BK \\ 0 & A - LC \end{bmatrix} \begin{bmatrix} \mathbf{x}_p \\ \mathbf{e} \end{bmatrix} + \begin{bmatrix} B \\ 0 \end{bmatrix} \mathbf{v} \qquad (11.90)$$

Since the composite state matrix is block diagonal, its poles are the poles of the plant with feedback and the observer error block. This result shows that under ideal conditions the poles of the plant and the poles of the error are independent of each other even though the error drives the plant. This is comforting even though we would have done the designs separately without this result.

■ Example 11.7

Since we have just examined an extensive performance design for the reactor in Example 11.6, let's now add an observer. I started by transforming $\mathbf{A}^{\mathsf{T}}, \mathbf{C}^{\mathsf{T}}$ to block-controllable form. The transformed state matrix is

$$\mathbf{A}_B = \begin{bmatrix} 0 & 1 & 0 & 0 \\ 20.97 & -5.297 & -2.643 & 0.2764 \\ 0 & 0 & 0 & 1 \\ 48.633 & 10.466 & -5.867 & -6.371 \end{bmatrix} \qquad (11.91)$$

The roots of $1, -\mathbf{A}_B[2,2], -\mathbf{A}_B[2,1]$ are -7.94 and 2.64. The roots of $1, -\mathbf{A}_B[4,4]$, $-\mathbf{A}_B[4,3]$ are -5.255 and -1.116. Again, these are somewhat different from the poles of $\mathbf{A}$ because the out-of-block entries are going to be canceled. I chose to test the effect of three different observers: one whose poles are all at -1, the second with poles all at -10, and the third with poles at -30. The -1 choice makes the observer about the same speed as the closed-loop system, and the other choices make it correspondingly faster. After choosing the $\mathbf{L}$ entries to the desired poles, transforming and transposing, they are

$$\mathbf{L}1 = \begin{bmatrix} 9.727 & -11.148 \\ 0.2764 & -4.371 \\ -10.764 & 22.568 \\ 2.26 & 0.9533 \end{bmatrix} \qquad (11.92)$$

$$\mathbf{L}10 = \begin{bmatrix} 13.896 & 10.14 \\ 0.2764 & 13.63 \\ 14.03 & 110.78 \\ 13.222 & 110.45 \end{bmatrix} \qquad (11.93)$$

$$\mathbf{L}30 = \begin{bmatrix} 149.91 & -137.65 \\ 0.2764 & 53.63 \\ 51.55 & 1193.1 \\ 146.76 & 1045 \end{bmatrix} \qquad (11.94)$$

You may have noticed in previous cases that the feedback gains are not necessarily less than unity magnitude. In these sets of gains, you can see the generally higher gains required for faster responses.

To test the effects of the observers, I used the following state description. The notation follows Fig. 11.34 and Example 11.6. The observer is applied to the servo system of that example.

$$\begin{bmatrix} \dot{\mathbf{x}}_p \\ \dot{\mathbf{x}}_o \end{bmatrix} = \begin{bmatrix} \mathbf{A} - \mathbf{BNiC} & -\mathbf{B(K + NiM)} \\ \mathbf{LC} - \mathbf{BNiC} & \mathbf{A]} - \mathbf{LC} - \mathbf{B(K + NiM)} \end{bmatrix} \begin{bmatrix} \mathbf{x}_p \\ \mathbf{x}_o \end{bmatrix} + \begin{bmatrix} \mathbf{BNi} \\ \mathbf{BNi} \end{bmatrix} \mathbf{r}$$

$$\mathbf{y}_p = [\mathbf{C} \quad 0] [\mathbf{x}_p{}^{\tau} \quad \mathbf{x}_o{}^{\tau}]^{\tau} \tag{11.95}$$

Since I would have to compose the state matrix three times for the different observers, I thought it would be useful to assign some of the constant submatrices to variable names. The top four rows don't change, so I assigned them to ASU. I assigned the state feedback $\mathbf{B(K + NiM)}$ to BF, and the output feedback term $\mathbf{BNiC}$ to BNC. I assumed perfect knowledge of the plant matrices, so there's no distinction between them and those for the observer. If the initial state is zero for both the plant and the observer, the observer error will start at, and remain at, zero. I tested the description with L1, the slowest observer, and a step in r_1. The responses were exactly the same as in Example 11.5, so I was reasonably confident I hadn't made any errors in putting the system description together.

Since the system's forced response is the same with and without the observer, the way to test different observers is to generate the response to an initial state which is different for the plant than for the observer. I chose to use $x_{p1} = 1$, all others being zero, as the initial condition for all the observers. This is meant to be illustrative, not exhaustive. For each case, I plotted the response for the plant state and then for the observer error $\mathbf{e} = \mathbf{x}_p - \mathbf{x}_o$. For the two faster observers I ran the response twice, once to get the plant response, which was always for 10 s, and again on a shorter time scale to spread out the error response.

Since all the poles of the compensated system are real, we should expect that, in the direct state feedback case the states should decay more or less smoothly to zero, with no large excursions. For L1, the slow observer, this is far from the case. The maximum magnitude is over 10, and it appears that it takes the state more than 10 s to settle down reasonably close to zero. Figures 11.35 and 11.36 show the plant state and observer error responses for the L1 observer. The observer error peaks at over 4, but it does settle faster than the plant state, which is a necessary sequence of events.

Our expectation is fulfilled with the L10 observer, as seen in Figs. 11.37 and 11.38. The plant variables never change sign and, after an initial hesitation, x_{p1} goes to zero about as fast as we would expect from the step responses in Example 11.5. The error variables also behave smoothly, approaching zero in about 0.6 s.

Figures 11.39 and 11.40 show the responses using the L30 observer. The plant variables behave even better with this observer, but the error variables have

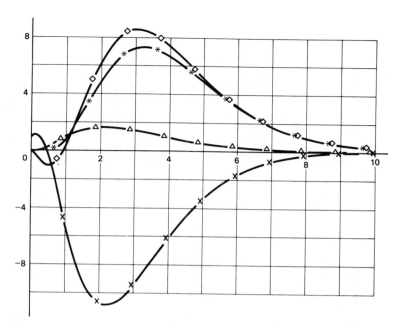

Figure 11.35 Initial condition response of the plant in Example 11.7. x, x_1; $\triangle$, x_2; $\diamond$, x_3; $\bigstar$, x_4.

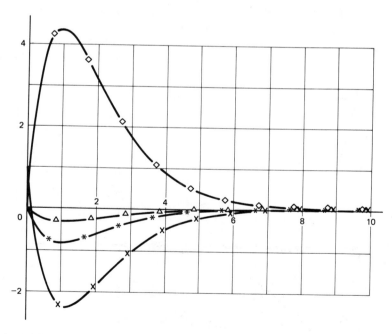

Figure 11.36 Observer error for Example 11.7, observer poles at -1. x--e_1; $\triangle$--e_2; $\diamond$--e_3; $\bigstar$--e_4.

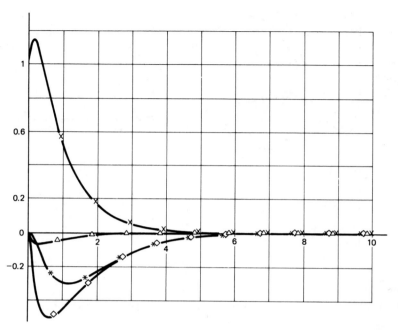

Figure 11.37   Initial condition response for the plant in Example 11.7, observer poles at -10. x, x_1; $\triangle$, x_2; $\diamondsuit$, x_3; $\bigstar$, x_4.

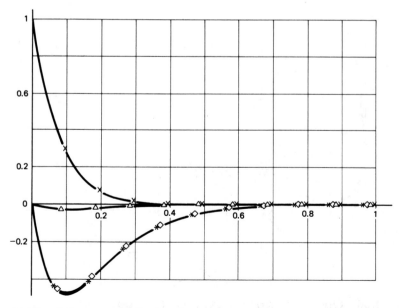

Figure 11.38   Observer error for Example 11.7, with observer poles at -10. Note the change in time scale. x, e_1; $\triangle$, e_2; $\diamondsuit$, e_3; $\bigstar$, e_4.

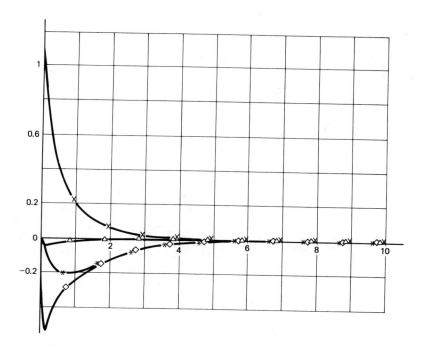

Figure 11.39 Plant response with observer poles at -30, Example 11.6. x, x_1; $\triangle$, x_2; $\diamond$, x_3; $\bigstar$, x_4.

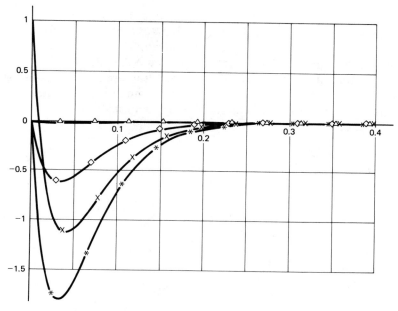

Figure 11.40 Observer error with the observer poles at -30, Example 11.7. Note the time scale change. x, e_1; $\triangle$, e_2; $\diamond$, e_3; $\bigstar$, e_4.

a little more amplitude swing than those for the L10 observer. One may conclude that the higher gains associated with faster poles give better control and smaller signals. Whether the difference between the L10 and L30 observers is worthwhile depends on the design performance requirements. Certainly either is preferable to the L1 observer, whatever the requirements.   ■

11.5   A SATELLITE SUBSYSTEM

In this section I will present an example which uses the analysis and design tools at your disposal on a model for a satellite subsystem. The subsystem consists of a solar panel, an antenna panel, and a boom connecting them. All three are free to rotate, and each has a torque actuator to drive it. A sketch which defines their angular positions with respect to an inertial reference is given in Fig. 11.41. This system has not, to my knowledge, ever been built. It was the subject of a U.S. Air Force study reported in 1986. The paper I took the model from uses an entirely different approach to analysis and design which is based on generalized Nyquist-Bode methods applied to transfer function matrices. The material I will present is exploratory—a look at things that can be done rather than a drive to a final design.

11.5.1   The Model

The main element in this system is a proposed large, flexible antenna. One can get an idea of the size involved from the fact that the unit for the torque inputs is Mft lbs and for the angle outputs it is μrad. The structure was analyzed by the finite element method with a program called NASTRAN, which is a commercial program for solving mechanical and structural stress distribution problems, and also has been used with a preprocessing program to solve magnetic field problems. The response modes have been simplified and normalized to generate the following state equations:

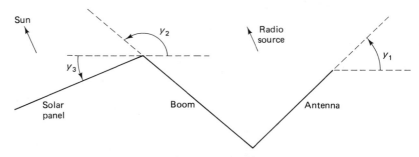

Figure 11.41   Edge view of satellite panel system.

$$\mathbf{A}_p = \begin{bmatrix} 0 & 0 & 0 & 1 & 0 & 0 \\ 0 & 0 & 0 & 0 & 1 & 0 \\ 0 & 0 & 0 & 0 & 0 & 1 \\ 0 & 0 & 0 & 0 & 0 & 0 \\ 0 & -0.02305 & 0 & 0 & -0.001518 & 0 \\ 0 & 0 & -0.08298 & 0 & 0 & -0.002881 \end{bmatrix}$$

$$\mathbf{B}_p = \begin{bmatrix} 0 & 0 & 0 \\ 0 & 0 & 0 \\ 0 & 0 & 0 \\ -9.557 & 0 & -6.758 \\ 5.517 & -10.5 & -0.3529 \\ 7.999 & -4.253 & -25.52 \end{bmatrix}$$

$$\mathbf{C}_p = \begin{bmatrix} -9.557 & 5.517 & 7.999 & 0 & 0 & 0 \\ 0 & -10.5 & -4.253 & 0 & 0 & 0 \\ -6.758 & -0.3529 & -25.52 & 0 & 0 & 0 \end{bmatrix} \qquad (11.96)$$

The nonzero portions of the input and output matrices are the transpose of each other, a feature of the normalization the authors used. These blocks show considerable cross-coupling in that each actuator has more drive on one state than the others, but the drive on the others isn't an order of magnitude smaller. The state matrix is very sparse. The upper right 3×3 block shows that x_1 is the integral of x_4, etc. If we think of the states x_1, x_2, x_3 as positions, then the nonzero entries in the bottom two rows show a high ratio of spring to damping forces. No state drives x_4; it is driven only by the input. The poles of $\mathbf{A}_p$ they are 0, 0, $-0.001441 \pm j0.2881$, $-0.000759 \pm j0.1518$. This confirms the inspection of the matrix; there are two lightly damped modes and a pole pair at the origin corresponding to the x_1 through x_4 system.

11.5.2 Decoupling

I have the option of treating this either as a regulator problem in which I add three integrators and achieve static decoupling, or as a servo problem in which I try dynamic decoupling. If the system decouples on one differentiation, I'm ahead on complexity. Choosing the servo approach, I first add state feedback to shift the poles. By using the block-controllable transformation, the state matrix transforms to three second-order blocks. I assign a pair of identical real poles to each block, -0.1212, -0.1398, and -0.2679. These are the square roots of the leftmost coefficient in each block, so that entry in $\mathbf{K}_B$ is zero. The feedback matrix is

$$\mathbf{K} = 10^{-3} \begin{bmatrix} -1.325 & 0.08121 & -1.639 & -21.86 & -2.353 & 5.752 \\ -0.9123 & 0.3535 & -0.8726 & -13.05 & -28.03 & 3.806 \\ -1.47 & 0.6702 & 0.38 & -10.97 & 7.348 & -18.1 \end{bmatrix}$$

$$(11.97)$$

The decoupling turns out to take two differentiations in each channel, so all the plant poles are turned into three double integrators. The cascade decoupling network is

$$\mathbf{Ni} = \begin{bmatrix} 8.799 & 5.519 & 0.8973 \\ 5.519 & 12.53 & -0.8974 \\ 0.8973 & -0.8974 & 1.761 \end{bmatrix} 10^{-3} \tag{11.98}$$

The combined state feedback network is $\mathbf{K} + \mathbf{NiM}$

$$= 10^{-4} \begin{bmatrix} 0 & 2.242 & -19.92 & 0 & 0.1476 & -0.6917 \\ 0 & 23.24 & -11.41 & 0 & 1.53 & -0.3963 \\ 0 & -3.17 & 28.17 & 0 & -0.2088 & 0.9782 \end{bmatrix} \tag{11.99}$$

11.5.3   Channel Compensation

Each channel is a double integrator, so a lead network as shown in Fig. 11.42 seems appropriate. Figure 11.43 shows the Bode plot design method. I choose to leave the low-frequency gain unchanged. Since the plant crosses unity gain at 1, I back up a decade and put in $z = 0.1$ rad/s. To provide reasonable damping, I put the pole at twice crossover, which makes $p = 20$. $k' = p/z = 200$. Now I need a state model for the compensator. Define

$$X_{ci} = \frac{1}{s + p} (R_i - \mathbf{C}_{pi}\mathbf{X}_p) \qquad V_i = k'(s + z)X_{ci} \tag{11.100}$$

Converting these to the time domain,

$$\dot{x}_{ci} = -px_{ci} + r_i - \mathbf{C}_{pi}\mathbf{x}_p$$

$$v_i = k'(\dot{x}_{ci} + zx_{ci})$$

$$= k'(r_i - \mathbf{C}_{pi}\mathbf{x}_p) - k'(p - z)x_{ci} \tag{11.101}$$

Define $k_c = k'(p - z)$. Then

$$\mathbf{v} = k'(\mathbf{r} - \mathbf{C}_p\mathbf{x}_p) - k_c\mathbf{x}_c \tag{11.102}$$

The presence of the zero requires a direct feedthrough from input to output in the time domain model. In the general case, a transfer function whose numerator has the same power as the denominator will have such a direct feed in its state-

$R_i - \mathbf{C}_i\mathbf{X}_p (s)$ ——— $\boxed{\dfrac{k'(s + z)}{s + p}}$ ——— $V_i(s)$ **Figure 11.42**   Channel lead network compensator.

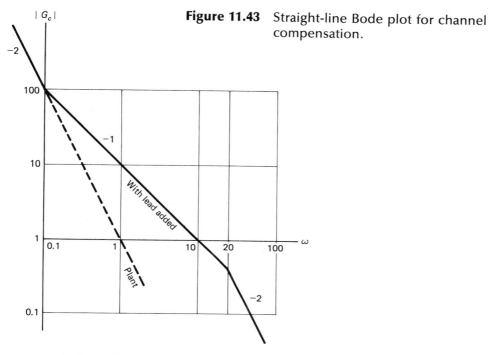

Figure 11.43   Straight-line Bode plot for channel compensation.

space equivalent. Figure 11.44 gives a block diagram for this model of the compensator. The servo system consisting of the plant and channel compensator now has the following closed-loop poles: three at -0.101 and three complex pairs at $-9.95 \pm j9.95$. The three real poles are nearly covered by the zeros at -0.1, so each channel will have responses typical of an underdamped quadratic.

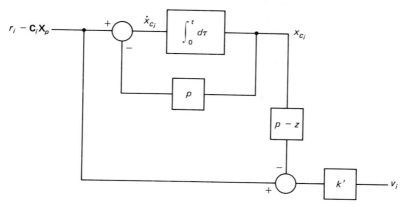

Figure 11.44   State-space version of the lead network showing input feed-through to produce the zero.

11.5.4 The Observer

Since the state variables are part of an approximate model of the structure, an observer is absolutely essential. Only the angle outputs and the torque inputs are accessible to the designer. The structure modes are all represented by poles between 0.1 and 0.3 rad/s in magnitude, but the closed-loop system poles are in the neighborhood of 14 rad/s so I chose -50 rad/s as the observer pole value. Following the procedure described in the last section, the output feedback for the observer is

$$\mathbf{L} = \begin{bmatrix} -9.015 & -4.668 & -2.048 \\ -0.9724 & -10.08 & 1.375 \\ 2.401 & 1.376 & -3.395 \\ -225.4 & -116.7 & -51.2 \\ -24.31 & -252 & 34.38 \\ 60.01 & 34.38 & -84.87 \end{bmatrix} \tag{11.103}$$

A check on the poles of $\mathbf{A}_o - \mathbf{LC}_o$ gives me six at -50.

11.5.5 The Composite Model

You may have noticed that I've been using subscripts on the matrices pertaining to the plant, controller, and observer. This is because I planned on keeping their identities separate in a composite model to test the performance of the entire system. By combining ideas and elements from Figs. 11.28, 11.34, and 11.44, I draw the block diagram for the present system as Fig. 11.45. This figure represents at least three levels of reality. The controller and observer can be implemented either in hardware or software, the plant model is entirely fictional, and the signals $\mathbf{r}$, $\mathbf{u}$, and $\mathbf{y}_p$ are physical quantities. The models for the observer and controller can be taken as the basic set of directions for the programmers or designers of those subsystems.

I have collected the equations describing the system in standard-form state equations in Fig. 11.46. It's important when one does this sort of thing to examine the terms in the result to see if they make sense. That is, do they represent the operations and signals one intends in the design? Is each group of states driven correctly according to the diagram? I have added labels to point out some of these terms in the equations.

In the course of testing the system, I will want to make changes in some of the model elements and generate some of the interior signals. To this end, I have created each model matrix in my workspace as a separate variable. I have written a function to compose, or assemble, them into the input, state, and output matrices for the composite system for the purpose of generating the responses of the entire

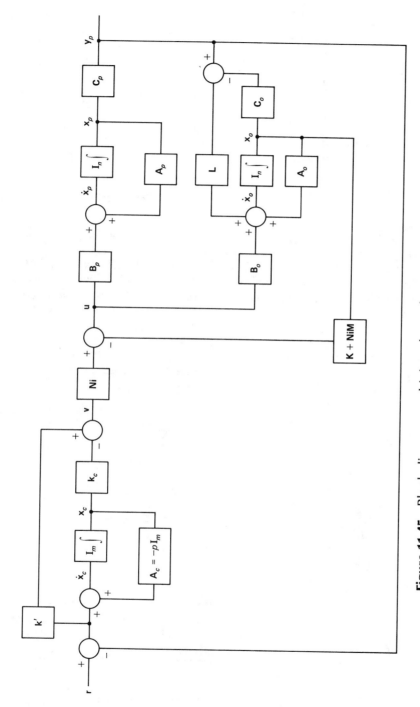

Figure 11.45 Block diagram combining plant, observer, and channel and decoupling compensation.

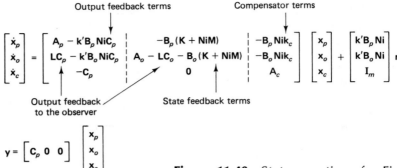

Figure 11.46   State equations for Fig. 11.45.

system to step or initial-condition inputs. The basic sequence I go through is the following: Create the individual matrices for the plant, the observer, the controller, the decoupling, and state feedback; run COMPEX to assemble them; run STATESTEP or STATIC to generate the responses. After that I can extract subsets of signals by combining the appropriate state responses and component matrices.

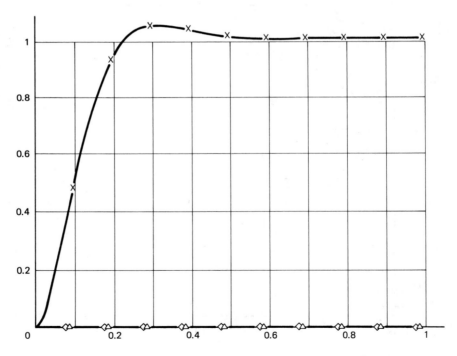

Figure 11.47   Output response to a step in r_1. x, y_1.

11.5.6   Performance

Figure 11.47 shows the output response to a unit step in r_1. We have perfect decoupling since only y_1 responded, and we have the expected second-order response. The overshoot is about 7%, and the final value is 1, as it should be. All the elements are apparently modeled, created or entered, and assembled correctly. The state vector now has 15 elements, so the complexity gives ample scope for mistakes.

The response to a unit initial condition in $x_1 = x_{p1}$ is given in Figs. 11.48– 11.50. x_1 hits a small negative peak (the response shape is like the step response) in about $\frac{1}{3}$ s, while the observer error is close to zero in about 0.1 s. The choice of -50 rad/s for the observer's poles seems to be adequate but not superior.

I want to see the behavior of **v** and **u** when a unit step is applied to an idea of the signal sizes and shapes. Equation (11.102) gives **v**. When I generate the state response to a step, I have $\mathbf{x}_p = \mathbf{x}[(1 \text{ through } 6),*]$ and $\mathbf{x}_c = \mathbf{x}[(13 \text{ through } 15),*]$. to use in that equation, and $\mathbf{u} = \mathbf{Niv} - (\mathbf{k} + \mathbf{NiM})\mathbf{x}_o$, with $\mathbf{x}_o = \mathbf{x}[7 \text{ through } 12,*]$. Doing these operations, I find the data presented in Figs. 11.51 and 11.52. Since the decoupling is perfect, only v_1 responds to the step, but all three torquers (u, u_2, u_3) have to act. On a request for 1 μrad displacement, the peak torque is 1.8 Mft-lb. The initial value is the peak in v_1 and, as expected, it is 200. The

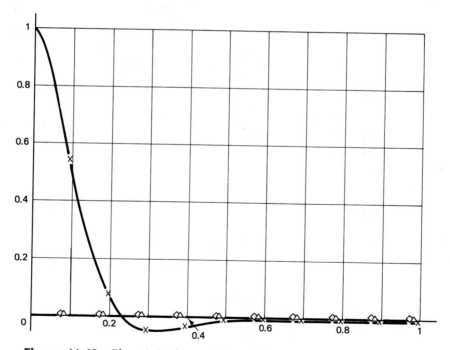

Figure 11.48   Plant initial condition response. x, x_1; $\triangle$, x_2; $\diamond$, x_3.

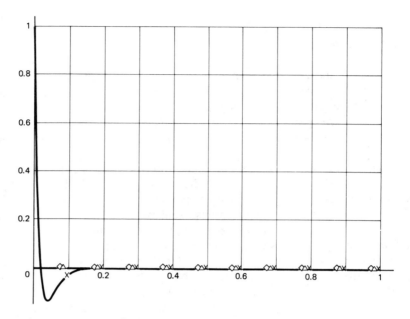

Figure 11.49 Observer error response. x, e_1; $\triangle$, e_2; $\diamond$, e_3.

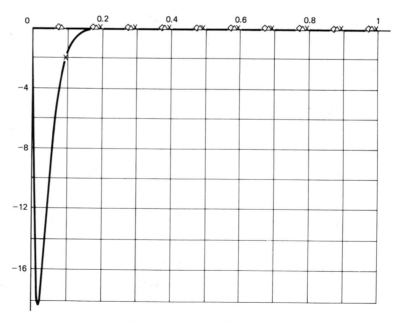

Figure 11.50 Observer error response. x, e_4; $\triangle$, e_5; $\diamond$, e_6.

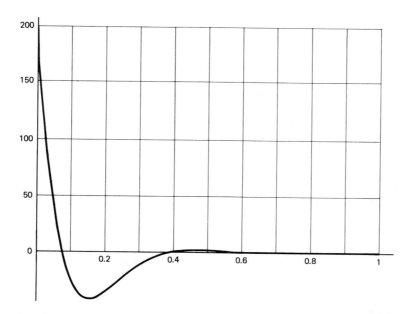

Figure 11.51 v_1 for a step in r_1. Channel compensator output.

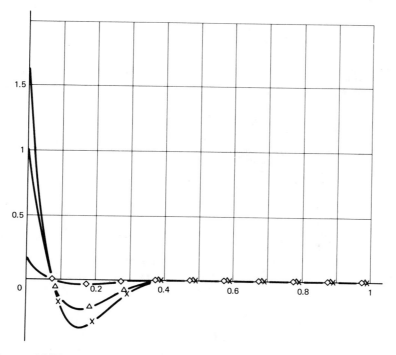

Figure 11.52 Actuator torque for a step in r_1. x, u_1; △, u_2; ◇, u_3.

controller output v_1 has some negative overshoot, which is necessary for braking the motion of the antenna, but it isn't very large. Likewise, the torques go through an acceleration and braking cycle, but the amplitude on the braking part is not large considering the very light damping of the structure model. Recall the torque signal for the motor-driven position system in earlier chapters.

11.5.7   Errors in Modeling

How will the system perform if the plant model is incorrect? There are two basic kinds of model errors, structural and numerical. Our models of real systems are generally approximations; usually we neglect some aspects which we think won't add to the model's ability to imitate or predict the system's behavior. This amounts to a deliberate introduction of error which we hope won't be significant but simplifies the model structure. In other cases the process or plant already exists, and we have to derive a model for it using some mixture of analysis and experiment. Then we can't help but be wrong to some extent in both structure and numerical values. Again, if the system is to be manufactured in quantity, there will be unit-to-unit variations in parameter values, as we have discussed in Chap. 9.

In the present case, I content myself with testing the effects of numerical differences between the plant model and the model of the plant used by the observer and which is the basis of the design procedure. What I do is specify a parameter percent tolerance e. I generate a matrix $\mathbf{E}$ of random numbers whose range limit is $\pm e$ and whose shape is the same as that of the matrix I want to vary. Then I add 1 to the elements of $\mathbf{E}$. This gives me a matrix whose elements are $1 + e_{ij}$. Finally I perform an element-by-element multiplication of $\mathbf{E}$ and the subject matrix. The result, for example, of doing this to the plant state matrix is that a_{pij} is replaced by $a_{pij}(1 + e_{ij})$. I did this for two values of e, 0.1 and 0.25, for all three plant model matrices. For the 10% range limit, the largest variation actually generated was about 7%. The output response to a step was identical to that in the previous case, but the observer error settled out to nonzero values as seen in Fig. 11.53. Remember that the step response is taken with an initial state, and therefore an initial error vector of zero. The output feedback to both the observer and the controller tends to force the output to match the input even if the internal values have to be in error.

For the 25% error range, the maximum that occurred for this test was 24%. Figure 11.54 shows that the output response has become somewhat coupled. The second and third outputs still tend to zero, but y_1 has a small final error. The observer error, shown in Figs. 11.55 and 11.56, has small initial oscillations before settling down to steady values similar to those for the previous case. The ability of a design to control the process in the face of various difficulties such as plant identification errors and various kinds of part failures is called in the literature *robust control*. A robust controller is one which has enough mental muscle in its design, and sometimes plenty of drive power, to overcome these system degeneracies. The present design appears to be pretty healthy with respect to plant parameter ignorance.

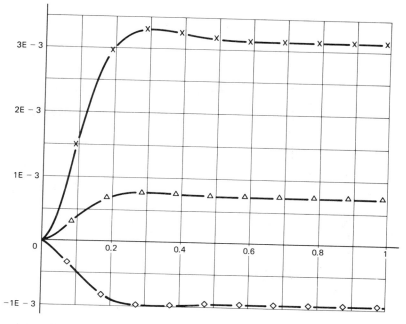

Figure 11.53   Observer error for a step in r_1 and 10% plant parameter uncertainty. x, e_1; △, e_2; ◇, e_3.

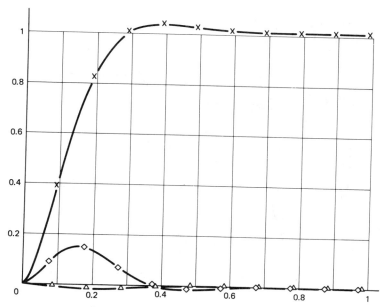

Figure 11.54   Output response to a step in r_1 and a 25% plant parameter uncertainty. x, y_1; △, y_2; ◇, y_3.

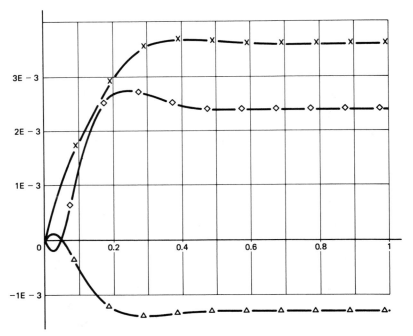

Figure 11.55 Observer error for a step in r_1 and 25% plant parameter uncertainty. x, e_1; △, e_2; ◇, e_3.

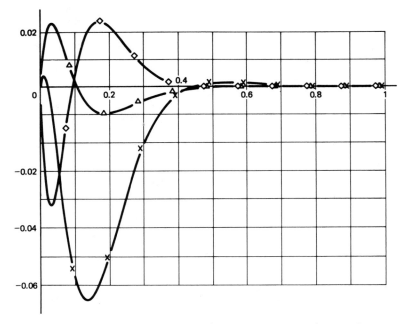

Figure 11.56 Observer error response for a step in r_1 and 25% plant parameter uncertainty. x, e_4; △, e_5; ◇, e_6.

11.5.8 Part Failures

In complex systems from power-generating stations to space stations we need to know what will happen if a part fails. One aspect of this knowledge is the dynamic behavior of the rest of the system. With the state-space model I can't tell (directly) if anything will go up in smoke, but I can predict the behavior of the signals in the rest of the system. As far as the model is concerned, the failure of a part means the signal it is supposed to generate goes to zero regardless of what the rest of the system is doing. In the case of an output sensor, this means that the row of the output matrix representing that signal should be replaced by zeros. In the case of an actuator, the column of the input matrix through which the actuator drives the states should be replaced by zeros. These are the easy ones. Other parts tend to be buried in model parameters, so their failure has to be traced through from the subsystem model or equations to the composite model to see what parameter(s) to take out.

To illustrate sensor failure, I first model a y_1 failure by zeroing out the first row of $\mathbf{C}_p$, running COMPEX to reconstruct the composite matrices, and applying a step to r_1. The output response appears in Fig. 11.57. Evidently, the controller can't stand this one, because it keeps applying torque to the structure and the

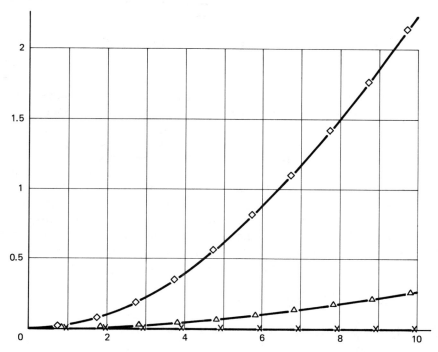

Figure 11.57 Output response to a step in r_1 and a failure in the sensor for y_1. x, y_1; $\triangle$, y_2; $\diamond$, y_3.

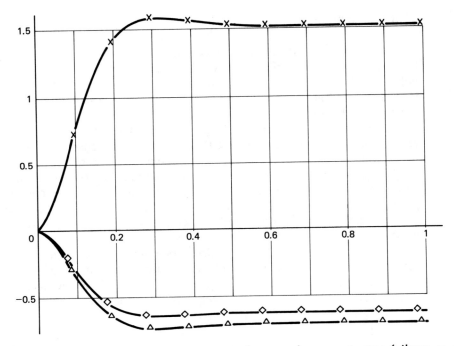

Figure 11.58 Ouput response to a step in r_1 and a u_2 actuator failure. x, y_1, $\triangle$, y_2; $\diamondsuit$, y_3.

error in channel 1 never changes. The fact that the errors in y_2 and y_3 are growing doesn't seem to help any. The system slowly proceeds to spin. On the other hand, a failure in the other output sensors, with y_1 working, is completely ignored when a step is applied to r_1. This is due to the perfect decoupling, which causes the expected values of the other outputs to be zero anyway.

Figures 11.58 and 11.59 show the output responses to a unit step in r_1 for two actuator failures. In both cases, the response is stable but very wrong. In fact, when u_1 fails, the angle of the antenna actually goes in the direction opposite that requested by the input. I have seen stability in the case of failure called "robust" behavior by some authors. If this system were implemented, steady-state error detection and alarms would have to be added to allow for correction or shutdown by the operating system or operators.

11.5.9 Conclusion

This is the end of both this example and this text. Again, the example is meant to be illustrative, not exhaustive. In practice, many runs of the model would be

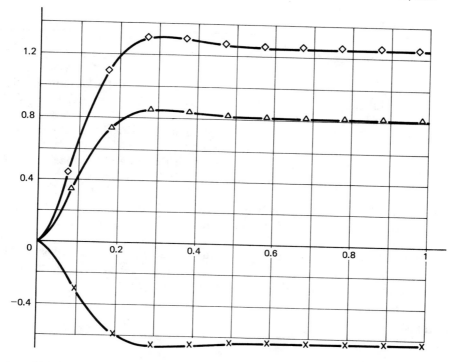

Figure 11.59   Output response for a step in r_1 and a u_1 acuator failure. x, y_1; △, y_2; ◇, y_3.

needed to test the system's behavior under various adverse conditions. Other controller designs might be tried, such as the regulator approach. Different pole selections for state and observer feedback may help in dealing with some problems.

The design process continues down to the hardware level through steps very much like those for a SISO system as described in Chapter 9. In this example, we have seen how to get the actuator signal amplitudes, and how to determine the effects of model errors and variations for MIMO case. The choices of amplifiers, sensors, and networks will be based on the same considerations as before. A new factor which appears in many MIMO systems is digital implementation of the controller. If digital implementation is to be used, interface circuits between the computer and the sensors and actuators will have to be designed or selected, algorithms which give the desired transfer functions may be written, etc. In any case, if you go on to practice control or systems engineering, you will have the time and opportunity to learn about and try many other options.

FURTHER READING

The decoupling problem is treated in the following papers.

Falb, P. L. and Wolovich, W. A., "Decoupling in the Design and Synthesis of Multivariable Control Systems", *IEEE Trans.* vol. AC-12 no. 6, December 1967.

Mufti, I. H., "A Note on the Decoupling of Multivariable Systems", *IEEE Trans.* vol. AC-14 no. 4, August 1969.

Christodoulou, M. A., "Decoupling in the Design and Synthesis of Singular Systems", *AUTOMATICA* vol. 22 no. 2, March 1986.

The chemical reactor came from Davison and Ferguson, listed at the end of Chapter 10. The large space antenna came from the following paper.

Enns, D. F. and Bugajski, D. J., "Multivariable Control Law Analysis for a Large Space Antenna", Honeywell System and Research Center, 3660 Marshall St. NE, Minn. MN 554118, March 1987.

PROBLEMS

Section 11.1.1

11.1   For the following system descriptions, find the plant's poles, then work through both the controllable-form transformation and Ackermann's formula to find the state feedback matrix which places the closed-loop poles at -1, -1.

(a) $\mathbf{A} = \begin{bmatrix} 0 & 0 \\ 0 & 0 \end{bmatrix}$ $\mathbf{B} = \begin{bmatrix} 1 \\ 1 \end{bmatrix}$

(b) $\mathbf{A} = \begin{bmatrix} 1 & 0 \\ 0 & -1 \end{bmatrix}$ $\mathbf{B} = \begin{bmatrix} 1 \\ 1 \end{bmatrix}$

(c) $\mathbf{A} = \begin{bmatrix} 0 & 1 \\ 0 & 0 \end{bmatrix}$ $\mathbf{B} = \begin{bmatrix} 0 \\ 1 \end{bmatrix}$

(d) $\mathbf{A} = \begin{bmatrix} 0 & 1 \\ 0 & -1 \end{bmatrix}$ $\mathbf{B} = \begin{bmatrix} 0 \\ 1 \end{bmatrix}$

11.2 For a general second-order system in controllable form, find the elements of Ackermann's formula. Use $P(s) = s^2 + d_1 s + d_2$.

11.3 Write a program that generates the matrix which transforms a state matrix to controllable form. You may find it convenient to divide the problem into parts which calculate $\mathbf{C}_t$ and $\mathbf{W}$. I assume you have a matrix inversion function or program available to test your program's result.

11.4 Write a program that implements Ackermann's formula. The inputs can be the controllability matrix and the coefficients of the desired system polynomial, as well as the system's state and input matrices.

Section 11.1.2

11.5 Let $\mathbf{B} = [a \quad b \quad c]^\top$. Use the $\mathbf{A}$ in Example 11.2 and show that the controllability matrix isn't invertible.

11.6 For the plant in Example 11.2, are there any $\mathbf{K}_c$ matrices with only two nonzero elements that will render the system single-input-controllable?

Section 11.1.3

11.7 Follow Example 11.3, but place the poles at $-(1 \pm j)/\sqrt{2}$ and -10.

11.8 Shift all the poles in Example 11.3 to zero.

11.9 Given

$$\mathbf{A} = \begin{bmatrix} -1 & 1 & 0 & 0 \\ 0 & -2 & 0 & 0 \\ 0 & 0 & -2 & 1 \\ 0 & 0 & 0 & -1 \end{bmatrix} \qquad \mathbf{B} = \begin{bmatrix} 0 & 0 \\ 1 & 0 \\ 0 & 0 \\ 0 & 1 \end{bmatrix}$$

Shift the poles in each block to 0 and -1.

Section 11.2

11.10 Derive (11.54).

11.11 Use (11.53) as the starting point to find values for the regulator matrices $\mathbf{K}_1$ and $\mathbf{K}_2$. Follow the procedure given for starting with (11.54).

11.12 Verify that $\mathbf{A}', \mathbf{B}'$ in (11.56) are controllable if $\mathbf{A}, \mathbf{B}$ are controllable.

11.13 Convert the block diagram in Fig. 11.25 to a schematic for $n = 3$. Use compensated operational amplifiers and practical R and C values. Design the filter as a 1-dB ripple Chebyshev with a passband of 0–1 kHz and a dc gain of 10.

11.14 Figure P11.14 shows a simple two-pole op-amp stage. Derive the state equations for the stage. Would this stage be suitable for a universal filter structure? If not, what modification would make it so?

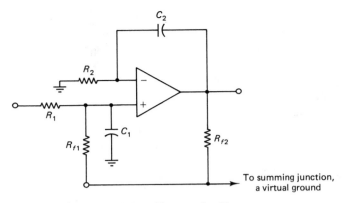

Figure P11.14   Two-pole filter stage.

11.15 Show that the transfer function for the filter in Example 11.5 has no zeros.

11.16 A plant has the description

$$A = \begin{bmatrix} -0.395 & 0.01145 \\ -0.011 & 0 \end{bmatrix} \qquad B = \begin{bmatrix} 0.03362 & 1.038 \\ 0.000966 & 0 \end{bmatrix}$$

$$C = \begin{bmatrix} 1 & 0 \\ 0 & 1 \end{bmatrix}$$

Design a unity-gain regulator with all its closed-loop poles at -0.1. Show the responses to step inputs and step output disturbances.

11.17 Suppose that disturbance signals drive the state derivative instead of the output. What is the regulator response in the steady state if the state equation is $\dot{x} = Ax + Bu + Dd$, with d a vector of step functions.

Section 11.3

11.18 Determine the decoupling matrices for the plant in Example 11.2.

11.19 Design tracking systems using the following plants. Assign all closed-loop poles to -1.

$$(a) \; A = \begin{bmatrix} 1 & -1 & -1 \\ 0 & 0 & 1 \\ -1 & 1 & -1 \end{bmatrix} \qquad B = \begin{bmatrix} 1 & 1 \\ 0 & 1 \\ -1 & -1 \end{bmatrix}$$

$$C = \begin{bmatrix} 1 & 1 & 0 \\ 0 & 1 & 1 \end{bmatrix}$$

$$(b)\ \mathbf{A} = \begin{bmatrix} 1 & -1 & 1 \\ -1 & -1 & 0 \\ -1 & -1 & 1 \end{bmatrix} \quad \mathbf{B} = \begin{bmatrix} 1 & 1 \\ 0 & -1 \\ 0 & 0 \end{bmatrix}$$

$$\mathbf{C} = \begin{bmatrix} 0 & 1 & 0 \\ -1 & 1 & 0 \end{bmatrix}$$

$$(c)\ \mathbf{A} = \begin{bmatrix} 1 & -1 & -1 & -1 \\ 0 & 1 & -1 & -1 \\ -1 & -1 & 1 & -1 \\ -1 & 0 & 0 & 0 \end{bmatrix} \quad \mathbf{B} = \begin{bmatrix} 0 & -1 \\ -1 & 1 \\ 1 & 0 \\ 1 & -1 \end{bmatrix}$$

$$\mathbf{C} = \begin{bmatrix} -1 & 1 & 0 & -1 \\ 1 & -1 & 0 & 0 \end{bmatrix}$$

Section 11.4

11.20 Derive (11.90), the result on separation of plant and observer poles.

11.21 Let $\mathbf{C} = [1\quad 1]$. Design observers for the plants in Prob. 11.1. If the plant is not observable with the given $\mathbf{C}$, is it observable through any single-output $\mathbf{C}$?

11.22 For the plants in Prob. 11.19, determine which are observable. Design observers for those that are. Prepare two designs for each case, placing the poles at -3 for one trial and at -10 for the next. Test and comment on the initial-condition responses.

Section 11.5

11.23 Given initially zero state, find the initial response of the channel compensator to a unit-step input in two ways:
    (a) Examine the state-space model in Fig. 11.44 and use physical reasoning.
    (b) Use the Laplace transform initial-value theorem on Fig. 11.42.

11.24 Redesign the system in Sect. 11.5 as a regulator. Place the dominant closed-loop poles in approximately the same places as the tracker design. Test your design for tolerance to parameter value errors and signal failures.

11.25 A Minimal Regulator Design Problem: For some purposes, the objective of the control design is to maintain the plant at a prescribed operating point. Even if the plant is basically nonlinear, its motion around the operating point can be linearized, as discussed in Chap. 2, and an appropriate linear controller can be designed. If the state variables are available for measurement, the minimum system that can perform this function is one in which the plant's variables are compared with the desired values and the difference signal applied to the plant's inputs through a gain network. Such

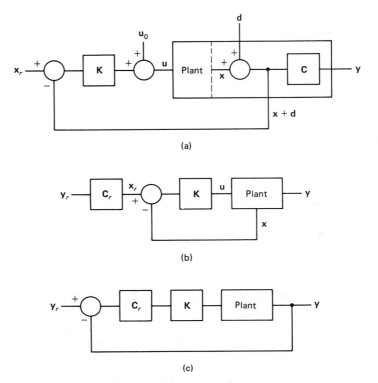

Figure P11.25 State regulator systems.

a design is shown in Fig. P11.25a. The input u_0 is that signal which will hold the plant at the nominal operating point in the absence of disturbances. One type of disturbance input is depicted as the signal d. If the plant is strongly nonlinear, it may have to be started up in the open-loop condition, closing the loop when the desired operating point is reached. For some plants, the u_0 signal may not be necessary, as the plant will come up smoothly in response to the reference state input. In any case, once the loop is closed, the equations that describe the linearized system are

$$\dot{\mathbf{x}} = (\mathbf{A} - \mathbf{BK})\mathbf{x} + \mathbf{BK}\mathbf{x}_r, \qquad \mathbf{y} = \mathbf{Cx} \qquad (11.105)$$

The steady-state response for constant reference inputs is

$$\mathbf{x}_{ss} = -(\mathbf{A} - \mathbf{BK})^{-1}\mathbf{BK}\mathbf{x}_r = (\mathbf{BK} - \mathbf{A})^{-1}\mathbf{BK}\mathbf{x}_r \qquad (11.106)$$

providing the inverse exists. The inverse exists if $\mathbf{A} - \mathbf{BK}$ has no poles at the origin. Evidently, if the numbers in $\mathbf{BK}$ are of the same order as those in $\mathbf{A}$, there will be a significant difference between the reference and output variables. This is a generalization of the single-loop system property that a low loop gain allows large error. To make this discussion more concrete, use the batch reactor in Example 11.5 as a case to study.

(a) Use the plant description and the stabilizing **K** found in Example 11.5. Define $\mathbf{x}_r = [x_{1r} \ \ x_{2r} \ \ x_{3r} \ \ x_{4r}]^T$. Find the steady-state values of the x_i for $\mathbf{x}_r = [1 \ \ 0 \ \ 0 \ \ 0]^T$ and $\mathbf{x}_r = [0 \ \ 1 \ \ 0 \ \ 0]^T$.

You will find that the error is quite large and that there is no particular correspondence between which input is nonzero and what happens at the output. In SISO systems, we saw that raising the gain is equivalent to raising the pole frequencies. Try this approach.

(b) Redesign **K** so that the closed-loop poles are all at -50. Then test the system again for the steady-state response to the two inputs given in (a).

You will find the response to x_{1r} is more satisfying, but that to x_{2r} is still poor. Although many of the numbers in **BK** are now large compared to those in **A**, the matrices are unstructured so that no clear correspondence exists between the inputs and the outputs. If you examine the corresponding problem for the block-controllable form of the system description, you will see why this happens. In order for each state derivative to be zero, the next lower state must be zero, except for the lowest one in the block. The lowest state variable in the block has zero derivative if the top state is a multiple of the input. This says that the only tidy equilibria are those for which only the top state in each block is nonzero.

Since each input in the block-controllable form drives a block, I have changed the question from, How do I design the system to get a desired state? to How do I design the system to get a desired constant output? This new goal is represented in Fig. P11.25b, where the reference or desired output is fed through a network, represented by a matrix $\mathbf{C}_r$, to generate an appropriate set of desired state signals. If the system were in the block-controllable form, I know that I want the reference outputs to correspond to nonzero values of, for the batch reactor, x_{1B} and x_{3B}, the top state variable in each block. Let's look at the block-controllable system and transform the results back to the original. The output equation is

$$\mathbf{y} = \mathbf{Cx} = \mathbf{CT}^{-1}\mathbf{x}_B = \mathbf{C}_B\mathbf{x}_B \qquad (11.107)$$

Given a desired $\mathbf{y}$, I want to know the needed $\mathbf{x}_B$. The elements of $\mathbf{x}_B$ that aren't top-of-block states are zero, so I can take out these elements and the corresponding columns in $\mathbf{C}_B$. Since the number of blocks is equal to the number of inputs and outputs, the new matrix is square. If it is invertible, then I have a solution. Call the columns of $\mathbf{C}_B$ we've kept $[C_{Bi}]$ and define

$$\mathbf{C}_M = [C_{Bi}]^{-1} \qquad (11.108)$$

Then the nonzero reference states are

$$[x_{iBr}] = \mathbf{C}_M \mathbf{y}_r \qquad (11.109)$$

To fill in the zero state variables, we need to insert zero rows below each row of $\mathbf{C}_M$. Define the result as

$$\mathbf{C}_{rB} = \begin{bmatrix} C_{M1} \\ 0 \\ 0 \\ C_{M2} \\ 0 \\ \vdots \\ 0 \end{bmatrix} \qquad (11.110)$$

Finally, the real desired state is $\mathbf{T}^{-1}\mathbf{x}_{rB}$, so

$$\mathbf{C}_r = \mathbf{T}^{-1}\mathbf{C}_{rB} \qquad (11.111)$$

(c) Find $\mathbf{C}_r$ for the batch reactor.

(d) With the $\mathbf{K}$ for the closed-loop poles at -50 and $\mathbf{C}_r$ from (c), find the steady-state output for a step in each component of $\mathbf{y}_r$.

(e) Plot the step responses for the design in (d).

(f) Determine the effects of errors in the plant parameter values on the steady-state responses.

(g) Design an observer for the system in (d).

(h) Again determine the effects of plant parameter misidentification for the system with the observer.

(i) Write the time and frequency domain equations for the output feedback system in Fig. P11.25c. Transform it to block-controllable form. By examining the form of the $\mathbf{K}_B \mathbf{C}_{rB} \mathbf{C}_B$ product, show that m^2 elements in $\mathbf{K}_B$ can be used for pole assignment, where m is the number of outputs.

(j) Redesign $\mathbf{K}$ for the batch reactor in the output feedback configuration with poles at -50.

(k) Determine the steady-state responses for the system in (j).

(l) Plot the step responses for the system in (j).

(m) Determine the effects of plant parameter value error on the results for (k).

11.26 A Minimal Zero-Final-Error Regulator Design: By converting to block-controllable form, we can design each block subsystem to have a pole at the origin. This puts an integrator into each input channel, even though the block outputs, the top states, are cross-coupled by the $\mathbf{C}_B$ matrix. Figure P11.26 shows an arrangement in which state feedback is used to convert the plant into integrating channels and output feedback is used to provide

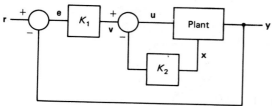

Figure P11.26 Another regulator design.

an error signal vector to drive the integrating channels. Use the batch reactor in Example 11.5 as a test system to explore this approach.

(*a*) Design K_2 so that K_{2B} places an origin pole in each block of the system
between **v** and **y**.

(*b*) Choose the elements of K_1 so that the closed-loop poles of the complete
system are real and in the range -0.1 to -10. You may find it easier
to work this problem entirely in the block-controllable form and transform the results back to the real signals at the end.

(*c*) Find the steady-state responses to constant inputs.

(*d*) Plot the step responses.

(*e*) Test the design's tolerance to variations in plant parameters.

EPILOGUE: WHERE DO YOU GO FROM HERE?

At this point in your learning, you have experienced a few ways of modeling a system or subsystem and practiced several methods which will help you to design systems that need feedback. Control system design, more than other kinds of design activity, is intimately concerned with the entire system from specification to prototype, so I have included elements and examples of the design process for most stages and levels. If you do not go on in control engineering, you will still have many occasions to use the methods and concepts of this book, both in the design of components and in thinking about larger systems and problems for which an appreciation of modeling and dynamics will give you a mental context.

If you do go on in control engineering, the fun has just begun. Your next course or book should be on digital (sampled-data) linear control systems because the mathematical tools are very close to the ones you have already acquired for continuous-time systems. Many, if not most, systems that were formerly aggregations of analog controllers and plants have been converted to digital forms, and most new designs for the MIMO systems are digital. In many cases, the digital design just emulates an equivalent analog design or performance requirement. In most cases, the plant or process to be controlled is continuous-time. For these reasons, continuous-time modeling and design are an essential prerequisite to the design of computer-controlled systems.

The next important area to explore is nonlinearity. While many systems are designed with deliberately chosen linear components, there are also a great many systems of high economic value which are essentially nonlinear. In some cases, these plants can be brought to a desired operating condition and maintained there by a linear controller. In many cases, such as aircraft flight control, the model of the system and the nature of the controller change with each mode of operation.

For example, maintaining a heading, making a lateral turn, climbing, and diving each require different models and possibly both controller parameter and structure changes. The simple action of a robot arm moving an object from one place to another means that the object's moment arm about one or more axes is changing over a wide range of values, so that the inertia being controlled by one or more motors is also changing over a wide range. These are just a few of the great variety of intriguing applications that await you in practice.

You might like to follow with an advanced study of linear systems, including such topics as MIMO analysis and design in the frequency domain, modeling for and design against noise effects, and optimization. Optimization methods are worth a course in themselves, though they are often included in graduate systems theory courses. The search procedures are useful in design problems other than those involving dynamic performance.

There are numerous refinements and extensions which I either have hinted at or not mentioned at all. An important example of this is the area of observer design. It is possible to design observers of lower complexity if some of the states are directly available as signals from the plant. This is a benefit. However, the observer structure also serves as a noise filter, so that there are cases where a full observer is preferable even if a reduced observer is possible. It is also possible to model the plant itself with fewer state variables than you might have expected just on physical grounds. This again leads to a simplification of the controller structure. You should pursue other books and past and current journal literature before trying these methods.

I have discussed, by example, design for system parameter variation and performance in the face of part failure. There are more sophisticated approaches available, ranging from numerical methods to choosing the optimum fixed-controller parameter values to the design of controllers which adapt their parameters and structures to follow changes in the plant. To use these methods without getting into too much trouble you will likely need a more sophisticated understanding of simulation than the material presented in this book. We are talking about complex, expensive systems that make these design procedures worthwhile, and computer testing of the design results is essential before you go to hardware.

You must maintain a continuing contact with professional journal literature. The engineering literature, without the supporting scientific and mathematical literature, is too vast for anyone to keep up with, even within one of the broad disciplines such as mechanical or electrical engineering. This means you must develop a policy on how to handle the necessity and impossibility of these two statements. An example you might consider is the following. Join the professional society for your discipline and read its magazine for a general awareness of ideas that have made it at least to the promising stage. Subscribe to the journal or transactions in your specialty, e.g., *IEEE Transactions on Automatic Control Systems* if you are a controls engineer with an electrical engineering background. Don't read every article. You don't have time to digest what's there, and you may wind up with mental indigestion. Read abstracts, read whatever is pertinent to your current project(s), and collect any numerical methods that are given either as the main point of a paper or a letter or that you happen to see mentioned in an abstract. When possible, seek work assignments that will lead you into areas which are either new or deeper than those you've studied in college courses. For such a project, always begin with a literature search and study. If one is available, use a computer search service to find potentially useful journal articles. Check current books in the area. In both cases, you may find the most valuable part of an article is its references. Again there's a danger that you may spend more time reading than you really should. You should always worry about the time spent searching for something as compared to the time it would take you to derive the result or design the part yourself. Page-flipping for the right formula or a canned design, like catalog engineering, isn't much fun and is a short path out of the profession if practiced exclusively. Again, awareness and judgment must be exercised to bring a balance between the desire for a better tool (more power), the cost of acquiring it, and an estimate of its real usefulness. The learned literature is replete with promises which, with our desires, create illusions of performance frequently beyond the rather narrow limits of valid or tested cases. Nevertheless, this phase of the project will provide you with the most enduring benefits, because you always find things you don't need immediately but which turn out to be foundations, analogies, or pathways that will help you in a context neither of us is aware of now.

FURTHER READING

Cox, K. J., and Hattic, P. D., "Shuttle Orbit Flight Control Lessons: Direction for Space Station," *Proc. IEEE*, vol. 75, no. 3, p. 336, March 1987. (A special issue on progress in space from shuttle to station.)

Dote, Y., "Application of Modern Control Techniques to Motor Control," *Proc. IEEE*, vol. 76, no. 4, p. 438, April 1988. (A survey paper with many references on specific applications.)

Hendricks, Russell et al., "Advanced Flight Control Development for Single-Pilot Attack Helicopters," Minneapolis, Minn., Honeywell, Inc., 1986.

Johnson, M. A., "Recent Trends in Linear Optimal Quadratic Multivariable Control System Design," *IEE (London) Proc.*, vol. 134, part D, no. 1, p. 53, January 1987.

Kailath, T., *Linear Systems*, Englewood Cliffs, N.J., Prentice Hall, 1980.

Special Issue on Multivariable Control Systems, *IEEE Trans.*, AC-26, no. 1, February 1981.

POLYNOMIAL TESTING

A.1   THE BASIC TESTS

As in Chaps. 3 and 4, we are interested here in learning something about the poles of a transfer function without having to actually factor the system polynomial. Let the system polynomial be given as

$$D(s) = a_0 s^n + a_1 s^{n-1} + a_2 s^{n-2} + \cdots + a_n \qquad \text{(A-1)}$$

where the coefficients are real numbers. The basic question for stability is, Are there any right-half-plane roots? From algebra, going all the way back to Descartes, if there is a sign change among the coefficients, there must be at least one such root. Therefore, a necessary but not sufficient condition is that all the coefficients have the same sign. Since the kth coefficient represents a sum of products of the roots taken k at a time, a zero coefficient indicates the presence of roots in both half-planes. A second necessary but not sufficient condition is that all the coefficients be nonzero.

As I mentioned in Chap. 1, in the late nineteenth century Routh and Hurwitz developed tests on coefficients that can detect the presence of any RHP roots. Since Routh's test is simpler, it is the one which is universally used. It is done by forming an array from the coefficients and then inspecting the signs of the first-column elements. There are as many RHP poles as there are sign changes in the first column.

The Routh array is formed as follows:

1. The first row is the even-indexed coefficients, a_0, a_2, a_4, etc.
2. The second row is the odd-indexed coefficients, a_1, a_3, a_5, etc.
3. The shorter of these rows is evened up with the longer by adding a zero element if necessary, to the right end.
4. Each succeeding row is formed from the two rows above it using 2×2 determinants. Let r_{ij} be the ijth element in the Routh array. Then

$$r_{ij} = \frac{r_{i-1,1} r_{i-2,j+1} - r_{i-2,1} r_{i-1,j+1}}{r_{i-1,1}} \tag{A-2}$$

If the number of elements in each row is m, then this calculation proceeds from $1 \leq j \leq m - 1$, and the last element is zero.

5. Continue the process until there are as many rows in the array as coefficients, that is, until $i = n + 1$.

This is not as tedious as it might appear because the zeros at the right ends of the third and fourth rows cause a zero in the next position to the left in the fifth row, and so on, so that the array is triangular.

■ Example A.1

$$D(s) = s^3 + 3s^2 + 3s + 1 = (s + 1)^3$$

All three poles are at -1, so there should be no sign changes. Steps 1–3 give the first two rows as

$$\begin{matrix} 1 & 3 \\ 3 & 1 \end{matrix}$$

Step 4 gives the first element of row 3 as $(3 \times 3 - 1 \times 1)/3 = \frac{8}{3}$. The second element is zero. Now we have

$$\begin{matrix} 1 & 3 \\ 3 & 1 \\ \frac{8}{3} & 0 \end{matrix}$$

Again, step 4 gives the first element of row 4 as $((\frac{8}{3}) \times 1 - 3 \times 0)/(\frac{8}{3}) = 1$. The second element is zero. Now the array is full:

$$1 \quad 3$$

$$3 \quad 1$$

$$\tfrac{8}{3} \quad 0$$

$$1 \quad 0$$

∎

■ Example A.2

$$D(s) = s^3 + 2s^2 + 5s + 52$$

This is a cubic whose roots are not obvious. The first two rows are

$$1 \quad 5$$

$$2 \quad 52$$

$r_{31} = (2 \times 5 - 1 \times 52)/2 = -21$, $r_{32} = 0$. Now we have

$$1 \quad 5$$

$$2 \quad 52$$

$$-21 \quad 0$$

$r_{41} = (-21 \times 52)/-21 = 52$, and again $r_{42} = 0$. The full array is

$$1 \quad 5$$

$$2 \quad 52$$

$$-21 \quad 0$$

$$52 \quad 0$$

There are two sign changes, so there are two poles in the RHP. You may verify this by factoring $D(s)$.

∎

A.1.1 Special Cases

There are two cases which require special treatment, one of which offers an opportunity to extract extra information. It is possible that a zero will occur in the first column before the last row. If this happens, it can, by step 4, cause an infinity in the next row. If all you need to know about is stability, you don't need to carry the test further because the zero entry indicates at least a pole on the imaginary axis, if not in the RHP. If you want to continue the test, there are at least three possibilities. The first is to replace the zero by an arbitrary small number and carry on. The second is to replace s by $1/x$ in D and form a new polynomial $D_i(x) = x^n D(1/x)$. The poles of D_i will be the reciprocal of those of D, but each function will have the same number in the right half-plane. Once $D_i(x)$ is formed, the Routh test is performed on it. The third is to multiply the original polynomial by an additional factor $s + a$ and start over again. Choosing a positive won't change the number of RHP poles, and almost any value will work. The zero replacement method is obviously the simplest, but reports say that it may give an incorrect count in the presence of pure imaginary poles. Examples of the occurrence of this borderline case will be seen in the next section.

The second special case is one in which a whole row goes to zero. In this case, the coefficients in the row above the zero row are used to form an auxiliary polynomial in s^2. The poles of the auxiliary function, which are also poles of D, are symmetrical about the imaginary axis and are the cause of the zero row.

Suppose the index of the zero row is t. Then let $a = n + 2 - t$. The auxiliary polynomial is

$$A(s) = r_{t-1,1}s^a + r_{t-1,2}s^{a-2} + \cdots + r_{t-1,1+a/2} \tag{A.3}$$

Since $A(s)$ has symmetrical poles, it is of even degree. It does not necessarily contain all the RHP poles of $D(s)$, but contains at least one. One can continue the test by replacing row t with the coefficients of dA/ds.

■ Example A.3

$$D(s) = s^7 + 11s^6 + 35s^5 + 25s^4 + 4s^3 + 44s^2 + 140s + 100$$

It is customary, when constructing a Routh array by hand, to label each row with the highest power of s for the corresponding auxiliary polynomial. This helps to keep track for a large array. The first five rows for this problem are

s^7	1	35	4	140
s^6	11	25	44	100
s^5	32.73	0	130.91	0
s^4	25	0	100	0
s^3	0	0	0	0

In this case, $t = 5$ and $n = 7$, so $a = 4$ just as the row label says. $A(s) = 25s^4 + 0s^2 + 100$. Its poles are at $\pm 1 \pm j$, and $dA/ds = 100s^3$. Replace the fifth row with 100   0   0   0. Then the sixth row is 0   100   0   0. Here's the other special case. Replace the first zero by ϵ so that the sixth row reads ϵ 100   0   0. Then the seventh row reads $-10^4/\epsilon$ 0   0   0 and the eighth row is 100   0   0   0. The full array is

s^7	1	35	4	140
s^6	11	25	44	100
s^5	32.73	0	130.91	0
s^4	25	0	100	0
s^3	100	0	0	0
s^2	ϵ	100	0	0
s	$-10^4/\epsilon$	0	0	0
1	100	0	0	0

Whether you take ϵ as a negative or positive number, there are two sign changes in the first column so there are two poles in the RHP. These must be the RHP poles of $A(s)$. Again, you may verify this by factoring D.   ■

A.2   STABILITY RANGE FOR A PARAMETER

Since even pocket calculators can find roots of polynomials these days, the real value of the Routh test is that one can analytically determine a parameter value range for which the system is stable. It won't tell anything about the dynamics, but that may not be important in some simple problems.

■ Example A.4

Suppose that the cubic in Example A.1 is part of a plant transfer in a unity-feedback system. That is,

$$G(s) = \frac{K}{s^3 + 3s^2 + 3s + 1}$$

What is the range of $K > 0$ for which the system is stable?

$$D(s) = s^3 + 3s^2 + 3s + 1 + K$$

The Routh array is

1	3
3	$1 + K$
$(8 - K)/3$	0
$1 + K$	0

r_{31} goes through zero at $K = 8$, so the range allowed is $0 \le K < 8$. A simple design might choose $K = 4$. ∎

■ Example A.5

The dc transfer for the system in the last example is only $\frac{4}{5}$, if one chooses $K = 4$. Suppose zero final error is required. Then we must add an integrator to the forward path.

$$G(s) = \frac{K}{s^4 + 3s^3 + 3s^2 + s}$$

and

$$D(s) = s^4 + 3s^3 + 3s^2 + s + K.$$

The Routh array is

1	3	K
3	1	0
$\frac{8}{3}$	K	0
$(8 - 9K)/8$	0	0
K	0	0

In this case, r_{41} goes through zero at $K = \frac{8}{9}$. Notice that when K exceeds the stable limit, two sign changes occur in the first column in both of these examples. This indicates the crossing of a complex conjugate pair of poles into the RHP. In the present case, when $r_{41} = 0$, we can use the auxiliary equation to find the values of s at the axis crossing.

$$A(s) = (\tfrac{8}{3})s^2 + K = (\tfrac{8}{3})s^2 + \tfrac{8}{9}$$

The poles cross at $s = \pm j/\sqrt{3}$.   ■

   The Routh test can be used to find the parameter ranges for any one- or two-parameter problem for which the algebra is tractable. The next example is a simple two-parameter problem.

■ Example A.6

$$G(s) = \frac{K}{s(1 + s)}$$

$$H(s) = \frac{1 - Ts}{1 + Ts}$$

The feedback is an all-pass model for a transducer delay. The question to be studied is, What are the combinations of gain and delay (K and T) for which the system is stable?

$$D(s) = Ts^3 + (1 + T)s^2 + (1 - KT)s + K$$

The Routh array is

$$
\begin{array}{ll}
T & 1 - KT \\
1 + T & K \\
\dfrac{1 + T - 2KT - KT^2}{1 + T} & 0 \\
K & 0
\end{array}
$$

The stability boundary is determined by the numerator of r_{31}. The condition is

$$1 + T - KT(2 + T) = 0$$

or

$$KT = \frac{1 + T}{2 + T}$$

The relationship between the maximum allowable gain and the delay is generally hyperbolic. For $T \ll 1$, $K_{max} \approx 1/(2T)$, and for $T \gg 2$, $K_{max} \approx 1/T$. A few points in the neighborhood of 1 and 2 suffice to define the curve for sketching.

T	0.2	0.5	1	2	5	10
K_{max}	2.73	1.2	$\frac{2}{3}$	$\frac{3}{8}$	0.171	0.09117

These results could serve as the basis for a cost-performance decision if the cost of reducing the delay is known and, for example, performance is defined by the steady-state velocity error with K chosen as a fixed fraction of K_{max}.   ∎

Extensions of the Routh test to stability margin and other performance properties can be found in the literature.

ALGORITHMS AND SOFTWARE PACKAGES

B.1   PROS AND CONS OF COMMERCIAL SOFTWARE

Before the advent of the personal computer, working engineers used either a mainframe computer or a tabletop system. In the former case, a few general-purpose software packages were available, but most of us just wrote our own programs, usually in FORTRAN. In the latter case, some software was available either free or at nominal cost from the machine's manufacturer. Again, most of us just wrote our own programs, either in a machine language or in BASIC. I've been told recently that there is a catalog listing over 3500 software packages for engineering use to run on PCs. Documentation has changed too. When I learned PL/1 in 1975, I did it with a manual about $\frac{1}{2}$ in. thick. Now, no self-respecting software publisher would sell a package with fewer than 300 pages of documentation, most of which you will never want or need to read. Even so, the amount of learning you will have to do is on the order of learning a new programming language. You should seriously consider the effort involved before committing money and time to a new software package. For a specialist who can foresee very frequent use of a package, an investment is very clearly justified. For a generalist, on the other hand, the investment is seldom rewarded in even the medium term. In my own case, I mainly use two products, a word processor and an APL interpreter.

I am a generalist, and you might conclude from the above that I have a bias against canned software. I have seen some products that create very nice displays and tried out some that claimed to be very user friendly, but I've always been put off by the size of the manual I would have to wade through if I bought it and the low use rate I know the package would actually have. Whatever language you prefer, the advantages of writing your own programs are

1. You control the models. You can make the compromise you want between accuracy and computation time. You determine what features are important, and you can readily incorporate new models changes as the situation warrants and as they appear in the literature.

2. You have already made a major learning investment in the language. Each application takes a relatively small investment in additional learning, and that is mostly about the problem itself.

3. You will develop a set of functions or subroutines which turn out to be common to many problems, such as complex arithmetic, two-port parameter manipulations, and plotting functions. These will form your own personal set of utilities from which you can draw to make solution of new problems easier.

To achieve these benefits though, you have to maintain sufficient documentation to be able to use a program 6 months after the last use. The documentation can be on paper or in comments in the source code. It is *absolutely essential*. Another very important point is this: Whatever you choose for a language, choose an implementation that allows pixel-level plotting. The most important contribution PCs have made is affordable computer power to support a large percentage of our daily professional needs. The second most important contribution is the general availability of good screen and printer plots. The graphical presentation of results with 200 or more points per curve is a tremendous aid to our understanding and presents opportunities that would have been missed with line-printer-style character plots or tables or a few performance figures.

I want to make a special case for learning APL, even though you probably have already learned at least one other programming language. The other languages I have learned or looked at all seem too me to be first- or second-level mnemonic covers for the computer's operation. Languages above Assembler are supposed to be architecture-independent, but one is very aware of storage requirements when declaring data types and dimensions. One is also very aware of the fact that operations are done on one datum at a time. Finally, the functions defined in the languages have one or two scalar arguments.

Such languages are like the machines themselves, essentially oriented to do arithmetic in either the numerical or logical sense. It's as if these languages never went to college. APL originated with an effort to standardize the use of functional symbols in mathematics. As such, its defined operators not only operate on scalars but also on vectors and arrays. Furthermore, the nature of an operand doesn't have to be predefined by you. The interpreter checks to see what kind of data you've assigned to the operand and allocates storage accordingly. No explicit data type or array dimension statements are required from you. Most importantly, a whole new class of operators is given to you for use on vectors and arrays. I like to impress people by telling them that matrix inverse is only one keystroke in

APL. Actually, I use matrix inverse rarely in comparison to the other array functions. An operator you ordinarily wouldn't think of is the compression operator /. The division symbol is ÷. A ← +/B means "A is assigned the sum of the elements in each row of **B**." The plus sign can be replaced by any arithmetic operator, and the result is as if the operator were inserted between each element in a row of **B** and the operation performed sequentially along the row. In addition, A ← C/B, where **C** is a vector of 1 and 0 elements, means "Pick out the elements in **B** that line up with the 1s in **C** and throw away the rest; then assign the result to **A**." Thus **C** is a selection vector. **C** can be formed as the result of a logical test on the elements of **B**. For example, C ← 2 ≤ B means "If an element in **B** is not less than 2 (the test is true) put a 1 in the corresponding element of **C**. If the test fails, put a 0 in that element of **C**." Thus, not only are there generalized versions of the operators you are familiar with, such as various kinds of transpose and rotation, taking and dropping of rows and columns, and inner and outer products, but there are operators for things that you do but don't have symbols for, such as reshaping an array, sorting, and assembling arrays from smaller arrays or vectors or scalars.

The operators have three major effects: They make the interpreted program go much faster than, for example, interpreted BASIC. Execution is not as fast as for a compiled program, but fast enough to interactive. The array operators give one new ways of thinking about how to solve problems. Development time is much shorter.

As an example of the short development time, a few months back I wanted to try out some ideas about multichannel filters on some simple filter structures. I didn't have a workspace for general network analysis, so I had to choose between learning to use a PC version of SPICE or writing a new workspace. I ran the sample SPICE program, discovered that I would need an ASCII editor to enter data into it, didn't like the character-plot output, didn't like the somewhat elaborate element description, and decided I'd make my own.

I chose to describe the networks in node equations with unknown voltages and current sources. Then I converted the equations to three matrices: a conductance, a capacitance, and an inverse inductance matrix. For n nodes I defined the input currents as a $2n$ vector—the first n values being the real parts and the last n values being the imaginary parts. I generated a vector of the frequencies for which I wanted the response and a selection vector to pick out the node voltages I wanted. The foregoing are all global variables. I wrote a function which uses them as follows to produce the frequency response of the network. Inside a loop which steps through the frequencies, it forms the imaginary matrix I ← (W[K]xC) + IL ÷ W[K], where W[K] is the kth radian frequency. The function then forms a $2n \times 2n$ matrix of the real and imaginary matrices, inverts it as described in the following section, and right-multiplies by the currents to find the node voltages. It's all just what you would do by hand, but much faster to program as a defined function in APL. The coding and debugging took me about an hour. Compare that with the time it would have taken to learn to use the ASCII editor and the SPICE program.

Another important aspect of APL is that it responds like a calculator. You don't have to write a program to find the product of A and B. You simply type $A \times B$, press ENTER, and the interpreter gives the result on the next line. Programs in primitive languages are user-defined functions (user-defined operators) in APL. To produce such a definition, one enters the function definition mode, types the routine, and returns to the immediate mode. The interpreter responds to the user-defined function, just as it does to a built-in (primitive) operator, immediately. The variables and functions defined in a working session are called a *workspace* and can be saved with a name. One can select variables and functions to copy from one workspace to another, making it easy to move utility routines from one application to the next.

It is so much faster to work in APL than in primitive languages, that I have come around to a new point of view. I believe that the only people who should study primitive languages are computer scientists and systems programmers. All the rest of us who just want to *use* a computer should learn only APL. The argument for FORTRAN has often been, "There's so much already invested in FORTRAN programs, people have to learn it." I propose that if a program has a good user interface, it isn't necessary to learn the underlying language. If the program is becoming obsolete, far less effort would be expended in rewriting its functions in APL than in going in and modifying the old program. APL is that good. A lot of time and money have been wasted by not using a college-level language in our professional practice.

B.2   ALGORITHMS

In this section, I present some of the algorithms I used in writing this book. They are presented both as APL functions and in an informal general style. The two versions of each algorithm do not always match, because I have written the general version to be codable in a primitive language requiring fixed-dimension arrays and lacking the facility to append values to arrays.

Some years ago, I wrote menus and a small manual for the workspaces that make the plots you've seen in this text. I call it the Linear Systems Package and sell it for a bit over cost. I do this partly because the purchaser will need access to an APL interpreter and partly as an incentive to learn APL. The menus allow one to choose to operate either by making simple responses or to escape to work directly in APL.

B.2.1   Complex Data Operations

I generally represent complex-valued signals as two-row arrays, or complex vectors, in which the first (top) row has the real parts and the second row (bottom) has the imaginary parts. Likewise, describing a transfer function in terms of poles and zeros means, for example, defining a complex vector GZ whose top row holds the real parts of the forward-path zeros and whose bottom row contains the imaginary parts. Although they are not used in the Linear Systems Package, I will also discuss complex matrices. A complex matrix is thought of as two matrices in separate planes, plane 1 being the real parts and plane 2 being the imaginary parts. Thus a complex matrix is three-dimensional, with the first dimension being 2. Vectors in polar form are also sometimes needed, e.g., for Bode plots. The usual form for these puts the magnitudes in the top row and the phase angles in the bottom row. Addition and subtraction of complex vectors is just a straight element-to-element operation in rectangular form, so they will not be discussed further.

Absolute Value

Z←CABS A
Z←(+ ≠ AxA)*0.5

A is a complex vector which is converted into a real vector Z. Each element of A is multiplied by itself (squared), and then the columns are added. This gives a real vector. Last, each element is raised to the 0.5 power (square-rooted).

Let B be a vector of the same dimensions as A.
For all i,j B[i,j] = A[i,j]xA[i,j].
For each j, Z[j] = B[1,j] + B[2,j].
Z[*] = Z[*]∧0.5.

Four-Quadrant Arctangent

Z←Y ATAN X
Z←(¯3O(Y ÷ (X + X = 0))x(X≠0)) + ((O ÷ 2)x(xY)x(1 − xX)) + ox((xX) = ¯1)x(Y = 0)

Y and X are real vectors of equal length, and Z will be a vector of the arctangents of $Y \div X$. To augment the basic two-quadrant function, one must fix the quadrant according to the signs of Y and X and take care of the special cases $Y = 0$ and $X = 0$. In the case of an element of X being 0, an error condition is avoided by testing the element first and adding the result to the element before the division is performed. If the element is zero, the value returned by atan( ) is multiplied by 0, and pi ÷ 2 times the sign of the Y element is given to Z. If the Y element is 0, and X is not, then (pi ÷ 2)(1 − signX) is given to Z. If both the elements X and Y are 0, 0 is given to Z.

For each k, YZ = (0 = Y[k]), XZ = (0 = X[k]),
z[k] = (NOT XZ)atan(Y[k] ÷ (XZ + X[k])) + (pi ÷ 2)(sign Y[k])(1 − sign X[k]),
Z[k] = Z[k] + pi(XZ = − 1)YZ.

Vector Multiply

Z←A CMUL B
Z←(− ≠ AxB),[0.5](+ ≠ AxθB)

A, B, and Z are complex vectors of the same length. The usual rule for complex multiplication in rectangular form is faster than conversion to and use of the polar form.

For each k,
Z[1,k] = A[1,k]B[1,k] − A[2,k]B[2,],
Z[2,k] = A[1,k]B[2,k] + A[2,k]B[1,k].

Vector Division

Z←A CDIV B
Z←(+ ⌿ AxB),[0.5] − ⌿ BxθA) ÷ (ρB)ρ + ⌿ BxB

Z is A divided by B. The method used is multiplication of the top and bottom of the fraction by the conjugate of B to get a real denominator. Again, this is faster than conversion to and use of the polar form.

Let D be a real vector of the same length as B.
For each k, D[k] = B[1,k]B[1,k] + B[2,k]B[2,k],
Z[1,k] = (A[1,k]B[1,k] + A[2,k]B[2,k]) ÷ D[k],
z[2,k] = (B[1,k]A[2,k] − B[2,k]A[1,k]) ÷ D[k].

Matrix Inverse

Z←CMINV1 M;T;R
Z←R,[0.5] − T + .xR← ⊞ M[1;;] + M[2;;] + .xT←(⊞ M[1;;]) + .xM[2;;]

Z←CMINV2 A;R;I;N
N←¯1 ↑ ρA ◇ R←A[1;;] ◇ I←A[2;;]
Z←(0, − N) ↓ ⊞ (R, − I),[1]I,R
Z←(2,N,N)ρZ

There are two ways to perform complex matrix inversion which I give here as method 1 and method 2. Method 1 involves two $n \times n$ real inversions, and method 2 uses one $2n \times 2n$ real inversion. Method 1 has the advantages of requiring about half the storage and less computation time than method 2. However, one of the $n \times n$ inversions in method 1 is of the real part. In an equation system which models a system with few dissipative elements, the real-part matrix may be too sparse to have an inverse by itself. If the physical system is properly modeled, method 2 always works. In presenting both methods, I assume you have real matrix multiplication and inversion functions or subroutines available.

Method 1:

Let the input matrix be M = Ri + jIi, the result matrix be Z = Ro + jIo, and T be an intermediate n×n matrix.

Ri = M[1,*,*], Ii = M[2,*.*].
T = (inverse Ri)Ii.
Ro = inverse (Ri + IiT), Io = TRo.
Z[1,*,*] = Ro, Z[2,*,*] = Io.

Method 2:

Let the input matrix be M = Ri + jIi and the output matrix be Z = Ro + jIo.
Ri = M[1,*,*], Ii = M[2,*,*].
The relation to be computed is

$$\begin{bmatrix} Ro & -Io \\ Io & Ro \end{bmatrix} = \text{inverse} \begin{bmatrix} Ri & -Ii \\ Ii & Ri \end{bmatrix}$$

Let T be the right-hand matrix. Then for i,j = 1 to n,
T[i,j] = T[n + i,n − j] = Ri[i,j].
For i = n + 1 to 2n, j = 1 to T[i,j] = Ii[i,j] T[i − n,j + n] = T[i − o].
T = inverse T.
For i,j = 1 to n, Z[1,i,j] = T[i,j], Z[2,i,j] = T[n + i,j].

The remaining three operations are just given in their APL forms because they are so straightward in any language. They are mainly listed for name reference in later functions.

To Polar Form

```
Z←POLAR X;N;XT
N←(÷2)×ρ,X ◊ XT←(2,N)ρX
Z←(ρX)ρ,(CABS XT),(XT[2;]ATAN XT[1;])
```

To Rectangular Form

```
Y←RECT Z;N
N←1↓ρZ
Y←(2,N)ρ(Z[1;]x2oZ[2;]),Z[1;]x1oZ[2;]
```

Vector of Square Roots

```
Y←CSQRT X;Z;N
Z←POLAR X
N←(÷2)×ρZ←,Z
Y←(N↑Z)*0.5
Y←(ρX)ρYx(22oZ),1o(Z←(÷2)×N↑Z)
```

B.2.2 Polynomial Operations

Polynomials are represented by vectors of their coefficients. For example, $s^2 + 2s + 2$ is represented by the vector 1 2 2. The polynomial coefficients are always listed from that for the highest power down to the constant.

Polynomial Multiplication

$Z \leftarrow A$ PMUL $B;N;M$
$M \leftarrow \rho A \ \diamond \ N \leftarrow \rho B$
$Z \leftarrow) \mathbb{Q} (1 - \iota M) \mathbb{Q} (M, ^-1 + M + N) \rho B, ((M-1)\rho 0)) + .xA$

If you write out the product of two general polynomials so that your result forms a table with each column having the coefficients of a given power, the result might look like this:

$$(a_0 s^m + a_1 s^{m-1} + a_2 s^{m-3} + \ldots)(b_0 s^n + b_1 s^{n-1} + b_2 s^{n-2} + \ldots) =$$

$$a_0 b_0 s^{n+m} + a_0 b_1 s^{n+m-1} + a_0 b_2 s^{n+m-2} + a_0 b_3 s^{n+m-3} + a_0 b_4 s^{n+m-4} + \ldots$$
$$+ a_1 b_0 s^{n+m-1} + a_1 b_1 s^{n+m-2} + a_1 b_2 s^{n+m-3} + a_1 b_3 s^{n+m-4} + \ldots$$
$$+ a_2 b_0 s^{n+m-2} + a_2 b_1 s^{n+m-3} + a_2 b_2 s^{n+m-4} + \ldots$$
$$+ a_3 b_0 s^{n+m-3} + a_3 b_1 s^{n+m-4} + \ldots$$

Looking at the terms in each column, we have a discrete convolution sum of the coefficients of the two polynomials. This leads to the following.

Let A and B be two input vectors of coefficients, Z be the output vector of coefficients.

M = no. of elements in A,
N = no. of elements in B.
Form a vector R of length $N + M - 1$, $R = [0]$.
$R[1$ to $N] = B$.
Duplicate R to form an M-row matrix T.
Rotate $T[i,*]$ right by i-1 positions. T = transpose T.
$z = T \ A$.

Polynomial Division

$C \leftarrow A$ PDIV $B;BT;N;M;L;K;J$
$N \leftarrow \rho A \ \diamond \ M \leftarrow \rho B \ \diamond \ L \leftarrow 1 + N - M \ \diamond \ C \leftarrow 0 \rho 0$
$BT \leftarrow M \uparrow A \ \diamond \ K \leftarrow 1$
RLOOP:$C \leftarrow C, BT[1] \div B[1]$
$J \leftarrow 1$
BLOOP:$BT[J] \leftarrow BT[J+1] - C[K]xB[J+1]$
$\rightarrow (M > J \leftarrow J + 1)/$BLOOP
$\rightarrow (K = L)/0$
$BT[M] \leftarrow A[K+M] \ \diamond \ K \leftarrow K + 1$
$\rightarrow$RLOOP

This algorithm is simply a straightforward implementation of synthetic division. It is meant to be used for polynomial deflation; that is, it is assumed that the polynomial represented by B is a factor of the one represented by A. Therefore, no remainder is returned.

A is a polynomial coefficient vector of length N, B is a polynomial coefficient vector of length M≤N, and C is the result of length

L = 1 + N − M. C = 0.
BT = A[1 to M], k = 1.
RL:C[k] = BT[1] ÷ B[1]
j = 1
BL:BT[j] = BT[j + 1] − C[k]B[j + 1]
j = j + 1, if M>j go to BL.
If k = L you're done.
BT[M] = A[k + M], k = k + 1
Go to RL.

Coefficients from Roots

A←COEFFS R;P;N;K
N←1 ↓ ρR ◇ →(N = 0)/ZERO
A← 2 1 ρ 1 0
P← 2 1 ρ 0 0
K←1
L:A←(A,P) − P,(A CMUL⍉(1 ↓ ρA),2)ρR[;K])
→(N≥K←K + 1)/L
A←A[1;]
→0
ZERO:A←,1

In this algorithm, the concept is that each root, a column of R, represents a first-order polynomial to be multiplied by the current polynomial. The current polynomial A is treated as a complex vector of coefficients, starting with an initial length of 1, value $1 + j0$. The current polynomial is shifted left once, with a zero appended on the right, to represent multiplication by the variable (e.g., s or x) to the first power. Then the root multiplies the current polynomial, and the product is subtracted from the shifted polynomial to form the next current result.

Let R be a complex vector of roots, of dimensions $2 \times N$. The result, A, is a vector of polynomial coefficients of dimension $2 \times N + 1$, until the last step where the imaginary row is stripped off. AT, RT are $2 \times N + 1$.

Initialize: A = AT = RT = 0. A[1,1] = 1, k = 1.
L:RT[*,2 to k + 1] = R[*,k].
AT[*,2 to k + 1] = A[*,1 to k].
A[*,1 to k + 1] = A[*,1 to k + 1] − (AT CMUL RT).
k = k + 1, if k≤N go to L.
A = A[1,*].

Evaluation for an Imaginary Argument

```
Y←A IMPOLY W;N;NR;NW;AP;WP
N←ρA ◇ NR←⌈(N+1)÷2
AP←(NR,2)ρ(⊖A),0,0,0
Y←(2,NW←ρW←,W)ρ0
K←1
WL:WP←(NR,2)ρ1,(W[K]*⍳N−1),0,0,0
Y[;K]←⍉−⌿ AP×WP
→(NW≥K←K+1)/WL
```

A straightforward way to evaluate a transfer function with the variable equal to a complex value is to convert the problem to polar form. The transfer function is represented by its factors, the argument value is put into each factor, and the results are converted to polar form and multiplied together. The polar conversion step is the slowest one in the process. It turns out to be faster, when using an imaginary argument, to work with the polynomials in the transfer function and treat the real and imaginary parts separately with real arithmetic. This is possible because the powers of the argument $j\omega$ are real for even index and imaginary for odd index. For example,

$$a_0s^4 + a_1s^3 + a_2s^2 + a_3s + a4 =$$
$$a_0\omega^4 - a_2\omega^2 + a_4 + j(-a_1\omega^3 + a_3\omega), \text{ for } s = j\omega.$$

To use this form then, the polynomial coefficients are placed in two rows of an array, the even and odd powers of the variable are placed in two rows in another array, the two arrays are multiplied together element by element, and the rows are compressed by alternate subtraction to find the real and imaginary parts of the result.

Let A be the polynomial coefficient vector, W be the scalar imaginary argument, N = length A, Y is the result, 2×1.

NR = round-up $((N+1) \div 2)$.
AP1 = WP1 = 2NR zeros.
For k = 1 to N, AP1[k] = A[N+1−k], WP1[k] = W∧k−1.
AP2 = WP2 = 2xNR array of zeros.
For k = 1 to NR, AP2[1,k] = AP1[2k−1], WP2[1,k] = WP1[2k−1],
AP2[2,k] = AP1[2k], WP2[2,k] = WP1[2k].
Y[1] = sum $((-1)∧(k-1))$AP2[1,k]WP2[1,k].
Y[2] = sum $((-1)∧(k-1))$AP2[2,k]WP2[2,k].

Evaluation for a Complex Argument

```
Y←A ACDPOLY X;N;P;PP;PDP;XT;XM;XA
N←−1−¯1↑ρA ◇ XT←2↑X,0
XT←POLAR XT
```

XM←θ1,(XT[1]*ιN)
XA←θ0,(XT[2]xιN)
XT←RECT XM,[0.5]XA
P←(2,1)ρ+/A CMUL XT
PP←(2,1)ρ+/((2,N)ρθιN)x(N TAKE A)CMUL(1 DROP XT)
PDP←(2,1)ρ+/((2,N−1)ρθ(ιN−1)×1+ιN−1)×((N−1)TAKE   A)CMUL(2 DROP XT)
Y←P,PP,PDP

This algorithm is used in the Laguerre root finder, so it assume that factors of the polynomial are not available. It also calculates the values of the first and second derivatives for the polynomial with a complex argument. The basic approach is to form a vector of the powers of the argument and then perform a complex multiplication with the coefficient vector. In APL, I use the polar form and the index generator to make the power vector. In languages restricted to scalar operation, looping is necessary whatever method is used, so a loop using complex multiplication in rectangular form is probably a faster way to generate the power vector. I do this in the following description.

Let A be a complex vector of polynomial coefficients with dimensions $2 \times N$. Let X be the complex argument with dimensions 2×1. P is the value of the polynomial, PP is the value of the first derivative, PDP is the value of the second derivative, all 2×1.

XT, PT, NT, are $2 \times N$ arrays of zeros.
XT[1,N] = 1.
For k = N − 1 to 1, XT[*,k] = XT[*,k + 1] CMUL X.
PT = A CMUL XT
P[1] = sum PT[1,*], P[2] = sum PT[2,*].
A[*,N] = XT[*,1] = 0.
Rotate XT one element left.
For k = 1 to N − 1, XT[k] = (N + 1 − k)XT[k].
PT = A CMUL XT.
pp[1] = sum PT[1,*], PP[2] = sum PT[2,*].
A[*,N − 1] = XT[*,1] = 0.
Rotate XT one element left.
For k = 1 to N − 2, XT[k] = (N − k)XT[k].
PT = A CMUL XT.
PDP[1] = sum PT[1,*], PDP[2] = sum PT[2,*].

Roots from Coefficients

One root:

R←A LG Z;N;K;Y;P;PP;PDP;DZ;D;AC
N←−1−¯1↑ρA ◇ R←2ρZ,0

```
AC←(2,N+1)ρ(,A),(N+1)ρ0 ◇ K←1
LOOP:Y←AC ACDPOLY R
P←Y[;1] ◇ PP←Y[;2] ◇ PDP←Y[;3]
→((CABS P)≤M)/0
D←CSQRT((N−1)x(((N−1)×PP CMUL PP)−N×P CMUL PDP)
DZ←−N×P CDIV PP+(xPP)×ID
R←R+DZ ◇ →(CNT≥K←K+1)/LOOP
'OUT OF COUNT' ◇ CNT,CABS P ◇ R
```

I use the Laguerre method as the basis of the root finder. It is an iterative method which uses the polynomial and its first two derivatives. It is expressed as follows: Let R be the current guess for the root, ΔR be the increment in the guess, and P, P', P'' be the polynomial of degree n, and its derivatives evaluated at R. All of these are complex numbers. In particular, $P' = P'r + jP'i$. Let $D = [(n - 1)((n - 1)P'P' - nPP'')]^{1/2} = Dr + jDi$, another complex number. Then $\Delta R = -nP/[P' + (\text{sign } P'r)|Dr| + j(\text{sign } P'i)|Di|]$. The next guess for the root is $R + \Delta R$. I won't attempt to justify the method but point out a few things about it. The use of the second derivative seems to move R to the neighborhood of a root very quickly. Close to a root, P and P'' are small and $\Delta R \to -P/P'$, a Newton's method expression. My experience has been that, if there is a real root, LG will find it first.

Let A be a $2 \times N + 1$ complex vector of polynomial coefficients.

Let Z be an initial guess for a root, a complex number.

M is the minimum magnitude tolerance for testing the polynomial at the trial root. CNT is the maximum number of trials before giving up.

```
DZ=DZD= 2×1 vector of zeros.
R=Z, k=1.
L:[P,PP,PDP]=A ACDPOLY R. (Polynomial evaluation routine given above.)
If M≥CABS P, it's done.
D=CSQRT[(N−1)((N−1)(PP CMUL PP)−N(P CMUL PDP))].
DZD[1,1]=PP[1,1]+(sign PP[1,1])(abs D[1,1]).
DZD[2,1]=PP[2,1]+(sign PP[2,1])(abs D[2,1]).
DZ=−NP CDIV DZD.
R=R+DZ. k=k+1. If k≤CNT go to L.
Print 'OUT OF COUNT', print CNT, CABS P, print R.
```

All the roots:

```
Z←LROOTS A;AR;N;NR;B;R;M;CNT
OVER:N←−1−ρA ◇ →(0≠A[1])/OK ◇ A←1↓A ◇ →OVER
OK:Z←(2,N)ρ0 ◇ AR←A÷A[1] ◇ CNT←100
```

```
NR←0 ◇ M←⌈/□CT,(1E⁻9)×AR[N+1]
ROOT:→(2=ρAR)/FIRST
→(3=ρAR)/SECOND
R←AR LG 0
→((|R[2])<|R[1]×1E⁻8)/REAL
NR←NR+2
Z[1;NR]←Z[1;NR−1]←R[1]
Z[2;NR]←−Z[2;NR−1]←R[2]
B←COEFFS⍉(2,2)ρZ[;NR],Z[;NR−1]
→NEXT
REAL:NR←NR+1 ◇ Z[1;NR]←R[1]
B←COEFFS(2,1)ρZ[;NR]
NEXT:AR←AR PDIV B ◇ →ROOT
FIRST:NR←NR+1
Z[1;NR]→−AR[2] ◇ →0
SECOND:NR←NR+2
R←−AR[3]−(AR[2]÷2)*2
→(R≥0)/REAL1
Z[1;NR]←Z[1;NR−1]←−AR[2]÷2
Z[2;NR]←−Z[2;NR−1]←(−R)*0.5 ◇ →0
REAL1:Z[1;NR]←(R←R*0.5)−AR[2]÷2
Z[1;NR−1]←−R+AR[2]÷2 ◇ →0
```

This function uses LG to extract one root at a time. It does the overhead of putting the root in a vector of roots, keeping track of how many have been found, and deflating (reducing by synthetic division) the polynomial each time a root is found. Since I am interested in polynomials with real coefficients, finding a complex root means that I have two roots—the one found and its conjugate. In that case, the corresponding quadratic is formed and used to deflate the polynomial. When the polynomial is reduced to first- or second-order, the function branches to appropriate finishing operations.

Let A be a real coefficient vector of length $N+1$. Let Z be the $2 \times N$ vector of roots, initially zero. Set values for M and CNT to pass to LG. The root counter is $NR=0$. Define vectors $B1=[1\ \ 0]$ and $B2=[1\ \ 0\ \ 0]$ for the polynomial corresponding to a found root (pair).

$AR = A \div A[1]$.
ROOT: If $N-NR=1$ go to FIRST.
If $N-NR=2$ go to SECOND.
$R = AR$ LG 0 (pass the reduced polynomial to LG with a guess of 0).
If $R[2,1] < R[1,1] \times 1E-8$ go to REAL.
$NR = NR+2$

$Z[1,NR] = Z[1,NR - 1] = R[1,1],$
$Z[2,NR] = - Z[2,NR - 1] = R[2,1].$
$B2[2] = - 2Z[1,NR], B2[3] = Z[1,NR]\wedge2 + Z[2,NR]\wedge2.$
$AR[NR + 1$ to $N + 1] = AR$ PDIV B2, $AR[NR] = AR[NR - 1] = 0$ (deflate). Go to ROOT.
REAL:$NR = NR + 1. Z[1,NR] = R[1,1]. B1[2] = - R[1,1].$
$AR[NR + 1$ to $N + 1] = AR$ PDIV B1, $AR[NR] = 0$ (deflate). Go to ROOT.
FIRST:$NR = NR + 1. Z[1,NR] = - AR[2].$ Done.
SECOND:$NR = NR + 2. RD = (AR[2] \div 2)\wedge2 - AR[3].$
If $RD \geq 0$ go to REAL.
$Z[1,NR] = Z[1,NR - 1] = - AR[2] \div 2,$
$Z[2,NR] = - Z[2,NR - 1] = (- RD)\wedge0.5.$ Done.
REAL1:$RD = RD\wedge0.5. Z[1,NR] = RD - AR[2] \div 2.$
$Z[1,NR - 1] = Z[1,NR] - 2RD.$ Done.

B.2.3 Function Minimization

More frequently than you might expect, I have found it necessary to numerically find a maximum or minimum or a zero of a function other than a polynomial. Newton's method is commonly used to find a zero, but it isn't suited for finding a minimum unless you know the function's value at the minimum or use its derivative so that you can convert the problem to a zero search. I prefer a method which searches for a function minimum so that I can use it also for zero finding by searching for the minimum magnitude of the function. Any search method requires that the function have only one local minimum in the search interval. If you're searching for a zero, the same is true: The magnitude must have only one local minimum in the search interval. The method I use is called "search by golden section."

```
Z←Y ZEROFIND FUNC;RES;K;X1;X2;X11;X21;F1;F2;D
RES←Y[1] ◇ X1←Y[2] ◇ X2←Y[3]
D←2 ÷ (3 + 5*0.5)
X11←X1 + DxX2 - X1
X21←X2 - DxX2 - X1
F1←⍕FUNC,' X11'
F2←⍕FUNC,' X21'
→(F1<F2)/CUTTOP
CUTBOTTOM:X1←X11 ◇ X11←X21
X21←X2 - DxX2 - X1
→((X2 - X1)<RES)/OUT1
F1←F2 ◇ F2←⍕FUNC,' X21'
→(F1>F2)/CUTBOTTOM
CUTTOP:X2←X21 ◇ X21←X11
```

```
X11←X1+DxX2−X1
→((X2−X1)<RES)/OUT2
F2←F1 ◇ F1←⏀FUNC,' X11'
→(F1>F2)/CUTTBOTTOM
→(F1≤F2)/CUTTOP
OUT1:Z←X21 ◇ → 0
OUT2:Z←X11
```

Given X_1 and X_2 as the left and right boundaries of the search interval, the idea is to find two interior values of X to test the function so that the search interval is shortened in an optimal way. Taking the test points at the middle of the interval is a fairly efficient method, as half of the interval is discarded after two function calls. The most efficient method I know is to take the test points in such a way that one of the current test points will also be a test point inside the reduced interval. This allows substantial reduction in the search interval after each function call. The optimum distance into the interval from each point is $2/(3 + \sqrt{5})$ times the interval length. This interval division has been called the "golden section" from the time of the classical Greeks.

Let RES be the resolution in location of the minimum value of FUNC(X). Let the starting interval be X1, X2. $D = 2 \div (3 + \text{sqrt}5)$.

```
X11 = X1 + D(X2 − X1), X21 = X2 − D(X2 − X1).
F1 = FUNC(X11), F2 = FUNC(X21).
If F1<F2 go to CUTTOP (drop the interval from X21 to X2).
CUTBOTTOM:X1 = X11, X11 = X21 (drop interval between X1 and X11).
X21 = X2 − D(X2 − X1).
If RES>X2 − X1 go to OUT1.
F1 = F2, F2 = FUNC(X21).
If F1>F2 go to CUTBOTTOM.
CUTTOP:X2 = X21, X21 = X11.
X11 = X1 + D(X2 − X1)
If RES>X2 − X1 to to OUT2.
F2 = F1, F1 = FUNC(X11).
If F1>F2 go to CUTBOTTOM.
If F1≤F2 go to CUTTOP.
OUT1:Z = X21. Done.
OUT2:Z = X11. Done.
```

FURTHER READING

Al-Khafaji, A. W., and Tooley, J. R., *Numerical Methods in Engineering Practice*. New York, Holt Rinehart and Winston, 1986.

Jennings, Alan, *Matrix Computation for Engineers and Scientists*, New York, John Wiley and Sons, New York, 1977.

Press, William H., et al., *Numerical Recipes; The Art of Scientific Computing*, Cambridge University Press, 1986.

Special Section on CAD of CONTROL SYSTEMS, *Proc. IEEE*, vol. 72, no. 12, December 1984.

THE PAPER DESIGN PROJECT

In recent years, I have included a paper design project as part of my linear control systems course. The class forms into two- or three-person groups, and about the middle of the semester I hand out instructions and a problem description to each group. I have been using five different basic problems, with additional numerical variations so that each group gets a unique problem. The final report is due during the last class week of the semester. I urge an instructor who wishes to try this to formulate a schedule for the students that fits the school calendar and gives the groups a task to accomplish each week.

C.1   INSTRUCTIONS AND EVALUATION

Each group is given a problem description which includes some information about the objects and variables to be controlled and some specifications to be met. The group's first task is to understand the problem statement, sketch the information given in terms of a block diagram, and then fill in the sketch with blocks to perform the remaining obvious functions. For example, if the group is given a speed control problem with known load dynamics, maximum ratings, and response specifications, it must choose a motor, amplifiers, and a speed sensor. These items are chosen for their ability to supply the maximum powers, speed, and estimated bandwidth required. The transfer functions of the elements and the system are then found, and the system analyzed for its dynamic response. It must be verified that the maximum excursion requirements can be met. The discrepancies between the actual dynamic response and the specifications must be corrected by design and insertion of an appropriate network. It may be that the specifications contain implicit conflicts which cannot be resolved. In that case, a compromise must be made. Here, as at other sticking points, the supervisor (or customer—in this case your instructor) should be consulted. When all requirements have been satisfied, the network must be designed. This means a circuit with appropriate parts, values, and ratings determined and specified. Then, the effect of component tolerances must be studied. Worst-case combinations of part values should be used to determine the variation in system responses. This may result either in specification of tight tolerances on some parts or in system redesign. The analysis-design-analysis loop may be traversed several times before the final design is achieved.

Once the design is complete, the report must be written. It should consist of the following sections, each of which is discussed: narrative text, figures, tables, and supporting calculations.

1. Narrative text: This part contains the problem description, a discussion of the most interesting or difficult parts of the design process, and a brief description of the final system. This portion may not exceed five pages and must be typed double-spaced. It may not include any equations or calculations but may include salient numerical values.

2. Figures: The figures must include, but are not limited to, the initial system block diagram, initial system responses, final system block diagram, final system responses, and schematics.

3. Tables: There must be at least one table—a list of all components and parts in the system. The columns must include quantity, item, value and tolerance, and maximum ratings.

4. Supporting calculations: This section must contain, at a minimum, calculations which verify that the components chosen will provide the maximum or minimum variable levels and swing rates required by the specifications.

Evaluation of the project report should be in two parts, each worth 100 points maximum. Technical merit is considered in the first part. The design is judged for accuracy, simplicity, and economy. Communication quality is rated in the second part. Language usage, spelling, and organization will be judged by the highest standards. All handwork, such as figure labels and supporting calculations, must be printed very neatly.

C.2 SPECIFICATION EXAMPLES

C.2.1 Paper Design Project 1: Tank Level Regulator

A mixing tank for a chemical process is to have its level maintained at 10 ft. It is supplied with two components from supply tanks, each inlet is through a valve, and the valves are connected together and driven by a single shaft. Mix is withdrawn from the tank at rates and times determined by the production schedule, so that the discharge amounts to a random disturbance. A float connected to a potentiometer bridge is used to sense the liquid level in the tank. The parameters of the system at this point are

- ☐ Valve load: 10 oz-in.-s^2 each
- ☐ Valve opening: zero to three-quarter turn
- ☐ Valve flow: 10 ft^3/s-rad
- ☐ Maximum discharge: (5 ft^2/s) × tank level in feet
- ☐ Tank cross section: 50 ft^2
- ☐ Sensor gain: 1 V/ft
- ☐ Sensor range: 8–12 ft.

The design requirements are for a maximum step discharge. The response should be critically damped, and the system should remain within the range limits of the valves and sensor. A limit circuit must be incorporated so that when the valve is at one end of its travel, current is blocked to keep the motor from further turning in that direction.

C.2.2 Paper Design Project 2: Tape Capstan Drive System

A capstan (wheel) drives a tape by friction at its edge. The loading of the tape may be modeled as an inertia connected to the shaft through a spring and a parallel damper. The constants are

- ☐ Inertia: 7.24 oz-in.-s^2
- ☐ Damper: 10 oz-in./(rad/s)
- ☐ Spring: 2900 oz-in./rad

The capstan inertia is negligible by comparison. The system driving the capstan must bring the speed to 100 rpm with zero final error. It should have a step response equivalent to a second-order system with a damping ratio of about 0.71. When it is turned on, it should bring the speed to within 1% of the final value in less than 0.1 s.

INDEX

A 8
B 9
C 0
D 1
E 2
F 3
G 4
H 5
I 6
J 7